Surface Properties and Surface Treatments of Wood and Wood-Based Composites

Surface Properties and Surface Treatments of Wood and Wood-Based Composites

Guest Editor

Marko Petric

Basel • Beijing • Wuhan • Barcelona • Belgrade • Novi Sad • Cluj • Manchester

Guest Editor
Marko Petric
University of Ljubljana
Ljubljana
Slovenia

Editorial Office
MDPI AG
Grosspeteranlage 5
4052 Basel, Switzerland

This is a reprint of the Special Issue, published open access by the journal *Coatings* (ISSN 2079-6412), freely accessible at: https://www.mdpi.com/journal/coatings/special_issues/wood_based_composites.

For citation purposes, cite each article independently as indicated on the article page online and as indicated below:

Lastname, A.A.; Lastname, B.B. Article Title. *Journal Name* **Year**, *Volume Number*, Page Range.

ISBN 978-3-7258-2673-5 (Hbk)
ISBN 978-3-7258-2674-2 (PDF)
https://doi.org/10.3390/books978-3-7258-2674-2

© 2024 by the authors. Articles in this book are Open Access and distributed under the Creative Commons Attribution (CC BY) license. The book as a whole is distributed by MDPI under the terms and conditions of the Creative Commons Attribution-NonCommercial-NoDerivs (CC BY-NC-ND) license (https://creativecommons.org/licenses/by-nc-nd/4.0/).

Contents

About the Editor vii

Preface ix

Jingyi Hang, Bo Zhang, Hongwei Fan, Xiaoxing Yan and Jun Li
Effect of Two Types of Chitosan Thermochromic Microcapsules Prepared with Syringaldehyde and Sodium Tripolyphosphate Crosslinking Agents on the Surface Coating Performance of Basswood Board
Reprinted from: *Coatings* **2024**, *14*, 1118, https://doi.org/10.3390/coatings14091118 1

Jinzhe Deng, Tingting Ding and Xiaoxing Yan
Effect of Two Types of Pomelo Peel Flavonoid Microcapsules on the Performance of Water-Based Coatings on the Surface of Fiberboard
Reprinted from: *Coatings* **2024**, *14*, 1032, https://doi.org/10.3390/coatings14081032 18

Ye Zhu, Ying Wang and Xiaoxing Yan
The Effects of Urea–Formaldehyde Resin-Coated *Toddalia asiatica* (L.) Lam Extract Microcapsules on the Properties of Surface Coatings for Poplar Wood
Reprinted from: *Coatings* **2024**, *14*, 1011, https://doi.org/10.3390/coatings14081011 35

Solène Pellerin, Fabienne Samyn, Sophie Duquesne and Véronic Landry
Preparation and Characterisation of UV-Curable Flame Retardant Wood Coating Containing a Phosphorus Acrylate Monomer
Reprinted from: *Coatings* **2022**, *12*, 1850, https://doi.org/10.3390/coatings12121850 56

Zuzana Vidholdová, Gabriela Slabejová and Mária Šmidriaková
Quality of Oil- and Wax-Based Surface Finishes on Thermally Modified Oak Wood
Reprinted from: *Coatings* **2021**, *11*, 143, https://doi.org/10.3390/coatings11020143 77

Sebastian Dahle, John Meuthen, René Gustus, Alexandra Prowald, Wolfgang Viöl and Wolfgang Maus-Friedrichs
Superhydrophilic Coating of Pine Wood by Plasma Functionalization of Self-Assembled Polystyrene Spheres
Reprinted from: *Coatings* **2021**, *11*, 114, https://doi.org/10.3390/coatings11020114 94

Matjaž Pavlič, Marko Petrič and Jure Žigon
Interactions of Coating and Wood Flooring Surface System Properties
Reprinted from: *Coatings* **2021**, *11*, 91, https://doi.org/10.3390/coatings11010091 107

Dace Cirule, Errj Sansonetti, Ingeborga Andersone, Edgars Kuka and Bruno Andersons
Enhancing Thermally Modified Wood Stability against Discoloration
Reprinted from: *Coatings* **2021**, *11*, 81, https://doi.org/10.3390/coatings11010081 120

Arnaud Maxime Cheumani Yona, Jure Žigon, Sebastian Dahle and Marko Petrič
Study of the Adhesion of Silicate-Based Coating Formulations on a Wood Substrate
Reprinted from: *Coatings* **2021**, *11*, 61, https://doi.org/10.3390/coatings11010061 133

Linda Makovicka Osvaldova, Patricia Kadlicova and Jozef Rychly
Fire Characteristics of Selected Tropical Woods without and with Fire Retardant
Reprinted from: *Coatings* **2020**, *10*, 527, https://doi.org/10.3390/coatings10060527 148

Jingyi Hang, Xiaoxing Yan and Jun Li
A Review on the Effect of Wood Surface Modification on Paint Film Adhesion Properties
Reprinted from: *Coatings* **2024**, *14*, 1313, https://doi.org/10.3390/coatings14101313 160

About the Editor

Marko Petric

Marko Petric received his PhD in chemistry in 1994 from the University of Ljubljana (UL) Slovenia. Currently, he holds the position of full professor in the field of "wood working and wood processing technologies" in the Department of Wood Science & Technology (UL, Biotechnical Faculty). He is the leader of the Surface Finishing Laboratory. Over the past few decades, he has held various leadership positions in the faculty (for example, Vice-Dean for the field of wood science and technology or Head of the Chair for Furniture; the Chairman of the Committee for Elections of the University of Ljubljana; since 1st October 2022, he has been the Chairman of the Committee for Doctoral Studies at the Biotechnical Faculty; etc.). He is or was professionally active in various committees; for example, he was the Leader of Section 3 (Wood Protecting Chemicals) of the International Research Group for Wood Protection (IRG/WP), Stockholm, Sweden, and a member of the Scientific Programme Committee (SPC) of the IRG/WP (until 2009), a member of the editorial boards of several established international scientific journals, a member of scientific programme committees of various international conferences, and many more. In 1996, he received the "Ron Cockroft Award" from the IRG/WP, and in 2018, he was awarded a Golden Plaquette from the University of Ljubljana for his research and pedagogic achievements in the field of wood science and technology for his contribution to the reputation of the UL.

His research and pedagogical activities include wood coatings for interiors and exteriors, their interactions with various substrates (wood, modified wood, and densified wood), wood liquefaction and preparation of new wood coatings on the basis of liquefied wood, and the applications of nanomaterials in wood coatings and in the treatment of wood. Over time, he has studied a considerable number of new, environmentally friendly wood preservatives.

Preface

Whenever products made of wood or wood-based composites are utilized, when their weathering is studied, or when the processes of wood deterioration due to exposure to biotic or abiotic factors are investigated, it is the surface of the object under consideration that must be taken into account first. Therefore, this Special Issue will examine the following topics: the surface finishing of wood and wood-based composites with different types of stains and film-forming coatings to obtain a fit-for-purpose finished product; the properties of wood surface systems, including resistance to weathering and biological deterioration parameters; plasma treatments; the wettability of wood and wood-based materials and their treated surfaces; and adsorption phenomena on untreated and surface finished lignocellulosic materials.

Marko Petric
Guest Editor

Article

Effect of Two Types of Chitosan Thermochromic Microcapsules Prepared with Syringaldehyde and Sodium Tripolyphosphate Crosslinking Agents on the Surface Coating Performance of Basswood Board

Jingyi Hang [1], Bo Zhang [2], Hongwei Fan [1,3], Xiaoxing Yan [1,3,*] and Jun Li [1,3]

1. Co-Innovation Center of Efficient Processing and Utilization of Forest Resources, Nanjing Forestry University, Nanjing 210037, China; hangjingyi@njfu.edu.cn (J.H.); fanhongwei@njfu.edu.cn (H.F.); lijun0099@njfu.edu.cn (J.L.)
2. School of Science, Nanjing University of Posts and Telecommunications, Nanjing 210023, China; zhangbo@njupt.edu.cn
3. College of Furnishings and Industrial Design, Nanjing Forestry University, Nanjing 210037, China
* Correspondence: yanxiaoxing@njfu.edu.cn

Abstract: In order to investigate the effect of thermochromic microcapsules on the surface coating performance of basswood board, two types of microcapsules prepared with syringaldehyde and sodium tripolyphosphate crosslinking agents were added to a UV primer and coated on the surface of basswood board. The color-change effect of the surface coating on basswood board with microcapsules added with syringaldehyde as the crosslinking agent was better than that with microcapsules added with sodium tripolyphosphate as the crosslinking agent, and the color difference varied more significantly with temperature. The effect of the two types of microcapsules on the glossiness of the surface coating on basswood board was relatively weak. The glossiness of the surface coating on basswood board with microcapsules containing syringaldehyde as the crosslinking agent showed an overall increasing trend with the increase in microcapsules, and the change trend was relatively gentle. The glossiness of the surface coating on basswood board with microcapsules containing sodium tripolyphosphate as the crosslinking agent increased first and then decreased as the amount of microcapsules added increased. The addition of microcapsules with syringaldehyde as the crosslinking agent had no significant effect on the reflectance in the visible light band of the surface coating on basswood board. Among the two groups of samples, the hardness increase in the surface coating on basswood board with syringaldehyde as the crosslinking agent was more significant. The adhesion level of the coating on the surface of the basswood board with the two microcapsules did not change. Neither of the microcapsules had a significant effect on the impact resistance of the surface on basswood board. In the comprehensive analysis, the surface coating on basswood board with microcapsules added with syringaldehyde as the crosslinking agent at a content of 4.0% had better comprehensive performance, better surface morphology, better color-change effect, and moderate mechanical properties. The color difference was found to be 21.0 at 25 °C, the reflectivity was found to be 57.06%, the hardness was found to be 3H, the adhesion was found to be five, and the impact resistance was found to be three.

Keywords: microcapsules; thermochromic; chitosan; coating properties

1. Introduction

Wood is the only renewable natural resource among the four major materials. Due to its unique texture and superior processing performance, it has always maintained a core position in the field of building decoration materials [1–7]. The improvement in people's own taste has led to higher requirements for wood in terms of a visual sense. Thermochromic wood can not only provide people with wonderful visual effects but can also meet users' personalized needs for decorative materials, so it has broad development potential in the field of furniture decoration [8–13]. Microencapsulation of color-changing materials can effectively expand the application range of thermochromic wood. In addition, it is necessary to treat the wood surface before use, and its color-changing characteristics mainly depend on the wood surface. Therefore, adding color-changing microcapsules into the coating of wood products is an efficient way of utilization [14,15]. Microcapsule technology can effectively prevent the core material from being affected by the outside world, improve the color-changing effect of the core material, and has become the main research object of color-changing coatings [16].

In microcapsule technology, choosing a suitable wall material is crucial, among which chitosan has attracted much attention due to its excellent performance. Chitosan is a light-yellow powder that is easily soluble in inorganic acids such as acetic acid and hydrochloric acid but not in water. It is derived from chitin, the second most abundant natural polymer in nature, making it a widely available and renewable biomass resource. By removing some acetyl groups from chitin, chitosan is produced, which can be considered an inexhaustible natural polymer material [17]. As a natural non-toxic material, chitosan is often employed in microcapsule technology, where it serves as a wall material to cover various materials and achieve specific functional outcomes. Furthermore, due to its excellent degradability and biocompatibility, chitosan is widely used in industries such as food, textiles, and medicine [18–20]. He et al. [21] synthesized photochromic microcapsules using spiropyran compounds as a core material and chitosan as a wall material to prepare photochromic hydrophobic fabrics. After 20 cycles of UV irradiation, it still had a photochromic effect, indicating that microencapsulation greatly improves the fatigue resistance of the photochromic dyes. Teng et al. [22] prepared microcapsules containing garlic essential oil (GEO) using chitosan grafted with gallic acid (GA), which enhanced the inhibition of nitrite-producing bacteria. Yin et al. [23] prepared chitosan microcapsules and seaweed salt solutions containing cinnamon essential oil, which can effectively inhibit the decrease in vitamin C content in mango and delay the appearance of mango respiratory peak. Thermochromic materials have the characteristic of having a temperature memory function and have great application potential in aerospace, military, anti-counterfeiting technology, construction, and other fields. In recent years, a variety of thermochromic materials have been prepared using different methods, with various discoloration mechanisms [24]. Zhu et al. [25] prepared thermochromic microcapsules using urea-formaldehyde resin as the wall material and thermochromic compounds as the core material by in situ polymerization and studied the thermochromic properties of the microcapsules in wood and wood coatings. The results showed that the microcapsules, when compounded with wood and coatings, also had good thermochromic properties and had broad application prospects in the preparation of intelligent materials. Hu et al. [26] developed an intelligent multifunctional wood material. During the coating process of medium-density fiberboard, thermochromic microcapsules were added to the coating. Microcapsules have a sensitive color-change phenomenon. The color of medium-density fiberboard can change between blue and brown at 20–29 °C.

Spray drying is a widely used and practical method for preparing microcapsules [27,28]. The drying process lasts for a short time, and the microcapsule emulsion can be directly dried into powder. The drying output is high, and the efficiency is high [29]. Previous studies have identified the optimal preparation conditions for the color-changing compound. Specifically, the optimal conditions were found to be a mass ratio of crystal violet lactone, bisphenol A, and decanol of 1:3:50, a reaction temperature of 50 °C, a reaction time of 1.5 h, and a stirring speed of 400 rpm. Under these conditions, the color-changing temperature of the compound closely matched real-life scenarios, the discoloration temperature range was moderate, and the solution appeared clearest when colorless. Based on the above results, the color-changing compound with this ratio was selected as the core material of the microcapsule. As one of the core components of the color-changing compound, bisphenol A plays a key role. However, since bisphenol A is widely considered to be a potentially toxic and carcinogenic compound, its safety in use deserves attention. Although the application of bisphenol A in coatings improves the color-changing performance, its potential health effects still need to be considered. In this experiment, bisphenol A was encapsulated in the microcapsule core material, and there was no free bisphenol A, thus ensuring the application safety of the microcapsules.

In the preparation process of microcapsules, a crosslinking agent, as a key chemical substance, can effectively connect polymers through their reaction, thereby forming a more stable material structure. In this experiment, chitosan was used as the wall material, and the color-changing compound was used as the core material. The chitosan-coated color-changing microcapsules were prepared by spray drying. The type of crosslinking agent was selected as the research variable to explore the effects of two crosslinking agents, syringaldehyde and sodium tripolyphosphate (STPP), on the performance of microcapsules. Thermochromic microcapsules with different mass fractions were added to UV coatings and coated on the surface of basswood boards. The color difference, optical and mechanical properties of the coatings were studied. The purpose was to make the coating containing microcapsules maintain the optical and mechanical properties of the original coating under the condition of obtaining a better color-changing effect. This provides a technical basis for the application of color-changing microcapsules and the preparation of color-changing coatings.

2. Materials and Methods

2.1. Test Materials

The materials required in this test are shown in Table 1. The size of the basswood board used in the test was 100 mm × 100 mm × 5 mm, and the UV primer was provided by Jiangsu Haitian Technology Co., Ltd., Jurong, China. The UV primer included epoxy acrylic resin, polyester acrylic resin, trihydroxy methacrylate, trimethyl methacrylate, leveling agent, photoinitiator 1173 (2-hydroxy-2-methylpropiophenone), defoamer, etc. The equipment used in this test is shown in Table 2.

Table 1. List of test raw materials.

Test Material	Molecular Formula	Manufacturer
Crystal violet lactone	$C_{26}H_{29}N_3O_2$	Wuhan Huaxiang Biotechnology Co., Ltd., Wuhan, China
Bisphenol A	$C_{15}H_{26}O_2$	Shanghai APB Chemical Reagent Co., Ltd., Shanghai, China
1-Decanol	$C_{10}H_{22}O$	Shanghai MacLean Biochemical Technology Co., Ltd., Shanghai, China
Chitosan	$(C_6H_{11}NO4)_n$	Shanghai Sinopharm Reagent Co., Ltd., Shanghai, China
Acetic acid	$C_2H_4O_2$	Shanghai Sinopharm Reagent Co., Ltd., Shanghai, China
(Z)-Sorbitan mono-9-octadecenoate (Span-80)	$C_{24}H_{44}O_6$	Shandong Yousuo Chemical Technology Co., Ltd., Linyi, China
NaOH	NaOH	Fuzhou Feijing Biotechnology Co., Ltd., Fuzhou, China
Syringaldehyde	$C_9H_{10}O_4$	Shanghai Een Chemical Technology Co., Ltd., Shanghai, China
STPP	$Na_5P_3O_{10}$	Shanghai MacLean Biochemical Technology Co., Ltd., Shanghai, China

2.2. Microcapsule Preparation Method

According to the parameters shown in Table 3, two crosslinking agents, syringaldehyde and STPP, were used to prepare chitosan thermochromic microcapsules, which were named 1# microcapsules and 2# microcapsules, respectively. "#" was the sample number unit. The amount of raw materials used to prepare chitosan thermochromic microcapsules is shown in Table 4.

Table 2. Experimental equipment.

Test Machine	Machine Model	Manufacturer
Electronic balance	JCS-W	Yongkang Huanyu Weighing Equipment Co., Ltd., Yongkang, China
Heat-collecting constant-temperature heating magnetic stirrer	DF-101S	Shenzhen Dingxinyi Experimental Equipment Co., Ltd., Shenzhen, China
Ultrasonic emulsifier disperser	TL-650CT	Jiangsu Tianling Instrument Co., Ltd., Yancheng, China
Small spray dryer	JA-PWGZ100	Shenyang Jingao Instrument Technology Co., Ltd., Shenyang, China
UV curing machine	CF-11	Bai De hong En Co., Ltd., Huzhou, China
Temperature measuring gun	TN400	Nomi Electronic Technology Co., Ltd., Changzhou, China
Portable color-difference meter	SC-10	Shanghai Hechen Energy Technology Co., Ltd., Shanghai, China
Gloss meter	X-rite ci60	Shenzhen Lai Te Instrument Equipment Co., Ltd., Shenzhen, China
Scanning electron microscopy	Quanta-200	Thermo Fisher Tech., Inc., Shanghai, China
Fourier transform infrared spectrometer	VERTEX 80V	Bruke Co., Ltd., Karlsruhe, Germany
UV spectrophotometer	U-3900	Hitachi Scientific Instruments (Beijing) Co., Ltd., Beijing, China
Universal mechanical testing machine	AGS-X	Shimazu Factory, Kyoto, Japan
Fine roughness tester	J8-4C	Shanghai Taiming Optical Instrument Co., Ltd., Shanghai, China
Pencil hardness tester	HT-6510P	Quzhou Aipu Measuring Instrument Co., Ltd., Quzhou, China
Paint film impactor	QCJ-40	Quzhou Aipu Measuring Instrument Co., Ltd., Quzhou, China
Paint film adhesion tester	QFH-A	Quzhou Aipu Measuring Instrument Co., Ltd., Quzhou, China

Table 3. Preparation parameters of the test.

Sample (#)	Crosslinking Agent	m(Crystal Violet Lactone)/m(Bisphenol A): m(1-Decanol)	m(Core Material)/m(Wall Material)	Temperature (°C)	Stirring Speed (rpm)
1	$C_9H_{10}O_4$	1:3:50	3:1	60	600
2	$Na_5P_3O_{10}$	1:3:50	3:1	60	600

Table 4. Detailed list of test raw materials.

Sample (#)	Chitosan (g)	1% Acetic Acid (mL)	Color-Changing Compound (g)	Span-80 (g)	Distilled Water (g)	Syringaldehyde (g)	Anhydrous Ethanol (mL)	STPP (g)	Distilled Water (g)
1	1.8	180	5.4	0.3	5.7	3.6	10	-	-
2	1.8	180	5.4	0.3	5.7	-	-	4.5	85.5

(1) Preparation of wall material: 1.8 g of chitosan was weighed and dissolved in 180 mL of 1% acetic acid solution to obtain the chitosan solution. A magnetic stirrer was added to the beaker, and the beaker was placed in the water bath at a set speed of 600 rpm and a temperature of 60 °C until the chitosan powder was completely dissolved to obtain the chitosan wall material solution.

(2) Preparation of the core material: First, the temperature of the water bath was raised to 30 °C, and 67.5 g of decanol was accurately weighed into the beaker, and the beaker was heated in the water bath to heat the decanol to a molten state. Then 4.05 g of bisphenol A and 1.35 g of crystal violet lactone were added to the beaker, and the solution in the beaker was evenly stirred with a magnetic stirrer. After stirring evenly, the water bath was

gradually heated to 50 °C and stirred at 400 rpm for 1.5 h to obtain the core emulsion. The core emulsion was cooled to room temperature, at which time the solution was colorless.

(3) Emulsification of the core material: 0.3 g of Span-80 and 5.7 g of distilled water were added in a beaker. After full stirring, an emulsifier solution with a mass fraction of 5% was obtained. The prepared emulsifier was dripped into 5.4 g of core material emulsion, the beaker was placed in the water bath, the temperature was adjusted to 60 °C, the rotating speed was 600 rpm, and the reaction time was carried out for 20 min. The stirred solution was placed in the ultrasonic emulsifier, and the ultrasonic treatment was carried out for 5 min to ensure that the emulsifier was evenly wrapped on the outer surface of the core material.

(4) Crosslinking reaction of microcapsules: The temperature of the water bath was set at 35 °C and the rotating speed was set at 600 rpm. The wall material solution was absorbed by a dropper and added to the core material emulsion. The core material and wall material solution were thoroughly mixed for 1 h. Then 0.5 mol/mL of NaOH solution was added to adjust the pH value of the solution to about 5, and the solution was placed in the ultrasonic emulsifier disperser. After 5 min of ultrasonic treatment, the crosslinking agent solution was added into the water bath for crosslinking reaction for 3 h. The temperature of the water bath was 60 °C, and the rotating speed was 600 rpm. Then, the obtained solution was spray-dried at the inlet temperature of 110 °C, the outlet temperature of 64 °C, and the feed rate of 100 mL/h. The dried powder was chitosan thermochromic microcapsule powder.

2.3. Preparation Method of the Surface Coating on Basswood Board

The coating process of the basswood board is the manual painting method. Before use, the basswood board was placed at room temperature with a relative humidity of 50.0% ± 5.0% for 7 d to make the moisture content reached about 14.5%. To sand the surface of the basswood until it was smooth and flat, 800 grit sandpaper was used, then a brush was used to remove loose powder. Microcapsules 1# and 2# were added to the UV primer at an addition amount of 1.0%, 2.0%, 3.0%, 4.0%, 5.0%, and 7.0%, respectively, among which the UV primer without microcapsules was the blank control group. The total mass of each group of coatings was controlled to be 3.00 g. The corresponding mass of microcapsules and UV primer were weighed according to Table 5. After mixing the microcapsules and UV primer evenly, the mixture was applied evenly on the surface of the basswood with a soft brush. Then the basswood board was placed on the conveyor belt of the UV curing machine, the conveying speed was adjusted to 0.05 m/s, and the curing time to 30 s. After the curing was completed, it stood for 1 min to allow the coating surface to cool to room temperature, and then the second sanding, coating, and curing were performed.

Table 5. List of materials for coatings.

Microcapsule Content (%)	Microcapsule Quantity (g)	UV Primer Quantity (g)
0	0.00	3.00
1	0.03	2.97
2	0.06	2.94
3	0.09	2.91
4	0.12	2.88
5	0.15	2.85
7	0.21	2.79

2.4. Testing and Characterization

2.4.1. Micromorphology Characterization

The microstructure of the paint film was characterized by optical microscopy and scanning electron microscopy [30].

2.4.2. Chemical Composition Test

The chemical composition of microcapsules and paint films was tested and characterized by infrared spectrometers. In the infrared test, the microcapsule powder was made into thin sheets by the powder press.

2.4.3. Optical Performance Test

(1) Color difference: According to the test method of GB/T11186.3-1989 [31], the portable color-difference meter was used for testing. After the colorimeter was calibrated, the test hole was placed on the coating to be tested, the test button was pressed, and the L, a, and b values were recorded. The L, a, and b values represent the values of the test sample after the temperature changes. L_0, a_0, and b_0 represent the values of the sample's starting test temperature (-5 °C). Each set of data was measured three times, and the average value was taken [32–34]. Formula (1) for the color difference (ΔE) between two points is as follows:

$$\Delta E = \left[(L - L_0)^2 + (a - a_0)^2 + ((b - b_0)^2 \right]^{\frac{1}{2}} \tag{1}$$

(2) Glossiness: The test was conducted in accordance with GB/T4893.6-2013 "Tests for physical and chemical properties of furniture surface paint films Part 6: Gloss determination method" [35]. A portable gloss meter was used to conduct the test. After calibrating the machine, the test hole was aligned with the coating to be tested, and the gloss value of the coating on the screen at incidence angles of 20, 60, and 85 was recorded [36–39].

(3) Reflectivity: An ultraviolet spectrophotometer was used to test the reflectivity of the paint film within the visible light wavelength range. Reflectivity refers to the ratio of the intensity of reflected light to the intensity of incident light when a beam of light shines on an object and is usually expressed as a percentage [40–42].

2.4.4. Mechanical Properties Test

(1) Hardness: The test was conducted in accordance with GB/T6739-2006 "Paints and varnishes-Determination of film hardness by pencil method" [43]. A pencil hardness tester was used to conduct the test. Several 6H-6B pencils were prepared, each with a flat lead core of about 4 mm exposed. The pencil was inserted into the hardness tester, and the tip of the pencil was pressed against the paint layer. The tip of the pencil was parallel to the paint film and pushed at a speed of approximately 0.5 mm/s over a distance of 1 cm.

(2) Adhesion: The test was conducted according to the standard GB/T4893.4-2013 "Tests for physical and chemical properties of furniture surface paint films Part 4: Adhesion cross-cutting method" [44]. First, a grid tool was used to draw a right-angle grid horizontally and vertically. Then a tape was stuck on the grid, pressed, and then torn off, and the shedding of the coating was observed on the tape. The adhesion grade of the coating is divided into 0, 1, 2, 3, 4, and 5, and the corresponding paint film shedding area is 5%, 15%, 35%, 55%, and more than 60%, respectively.

(3) Impact resistance: According to GB/T4893.9-1992 "Furniture surface paint film impact resistance test" [45], a paint film impactor was used to conduct the test. A sample was placed in the horizontal base, the 1.0 kg impact block was dropped freely from a certain height to impact the sample, and the coating was observed for cracks and spalling. Before the experiment, the coating without microcapsules was tested, and the impact height was 10 mm, 25 mm, 50 mm, 100 mm, 200 mm, 400 mm. Each height was impacted five times from low to high until the impact pattern caused grade 5 cracks to stop. After testing, it was found that in the coating without microcapsules grade 5 cracks appeared when the height was 50 mm, so the height was set to 50 mm to test the impact resistance of each coating at this height. Each coating was impacted five times. By comparing the data, as in Table 6, the average of the five impact levels was rounded to an integer. The average was the impact resistance level.

Table 6. Grade of impact site.

Impact Resistance Level	Variation in Paint Film Surface
1	No visible changes.
2	There are no cracks on the paint film surface, but impact marks are visible.
3	There are slight cracks on the paint film surface, usually 1–2 ring cracks or arc cracks.
4	There are moderate to severe cracks on the paint film surface, usually 3–4 ring cracks or arc cracks.
5	The paint film surface is severely damaged, usually with more than 5 ring cracks, arc cracks, or paint film falling off.

3. Results and Discussion

3.1. Morphological Analysis

3.1.1. Macroscopic Morphology Analysis

The color change of 1# microcapsule was shown in Figure 1. The color of the microcapsule after freezing (−5 °C) was dark green, changing the blue color of the original discoloration complex, which was preliminarily judged to be related to the crosslinking agent added. Syringaldehyde is a brownish-yellow substance, which exhibited a dark-green color when mixed with the discoloration compound. When the microcapsules were cooled to room temperature, the color change was still not obvious, so the heating measures were taken. When placed in an oven and heated to 35 °C, it gradually turned yellow-green. When heated to 45 °C, the yellow-green became lighter. After 55 °C, the color of the microcapsules no longer changed. When syringaldehyde was used as a crosslinking agent to react with the color-changing compound, the brown color of syringaldehyde may have masked the original color change of the compound or formed a stable color state during the discoloration process, rendering the color change insignificant.

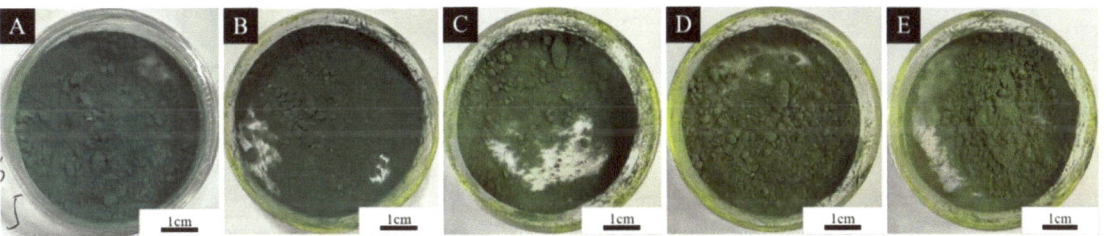

Figure 1. Color change of 1# microcapsules at different temperatures: (**A**) −5 °C, (**B**) 25 °C, (**C**) 35 °C, (**D**) 45 °C, (**E**) 55 °C.

The color change of 2# microcapsule is shown in Figure 2. After freezing (−5 °C), the microcapsules turned light yellow, changing the blue color of the original color-changing compound. Preliminary judgment was that this was related to the added crosslinking agent. STPP is a white substance, which appeared light yellow when mixed with a color-changing compound. When the microcapsules were cooled to room temperature, the color did not change. The temperature was raised by placing the microcapsules in an oven and heating them to 55 °C. During the entire heating process, the color of the microcapsules did not change significantly, which may be because the crosslinking agent STPP used was less sensitive to temperature changes. In addition, STPP may have formed a relatively stable chemical structure with chitosan, further inhibiting the effect of temperature changes on the color of the microcapsules.

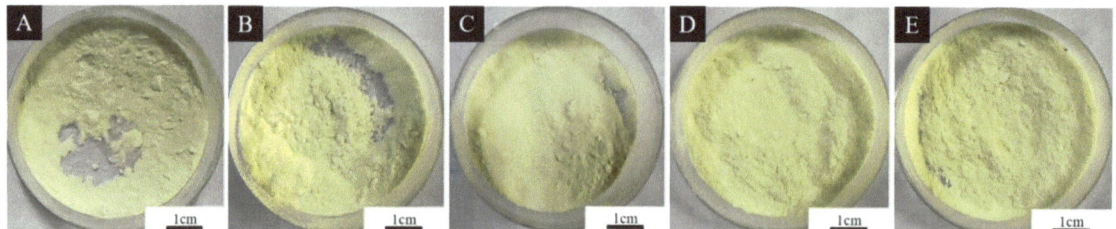

Figure 2. Color change of 2# microcapsules at different temperatures: (**A**) -5 °C, (**B**) 25 °C, (**C**) 35 °C, (**D**) 45 °C, (**E**) 55 °C.

3.1.2. Microtopography Analysis

Figure 3 shows SEM images of 1# and 2# microcapsules. Spherical microcapsules with relatively uniform particle size and high yield were successfully produced by spray drying. Upon comparing the two SEM images, it can be clearly seen that there are large pieces of adhered and irregularly shaped materials in the 1# microcapsule, and the particle size distribution varies greatly. Although the microcapsules also appear spherical, there is obvious adhesion and agglomeration. The particle size distribution of 2# microcapsules is relatively uniform. The microcapsules are spherical, but there is an obvious agglomeration phenomenon, and the surface of the microcapsules is loose and porous. Both microcapsules showed a severe aggregation, which may be related to the wall material and the crosslinking between chitosan and aldehyde groups in aqueous medium.

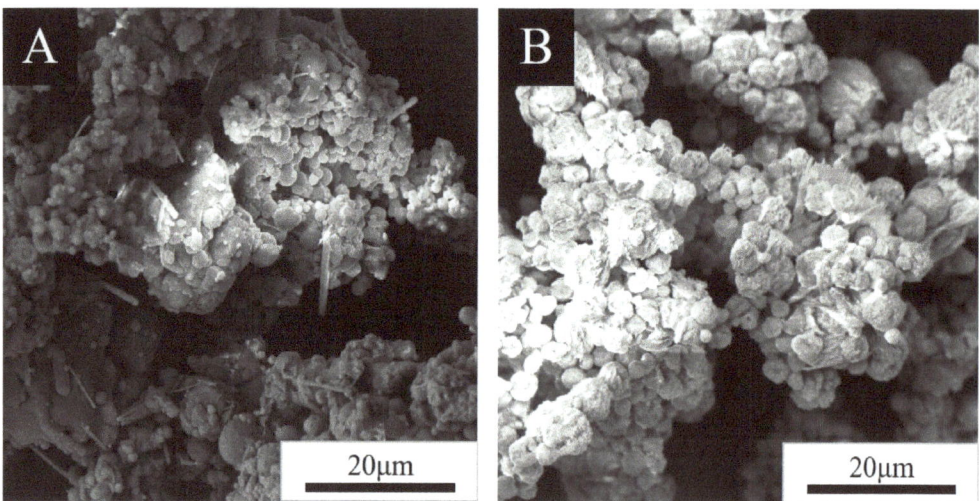

Figure 3. SEM microscopy images of the different microcapsules: (**A**) 1# microcapsule, (**B**) 2# microcapsule.

The microscopic morphology of the paint film with added 1# microcapsules is shown in Figure 4. When the content of the microcapsules was 0%, the surface of the paint film was relatively flat. When the content of microcapsules was 1.0%, the surface of the paint film had a slightly raised granular sense. When the content was 4.0%, the particle sense of the paint film surface increased. When the content reached 7.0%, due to the aggregation of a small amount of microcapsules in the paint, the paint film presented a small area of raised wrinkles. This showed that when the content of the 1# microcapsules was low, there was little effect on the smoothness of the coating. The micromorphology of the paint film with the added 2# microcapsule is shown in Figure 5. When the content of 2# microcapsule was

0%, the surface of the paint film was relatively flat. When the content of 2# microcapsules was 1.0%, the convexity on the paint film surface was small. When the content of 2# microcapsule was 4.0% and 7.0%, the wrinkles and granularity on the surface of the paint film were stronger. This was because the 2# microcapsule was unevenly dispersed in the paint, forming agglomerates, resulting in larger particles appearing on the surface of the paint film. When the content of 2# microcapsule in the paint film was 4.0% or above, it would have a negative impact on the surface morphology of the paint film.

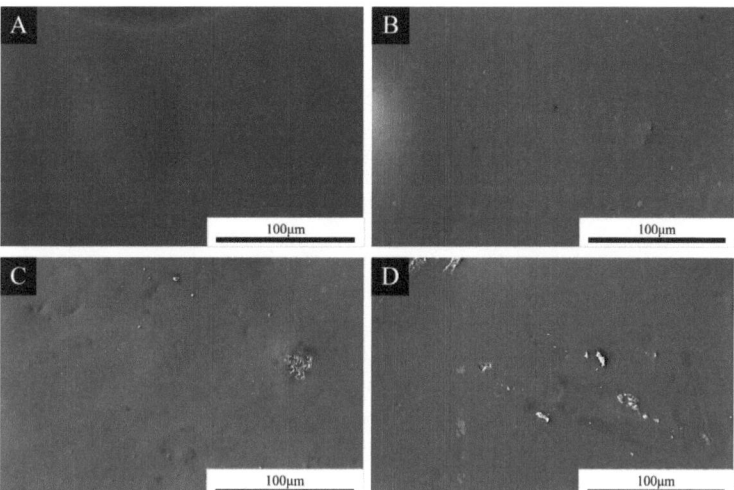

Figure 4. SEM images of paint films with different 1# microcapsule contents: (**A**) 0%, (**B**) 1.0%, (**C**) 4.0%, (**D**) 7.0%.

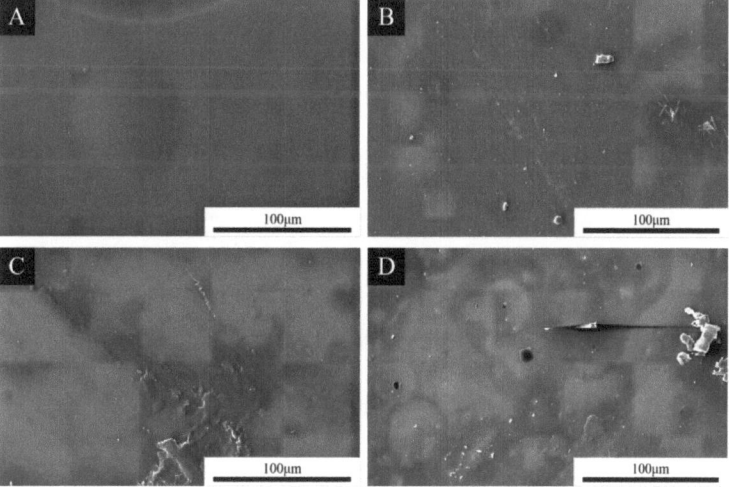

Figure 5. SEM images of paint films with different 2# microcapsule contents: (**A**) 0%, (**B**) 1.0%, (**C**) 4.0%, (**D**) 7.0%.

3.2. Chemical Composition Analysis

Figure 6 shows the infrared spectra of the blank paint film and the paint film with two kinds of microcapsules added. The absorption peak at 1721 cm^{-1} was the absorption peak of C=O in the paint [46]. Due to the strengthening of intermolecular and intramolecular hydrogen bonds, the vibration frequency moved to a lower wave number, the peak area became larger, and the broad peak at 3449 cm^{-1} was attributed to the overlapping peaks of –OH and N–H, indicating the formation of hydrogen bonds between chitosan-syringaldehyde microcapsules. The characteristic absorption peak of –RC=N– appeared at 1721 cm^{-1}, indicating that the amino group of chitosan and the aldehyde group of syringaldehyde form a Schiff base [47,48]. The characteristic peaks of the thermochromic complex also appeared in the infrared spectrum of the microcapsule. The 2967 cm^{-1} was attributed to the stretching vibration peak of C–CH$_3$ in bisphenol A. The ester carbonyl C=O absorption peak of the non-lactone ring structure appeared at 1610 cm^{-1}, and 1183 cm^{-1} corresponded to the symmetrical stretching absorption peak of the carboxylate, proving that the lactone ring in the molecule is open to form a conjugated chromogenic structure [49]. This proves that after adding the prepared 1# microcapsule to the UV primer, the wall material and core material components belonging to the 1# microcapsule still exist, and there is no chemical reaction between the 1# microcapsule and the UV primer. Although the chitosan content in the two formulations was the same, there was no obvious signal in the FTIR spectrum of the chitosan coated with 2# microcapsules. This phenomenon may have occurred due to the thick coating layer of the 2# microcapsules, which obscured the FTIR characteristic peak of chitosan. Additionally, STPP, used as a crosslinking agent, may have interfered with the FTIR spectrum, affecting the infrared absorption characteristics of chitosan.

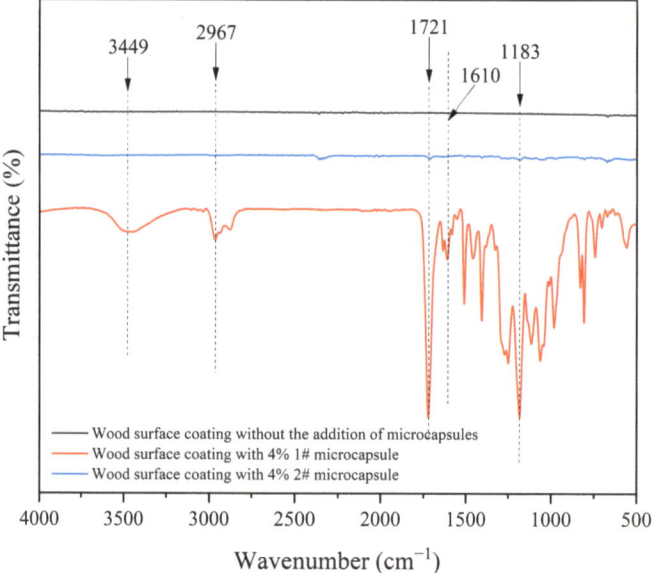

Figure 6. Infrared spectrum of paint film.

3.3. Analysis of the Influence of Microcapsule Content on the Optical Properties of Basswood Surface Coating

3.3.1. Effect of *Microcapsule* Content on Color Difference of Surface Coating on Basswood Board

The effects of different contents of microcapsules on the color difference of the surface coating on basswood board are shown in Tables 7 and 8. The L value represents the lightness or darkness of the measured coating color, and the larger the L value, the brighter the color. The a value represents the red-green value. A positive a value means the color is reddish, and a negative a value means the color is greenish. The b value represents the yellow-blue value. A positive b value represents a yellowish color, and a negative b value represents a bluish color. Since the 1# microcapsules prepared by the addition of syringaldehyde crosslinking agent were yellow-green, the b value representing yellowness tends to increase with the increase of microcapsule content. The L value representing lightness and darkness, and the a value representing red and green values, changed less. The surface coating on basswood board with 1# microcapsules mostly changed color significantly at around 25 °C, which meant it started to change color. When the mass fraction of 1# microcapsule added was 4%, the color difference changed most significantly with the change of temperature, and the color difference was 21.0 at a temperature of 25 °C. As the temperature rose, the color-difference value of adding 2# microcapsules at the same mass fraction fluctuated within a certain range without obvious changes. Because the color of 2# microcapsules prepared by adding the STPP crosslinking agent was light yellow and lighter in color, it had little effect on the color of the surface coating on basswood board. When the mass fraction of 2# microcapsules was 2%, the color difference changed most significantly with the change of temperature, and the color difference was 10.7 at a temperature of 50 °C. The individual color-difference values in Table 8 were relatively large, probably because the paint film was transparent at the experimental temperature (−5 °C–55 °C), and the original wood color of the basswood board affected the experimental data. Moreover, the color of different parts of the same wooden board was uneven, which caused the measured individual color-difference values to be too large compared to other color-difference values. In previous studies, the application of different microcapsule materials in thermochromic coatings showed various color-difference effects. For example, Li et al. [50] studied color-changing microcapsules using urea-formaldehyde resin as wall material and found that the color-difference change was most significant at 80 °C, with the maximum color difference reaching 15.80. However, in this study, the color difference of the 1# microcapsules using chitosan as the wall material reached 21.0 at 25 °C, which shows that chitosan, as a green microcapsule wall material, has higher discoloration sensitivity at relatively low temperatures and is especially suitable for application scenarios near room temperature.

Table 7. The color difference of the surface coating on basswood board with 1# microcapsule at different temperatures.

Microcapsule Content (%)	Colorimetric Parameters	−5 °C	0 °C	5 °C	10 °C	15 °C	20 °C	25 °C	30 °C	35 °C	40 °C	45 °C	50 °C	55 °C
0.0	L	81.8	81.9	82.2	82.4	81.9	81.8	81.9	82.0	82.8	82.4	82.6	82.8	81.8
	a	6.8	6.9	6.9	7.0	7.1	7.0	7.0	6.8	7.0	6.9	6.9	7.0	6.9
	b	29.7	29.8	29.9	30.1	29.9	29.8	29.8	30.1	30.1	29.9	29.9	30.0	30.5
	ΔE	-	0.2	0.5	0.8	0.4	0.2	0.2	0.5	1.1	0.6	0.8	1.1	0.8
1.0	L	82.8	82.6	81.3	81.0	81.7	81.2	79.3	80.3	83.2	81.1	81.2	81.4	82.4
	a	6.2	6.0	5.7	6.4	6.1	5.6	7.2	7.1	5.6	5.8	6.3	7.0	6.0
	b	41.6	41.6	41.4	40.4	42.5	41.7	38.9	38.7	39.5	39.9	41.0	42.1	40.5
	ΔE	-	0.3	1.6	2.2	1.6	1.7	4.5	3.9	2.1	2.5	1.7	1.7	1.2
2.0	L	78.2	80.4	71.5	72.8	71.1	78.7	72.2	71.7	71.2	71.3	71.9	71.7	73.0
	a	8.8	7.2	10.2	9.2	9.6	7.9	10.1	8.2	9.2	9.2	10.6	8.7	9.2
	b	42.1	43.9	38.4	37.0	40.8	45.3	37.2	36.3	36.4	35.2	37.9	38.0	37.9
	ΔE	-	3.3	6.9	7.4	7.3	3.4	7.3	8.7	9.0	9.8	7.8	7.7	6.7
3.0	L	77.9	78.5	77.7	72.7	71.8	71.4	69.4	68.8	69.9	71.9	68.4	68.4	69.3
	a	5.4	5.6	4.9	6.9	6.8	7.3	9.4	10.4	11.8	6.4	6.7	7.3	8.3
	b	48.8	48.7	48.9	42.4	44.9	46.9	41.8	42.0	40.8	40.2	40.7	40.3	42.5
	ΔE	-	0.6	0.6	8.4	7.4	7.0	12.5	8.5	13.0	10.5	12.6	12.9	11.1
4.0	L	82.0	82.1	74.8	74.4	80.6	73.0	67.9	68.2	68.5	69.0	67.9	72.5	72.8
	a	5.9	5.3	5.7	5.7	5.4	7.7	7.0	6.9	9.3	11.9	7.2	7.1	5.7
	b	54.9	52.7	54.3	53.9	54.1	47.2	39.6	38.2	39.7	40.2	39.6	42.3	42.1
	ΔE	-	2.3	7.2	7.7	1.7	12.0	21.0	21.7	17.0	19.8	20.9	15.8	15.8
5.0	L	77.8	76.2	78.7	73.9	74.5	81.7	75.4	75.6	75.2	76.6	75.4	78.0	76.1
	a	4.8	4.9	4.4	5.6	6.7	4.2	7.3	7.0	7.3	7.2	7.8	8.3	12.8
	b	56.6	56.6	56.7	55.5	55.2	59.6	56.3	56.3	55.2	58.7	57.2	59.6	58.9
	ΔE	-	1.6	1.5	4.2	4.1	5.0	6.5	3.1	3.9	3.4	3.9	4.8	8.5
7.0	L	77.1	76.7	74.8	71.4	71.5	70.3	69.1	67.6	67.4	68.4	67.5	67.3	68.6
	a	7.2	6.9	7.5	8.2	8.2	9.3	7.3	8.7	8.3	8.0	9.5	7.9	9.1
	b	61.1	61.0	60.9	55.1	54.8	58.4	54.1	52.7	52.9	54.4	52.4	52.8	54.8
	ΔE	-	0.5	2.3	8.3	8.5	7.6	10.6	12.8	12.8	11.0	13.1	12.6	10.8

Table 8. The color difference of the surface coating on basswood board with 2# microcapsule at different temperatures.

Microcapsule Content (%)	Colorimetric Parameters	−5 °C	0 °C	5 °C	10 °C	15 °C	20 °C	25 °C	30 °C	35 °C	40 °C	45 °C	50 °C	55 °C
0.0	L	81.8	81.9	82.2	82.4	81.9	81.8	81.9	82.0	82.8	82.4	82.6	82.8	81.8
	a	6.8	6.9	6.9	7.0	7.1	7.0	7.0	6.8	7.0	6.9	6.9	7.0	6.9
	b	29.7	29.8	29.9	30.1	29.9	29.8	29.8	30.1	30.1	29.9	29.9	30.0	30.5
	ΔE	-	0.2	0.5	0.8	0.4	0.2	0.2	0.5	1.1	0.6	0.8	1.1	0.8
1.0	L	73.0	74.6	69.0	73.0	73.8	69.5	69.8	68.3	68.8	68.7	69.4	71.0	72.1
	a	8.1	8.0	9.0	8.3	8.7	8.4	8.0	8.1	9.2	9.4	6.8	8.4	9.4
	b	21.3	21.5	21.2	22.2	23.1	19.3	20.9	19.7	18.6	20.3	20.8	21.6	21.0
	ΔE	-	1.6	4.1	0.9	1.0	4.0	3.5	5.0	5.1	4.6	3.9	2.0	1.7
2.0	L	74.9	75.3	74.7	74.8	73.8	82.7	78.4	79.3	78.1	78.5	78.7	84.6	78.6
	a	8.6	8.7	8.3	8.1	9.8	6.3	7.1	6.5	5.7	6.4	5.9	5.9	7.2
	b	28.8	28.8	28.0	33.1	28.0	32.7	29.6	30.1	28.8	26.9	29.1	32.4	28.9
	ΔE	-	0.4	1.0	4.3	1.7	9.0	3.9	5.1	4.3	4.6	2.8	10.7	3.8
3.0	L	72.1	71.8	70.7	73.9	74.5	74.2	70.9	71.7	71.1	71.5	71.9	70.7	71.1
	a	10.9	13.8	10.1	9.2	9.2	9.5	10.3	9.8	10.7	14.8	12.1	9.0	10.0
	b	22.0	23.1	20.9	20.4	23.1	24.0	21.4	21.1	21.8	22.7	22.8	19.6	21.1
	ΔE	-	3.1	2.0	3.0	3.1	3.2	1.5	1.5	0.5	4.0	1.5	3.4	1.6
4.0	L	76.1	80.8	82.0	74.1	80.2	81.3	78.1	70.7	71.8	74.1	73.7	73.0	73.7
	a	6.3	6.2	5.3	8.7	6.9	5.4	8.1	7.6	7.7	8.1	8.9	8.4	9.4
	b	31.1	30.9	32.3	25.7	31.6	32.3	28.0	25.6	28.8	24.0	23.3	23.2	24.4
	ΔE	-	1.8	6.1	6.2	4.1	5.4	4.1	7.8	5.1	7.6	5.6	8.8	7.8
5.0	L	76.6	84.2	76.2	76.4	77.1	83.8	72.6	69.8	71.5	75.3	70.0	71.4	75.3
	a	6.9	5.5	7.7	7.8	7.9	8.8	8.2	8.9	10.7	8.7	11.0	7.1	9.0
	b	22.6	21.5	20.7	23.6	25.7	21.6	23.7	21.4	25.6	25.5	23.0	22.7	25.6
	ΔE	-	7.7	2.2	1.4	3.3	7.5	4.4	7.2	7.3	3.7	7.8	5.2	3.9
7.0	L	74.6	73.0	74.8	78.3	71.8	71.8	73.9	72.7	74.2	77.2	76.9	73.1	78.9
	a	9.1	10.2	8.4	8.6	7.6	10.2	15.4	9.0	13.0	7.9	6.2	6.4	4.6
	b	34.1	32.6	29.4	35.4	34.6	31.2	33.1	31.1	33.3	33.8	40.0	31.8	36.1
	ΔE	-	2.5	4.8	4.0	3.2	4.2	6.4	3.7	4.0	3.4	7.0	3.9	6.5

Significance analysis was performed on the data in Tables 7 and 8. Based on the coating-color-difference data obtained at a temperature of 25 °C, the non-repeated two-way ANOVA method was used for significance analysis. Three values of the F, the p-value, and the F_{crit} were obtained, respectively. It should be noted that the F represents the test statistic, a statistical measure used in hypothesis testing calculations. The p-value represents the significance level, which evaluates the range and interval of the overall parameter and evaluates the probability that the experiment may occur. The F_{crit} is the critical value of F at the corresponding significance level. Among them, $F > F_{crit}$ indicates that there is a difference between the two sets of data. $F < F_{crit}$ means there is no difference between the two sets of data. The standard for determining significant differences is $0.01 < p\text{-value} < 0.05$, indicating a significant difference. A $p\text{-value} \leq 0.01$ means the difference is extremely significant. $p\text{-value} > 0.05$ means not significant. As shown in Table 9, the results obtained according to the above method are $F < F_{crit}$ and $p\text{-value} > 0.05$, indicating that the content of microcapsules in the paint film has no significant impact on the color difference of the coating. This may be because the change in microcapsule content has not reached the critical point of producing significant color difference. Although the change in content has a certain effect, it is not enough to show a statistically significant difference. Although the crosslinking agent species did not reach the threshold of $p\text{-value} < 0.05$ in the significance analysis, the p-value close to 0.05 indicates that it may have an impact on the coating performance.

Table 9. Significance analysis of color difference.

Difference Source	SS	d_f	MS	F	p-Value	F_{crit}
content	176.0271429	6	29.33785714	1.541605675	0.306188138	4.283865714
crosslinking agent	106.4257143	1	106.4257143	5.592313178	0.055915742	5.987377607
error	114.1842857	6	19.03071429			
total	396.6371429	13				

3.3.2. Effect of Microcapsule Content on the Glossiness of Surface Coating on Basswood Board

The changes in the glossiness of the surface coating on basswood board with different addition amounts of 1# microcapsules and 2# microcapsules added to the UV primer are shown in Table 10. The glossiness of the surface coating on basswood board with 1# microcapsule added increased as the amount of microcapsule added increased, and the change trend was relatively gentle. The glossiness of the surface coating on basswood board with 2# microcapsules added showed an overall trend of first increasing and then decreasing as the amount of microcapsules added increased. When the microcapsule addition amount was 3%, the glossiness of the coating was the highest. This was due to the microcapsule powder was relatively coarse. When the microcapsule content was between 3% and 7%, as the microcapsule content increased, the particle size of the coating increased, the diffuse reflection of the coating surface increased, and the glossiness of the coating was reduced. Cellulose was unevenly distributed in the microcapsule wall material and was more prone to agglomeration. However, because it was added to the UV primer, its impact on the overall paint film was relatively small. In summary, the effects of 1# and 2# microcapsules on the gloss of the coating were relatively weak.

Table 10. Gloss of the coating for different microcapsule content.

Sample	Microcapsule Content (%)	Gloss at 20° (GU)	Gloss at 60° (GU)	Gloss at 85° (GU)
Basswood surface coating with 1# microcapsules added	0.0	9.3	37.7	34.8
	1.0	8.6	36.0	38.3
	2.0	9.6	38.7	45.9
	3.0	14.8	46.5	43.5
	4.0	12.7	45	44.1
	5.0	10.4	40.2	43.8
	7.0	17.1	49.2	55.5
Basswood surface coating with 2# microcapsules added	0.0	9.3	37.7	34.8
	1.0	25.3	64.9	68.2
	2.0	36.0	76.9	81.2
	3.0	32.0	66.2	69.3
	4.0	22.9	61.4	60.0
	5.0	17.9	60.0	63.5
	7.0	16.3	53.2	67.4

3.3.3. Effect of Microcapsule Content on the Reflectivity of the Surface Coating on Basswood Board

The reflectivity values and reflectivity curves in the visible light band on the surface of basswood with different contents of 1# microcapsules added are shown in Table 11 and Figure 7, respectively. The reflectivity of the visible light band on the surface on basswood board with different contents of 1# microcapsules added slightly decreased with the increase in microcapsule content. However, the spacing between each group of curves was very close, and the numerical differences were small. This showed that the influence of 1# microcapsules on the visible light band reflectivity of the surface coating on basswood board was not significant.

Table 11. The effect of adding 1# microcapsule on the surface of basswood board.

Sample	Microcapsule Content (%)	Visible Light Reflectance (%)
Basswood surface coating with 1# microcapsules added	0	64.57
	1.0	60.33
	2.0	58.66
	4.0	57.06
	5.0	59.45
	7.0	54.64

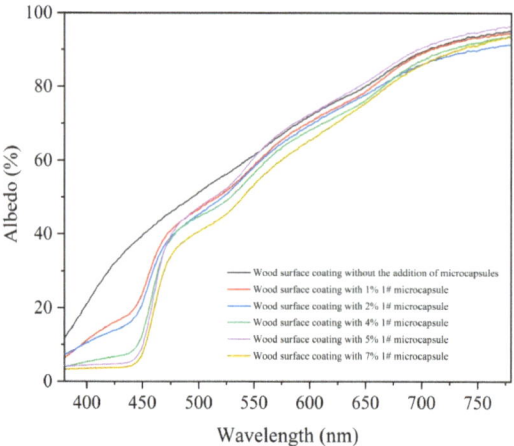

Figure 7. The reflectance of the visible light band of the 1# microcapsule basswood board was added.

3.4. Analysis of the Effect of Microcapsule Content on the Mechanical Properties of Surface Coating on Basswood Board

The effect of microcapsule content on the mechanical properties of surface coating on basswood board is shown in Table 12. Among the two groups of samples, the hardness of the surface coating on basswood board with 1# microcapsules added increased most significantly. When the addition amount was 3.0%, the coating hardness reached 4H. When the microcapsule content was 7.0%, the hardness of both coatings decreased slightly, and the hardness was H. According to the data in the Table 12, the adhesion level of the surface coating on basswood board with different contents of the two microcapsules added was 5, which was the same as the adhesion level of the surface coating on basswood board without adding microcapsules, indicating that the adhesion of the coating had not changed. This showed that when the content of the two microcapsules was 7.0% or less, it would not affect the adhesion of the coating on the surface of basswood boards and could be used in practice. The impact resistance level of the surface of basswood without microcapsules was level 3, which means that there were one to two circles of slight cracks on the coating surface. The impact resistance level of the surface coating on basswood board with the addition of the two types of microcapsules remained at levels 3 and 4, which showed that neither of the two microcapsules would have a significant impact on the impact resistance of the basswood board surface.

Table 12. Effect of microcapsule content on mechanical properties of coating surface.

Sample	Microcapsule Content (%)	Hardness	Adhesion Level (Level)	Impact Resistance Level (Level)
Basswood surface coating with 1# microcapsules added	0.0	2H	5	3
	1.0	2H	5	4
	2.0	2H	5	4
	3.0	4H	5	3
	4.0	3H	5	3
	5.0	H	5	4
	7.0	H	5	4
Basswood surface coating with 2# microcapsules added	0.0	2H	5	3
	1.0	4H	5	3
	2.0	3B	5	4
	3.0	H	5	3
	4.0	4H	5	3
	5.0	2H	5	4
	7.0	H	5	4

4. Conclusions

Two kinds of chitosan thermochromic microcapsules prepared with syringaldehyde and STPP crosslinking agent were added into UV primer and applied on the surface of basswood board. The morphology, chemical composition, optical properties, and mechanical properties of the surface coating on basswood boards were tested. The results showed that the 1# microcapsule coating prepared with syringaldehyde had a better color-changing effect than the 2# microcapsule coating prepared with STPP, and the color difference changed more obviously with temperature. When the content of microcapsules with syringaldehyde as the crosslinking agent was 4.0% and the temperature was 25 °C, the color difference of the surface coating on basswood board was the largest, reaching 21.0. With the increase in microcapsule content, the surface gloss of the basswood surface coating with microcapsules using syringaldehyde as the crosslinking agent showed an overall upward trend, and the change trend was relatively gentle. The reflectance of the visible light band of the coating decreased slightly. The hardness of the coating was significantly improved, with little change in adhesion level and impact resistance. With the increase in microcapsule content, the surface glossiness of the surface coating on basswood board with microcapsules using STPP as the crosslinking agent showed an overall trend of first increasing and then decreasing. The adhesion level and impact resistance of the coating changed little. This study demonstrated the potential of chitosan as a green and environmentally friendly material in thermochromic microcapsules, providing a new perspective for the application of eco-friendly materials in the field of coatings. Future research should be the further optimization of the preparation process conditions of microcapsules to enhance their color-changing effect and overall performance, thereby expanding their potential applications in coatings and other related fields.

Author Contributions: Conceptualization and methodology, writing—review and editing, J.H.; validation, B.Z.; resources, data management, H.F.; formal analysis, J.L.; investigation and supervision, X.Y. All authors have read and agreed to the published version of the manuscript.

Funding: This project was partly supported by the Natural Science Foundation of Jiangsu Province (BK20201386).

Institutional Review Board Statement: Not applicable.

Informed Consent Statement: Not applicable.

Data Availability Statement: Data are contained within the article.

Conflicts of Interest: The authors declare that there are no conflicts of interest.

References

1. Mai, C.; Schmitt, U.; Niemz, P. A brief overview on the development of wood research. *Holzforschung* **2022**, *76*, 102–119. [CrossRef]
2. Zhang, X.Y.; Xu, W.; Li, R.R.; Zhou, J.C.; Luo, Z.Y. Study on sustainable lightweight design of airport waiting chair frame structure based on ANSYS workbench. *Sustainability* **2024**, *16*, 5350. [CrossRef]
3. Hu, W.G.; Yu, R.Z. Mechanical and acoustic characteristics of four wood species subjected to bending load. *Maderas-Cienc. Tecnol.* **2023**, *25*, 39. [CrossRef]
4. Chen, Y.T.; Sun, C.S.; Ren, Z.R.; Na, B. Review of the Current State of Application of Wood Defect Recognition Technology. *BioResources* **2023**, *18*, 49. [CrossRef]
5. Zhu, J.G.; Niu, J.Y. Green Material Characteristics Applied to Office Desk Furniture. *Bioresources* **2022**, *17*, 2228–2242. [CrossRef]
6. Hu, W.G.; Luo, M.Y.; Hao, M.M.; Tang, B.; Wan, C. Study on the effects of selected factors on the diagonal tensile strength of oblique corner furniture joints constructed by wood dowel. *Forests* **2023**, *14*, 1149. [CrossRef]
7. Tao, C.L.; Gao, Z.X.; Cheng, B.D.; Chen, F.W.; Yu, C. Enhancing wood resource efficiency through spatial agglomeration: Insights from China's wood-processing industry. *Resour. Conserv. Recycl.* **2024**, *203*, 107453. [CrossRef]
8. Liu, Y.; Hu, W.G.; Kasal, A.; Erdil, Y.Z. The State of the Art of Biomechanics Applied in Ergonomic Furniture Design. *Appl. Sci.* **2023**, *13*, 12120. [CrossRef]
9. Hu, W.G.; Luo, M.Y.; Liu, Y.Q.; Xu, W.; Konukcu, A.C. Experimental and numerical studies on the mechanical properties and behaviors of a novel wood dowel reinforced dovetail joint. *Eng. Fail. Anal.* **2023**, *152*, 107440. [CrossRef]
10. Hu, W.G.; Fu, W.J.; Zhao, Y. Optimal design of the traditional Chinese wood furniture joint based on experimental and numerical method. *Wood Res.* **2024**, *69*, 50–59. [CrossRef]

11. Qin, Y.Z.; Yan, X.X. Preparation of Healable Shellac Microcapsules and Color-Changing Microcapsules and Their Effect on Properties of Surface Coatings on Hard Broad-Leaved Wood Substrates. *Coatings* **2022**, *12*, 991. [CrossRef]
12. Zhang, N.; Xu, W.; Tan, Y. Multi-attribute hierarchical clustering for product family division of customized wooden doors. *Bioresources* **2023**, *18*, 7889–7904. [CrossRef]
13. Wang, C.; Zhang, C.Y.; Zhu, Y. Reverse design and additive manufacturing of furniture protective foot covers. *Bioresources* **2024**, *19*, 4670–4678. [CrossRef]
14. Yan, X.X.; Chang, Y.J. Investigation of waterborne thermochromic topcoat film with color-changing microcapsules on Chinese fir surface. *Prog. Org. Coat.* **2019**, *136*, 105262. [CrossRef]
15. Jia, Z.; Bao, W.H.; Tao, C.Y.; Song, W.L. Reversibly photochromic wood constructed by depositing microencapsulated/polydimethylsiloxane composite coating. *J. Forestry Res.* **2022**, *33*, 1409–1418. [CrossRef]
16. Wang, X.M.; Wang, X.X.; Cui, Y.F. Methyl red modified crystal violet lactone microcapsules for natural and composite fabrics producing a violet to orange-red effect at low temperature. *Mater. Chem. Phys.* **2023**, *305*, 127901. [CrossRef]
17. Meng, Q.Y.; Zhong, S.L.; Wang, J.; Gao, Y.; Cui, X.J. Advances in chitosan-based microcapsules and their applications. *Carbohyd. Polym.* **2023**, *300*, 120265. [CrossRef]
18. Valle, J.A.B.; Valle, R.D.S.C.; Bierhalz, A.C.K.; Bezerra, F.M.; Hernandez, A.L.; Arias, M.J.L. Chitosan microcapsules: Methods of the production and use in the textile finishing. *J. Appl. Polym.* **2021**, *138*, e50482. [CrossRef]
19. Xiao, Z.B.; Ma, X.J.; Sun, P.L.; Kang, Y.X.; Niu, Y.W.; She, Y.B.; Zhao, D. Tomato preservation with essential oil microcapsules-chitosan coating. *J. Food Meas. Charact.* **2024**, *18*, 5928–5944. [CrossRef]
20. Chen, Z.; Zhang, W.W.; Yuan, Z.H.; Wang, Z.Y.; Ma, R.J.; Chen, K.L. Preparation of strawberry chitosan composite microcapsules and their application in textiles. *Colloids Surfaces A* **2022**, *652*, 129845. [CrossRef]
21. He, Z.C.; Bao, B.W.; Fan, J.; Wang, W.; Yu, D. Photochromie cotton fabric based on microcapsule technology with anti-fouling properties. *Colloids Surfaces A* **2020**, *594*, 124661. [CrossRef]
22. Teng, X.X.; Zhang, M.; Mujumdar, A.S.; Wang, H.Q. Garlic essential oil microcapsules prepared using gallic acid grafted chitosan: Effect on nitrite control of prepared vegetable dishes during storage. *Food Chem.* **2022**, *388*, 132945. [CrossRef] [PubMed]
23. Yin, C.; Huang, C.X.; Wang, J.; Liu, Y.; Lu, P.; Huang, L.J. Effect of Chitosan- and Alginate-Based Coatings Enriched with Cinnamon Essential Oil Microcapsules to Improve the Postharvest Quality of Mangoes. *Materials* **2019**, *12*, 2039. [CrossRef]
24. Cheng, Y.L.; Zhang, X.Q.; Fang, C.Q.; Chen, J.; Wang, Z. Discoloration mechanism, structures and recent applications of thermochromic materials via different methods: A review. *J. Mater. Sci. Technol.* **2018**, *34*, 2225–2234. [CrossRef]
25. Zhu, X.D.; Liu, Y.; Li, Z.; Wang, W.C. Thermochromic microcapsules with highly transparent shells obtained through *in-situ* polymerization of urea formaldehyde around thermochromic cores for smart wood coatings. *Sci. Rep.* **2018**, *8*, 4015. [CrossRef]
26. Hu, L.; Lyu, S.Y.; Fu, F.; Huang, J.D.; Wang, S.Q. Preparation and properties of multifunctional thermochromic energy-storage wood materials. *J. Mater. Sci.* **2016**, *51*, 2716–2726. [CrossRef]
27. Jiang, J.Y.; Ma, C.; Song, X.N.; Zeng, J.H.; Zhang, L.W.; Gong, P.M. Spray drying co-encapsulation of lactic acid bacteria and lipids: A review. *Trends Food Sci. Technol.* **2022**, *129*, 134–143. [CrossRef]
28. Ren, W.B.; Tian, G.F.; Zhao, S.J.; Yang, Y.; Gao, W.; Zhao, C.Y.; Zhang, H.J.; Lian, Y.H.; Wang, F.Z.; Du, H.J.; et al. Effects of spray-drying temperature on the physicochemical properties and polymethoxyflavone loading efficiency of citrus oil microcapsules. *LWT-Food Sci. Technol.* **2020**, *133*, 109954. [CrossRef]
29. Wang, B.; Adhikari, B.; Mathesh, M.; Yang, W.R.; Barrow, C.J. Anchovy oil microcapsule powders prepared using two-step complex coacervation between gelatin and sodium hexametaphosphate followed by spray drying. *Powder Technol.* **2019**, *358*, 68–78. [CrossRef]
30. Wu, S.S.; Zhou, J.C.; Xu, W. A convenient approach to manufacturing lightweight and high-sound-insulation plywood using furfuryl alcohol/multilayer graphene oxide as a shielding layer. *Wood Mater. Sci. Eng.* **2024**. [CrossRef]
31. *GB/T 11186.3-1989*; Methods for Measuring the Colour of Paint Films. Part III: Calculation of Colour Differences. Standardization Administration of the People's Republic of China: Beijing, China, 1990.
32. Wang, C.; Zhou, Z.Y. Optical properties and lampshade design applications of PLA 3D printing materials. *Bioresources* **2023**, *18*, 1545–1553. [CrossRef]
33. Hu, J.; Liu, Y.; Wang, J.X.; Xu, W. Study of selective modification effect of constructed structural color layers on European beech wood surfaces. *Forests* **2024**, *15*, 261. [CrossRef]
34. Hu, J.; Liu, Y.; Xu, W. Influence of Cell Characteristics on the Construction of Structural Color Layers on Wood Surfaces. *Forests* **2024**, *15*, 676. [CrossRef]
35. *GB/T 4893.6-2013*; Test of Surface Coatings of Furniture. Part VI: Determination of Gloss Value. Standardization Administration of the People's Republic of China: Beijing, China, 2013.
36. Hu, J.; Liu, Y.; Xu, W. Impact of cellular structure on the thickness and light reflection properties of structural color layers on diverse wood surfaces. *Wood Mater. Sci. Eng.* **2024**. [CrossRef]
37. Wang, C.; Yu, J.H.; Jiang, M.H.; Li, J.Y. Effect of slicing parameters on the light transmittance of 3D-printed polyethylene terephthalate glycol products. *Bioresources* **2024**, *19*, 500–509. [CrossRef]
38. Hu, W.G.; Luo, M.Y.; Yu, R.Z.; Zhao, Y. Effects of the selected factors on cyclic load performance of T-shaped mortise-and-tenon furniture joints. *Wood Mater. Sci. Eng.* **2024**. [CrossRef]

39. Wang, C.; Yu, J.H.; Jiang, M.H.; Li, J.Y. Effect of selective enhancement on the bending performance of fused deposition methods 3D-printed PLA models. *Bioresources* **2024**, *19*, 2660–2669. [CrossRef]
40. Hu, W.G.; Yu, R.Z. Study on the strength mechanism of the wooden round-end mortise-and-tenon joint using the digital image correlation method. *Holzforschung* **2024**. [CrossRef]
41. Wu, W.; Xu, W.; Wu, S.S. Mechanical performance analysis of double-dovetail joint applied to furniture T-shaped components. *BioResources* **2024**, *19*, 5862–5879. [CrossRef]
42. Wang, C.; Zhang, C.Y.; Ding, K.Q.; Jiang, M.H. Immersion polishing post-treatment of PLA 3D printed formed parts on its surface and mechanical performance. *BioResources* **2023**, *18*, 7995–8006. [CrossRef]
43. GB/T 6739-2006; Paint and Varnishes-Determination of Film Hardness by Pencil Test. Standardization Administration of the People's Republic of China: Beijing, China, 1998.
44. GB/T 4893.4-2013; Test of Surface Coatings of Furniture. Part IV: Determination of Adhesion-Cross Cut. Standardization Administration of the People's Republic of China: Beijing, China, 2013.
45. GB/T 4893.9-1992; Furniture-Test for surfaces-Assessment of resistance to impact. Standardization Administration of the People's Republic of China: Beijing, China, 1992.
46. Li, D.W.; Cui, H.P.; Hayat, K.; Zhang, X.M.; Ho, C.T. Superior environmental stability of gelatin/CMC complex coacervated microcapsules via chitosan electrostatic modification. *Food Hydrocoll.* **2022**, *124*, 107341. [CrossRef]
47. Sripetthong, S.; Nalinbenjapun, S.; Basit, A.; Surassmo, S.; Sajomsang, W.; Ovatlarnporn, C. Preparation of Self-Assembled, Curcumin-Loaded Nano-Micelles Using Quarternized Chitosan-Vanillin Imine (QCS-Vani Imine) Conjugate and Evaluation of Synergistic Anticancer Effect with Cisplatin. *J. Funct. Biomater.* **2023**, *14*, 525. [CrossRef] [PubMed]
48. Mao, S.; Zhang, L.L.; Feng, J.Y.; Han, P.; Lu, C.W.; Zhang, T.H. Development of pH-responsive intelligent and active films based on pectin incorporating Schiff base (Phenylalanine/syringaldehyde) for monitoring and preservation of fruits. *Food Chem.* **2024**, *435*, 137626. [CrossRef]
49. Liu, H.S.; Deng, Y.H.; Ye, Y.; Liu, X.Q. Reversible Thermochromic Microcapsules and Their Applications in Anticounterfeiting. *Materials* **2023**, *16*, 5150. [CrossRef] [PubMed]
50. Li, W.B.; Yan, X.X.; Zhao, W.T. Preparation of Crystal Violet Lactone Complex and Its Effect on Discoloration of Metal Surface Coating. *Polymers* **2022**, *14*, 4443. [CrossRef] [PubMed]

Disclaimer/Publisher's Note: The statements, opinions and data contained in all publications are solely those of the individual author(s) and contributor(s) and not of MDPI and/or the editor(s). MDPI and/or the editor(s) disclaim responsibility for any injury to people or property resulting from any ideas, methods, instructions or products referred to in the content.

Article

Effect of Two Types of Pomelo Peel Flavonoid Microcapsules on the Performance of Water-Based Coatings on the Surface of Fiberboard

Jinzhe Deng [1,2], Tingting Ding [1,2] and Xiaoxing Yan [1,2,*]

[1] Co-Innovation Center of Efficient Processing and Utilization of Forest Resources, Nanjing Forestry University, Nanjing 210037, China; dengjinzhe@njfu.edu.cn (J.D.); dingtingting@njfu.edu.cn (T.D.)
[2] College of Furnishings and Industrial Design, Nanjing Forestry University, Nanjing 210037, China
* Correspondence: yanxiaoxing@njfu.edu.cn

Abstract: In order to achieve antibacterial properties in water-based coatings, two types of antibacterial pomelo peel flavonoid microcapsules were added to water-based coatings and decorated on the surface of fiberboard. The surface coatings of the substrates were tested and analyzed. The antibacterial rate of the surface coatings of the two groups of fiberboards gradually increased with the increase in the content of the microcapsules. The color difference of the surface coatings of both groups increased slightly, the glossiness decreased, the gloss loss rate increased greatly, and the reflectivity increased slightly. The adhesion of the surface coatings of the two groups of fiberboards did not change significantly, the roughness gradually increased, the hardness of the melamine-resin-coated pomelo peel flavonoid microcapsules gradually increased, and the impact resistance slightly improved. Compared with the antibacterial results of the coating without substrate at the same content, the antibacterial effect of the fiberboard surface coating was slightly decreased. Overall, the surface coating on the fiberboard with 9.0% chitosan-coated pomelo peel flavonoid microcapsules demonstrated superior performance, superior coating morphology, and enhanced antibacterial properties. The antibacterial rate was 73.7% against *Escherichia coli*, and the antibacterial rate was 77.4% against *Staphylococcus aureus*. The color difference was 3.85, the gloss loss rate was 90.0%, and the reflectivity was 20.19%. The hardness was HB, the adhesion was level 1, the impact resistance level was 3, and the roughness was 1.94 µm. This study explored the effect of antibacterial microcapsules on coating performance, providing a technical basis for the application of the antibacterial microcapsules.

Keywords: microcapsules; antibacterial; pomelo peel flavonoids; wooden material; water-based coatings

Citation: Deng, J.; Ding, T.; Yan, X. Effect of Two Types of Pomelo Peel Flavonoid Microcapsules on the Performance of Water-Based Coatings on the Surface of Fiberboard. *Coatings* 2024, 14, 1032. https://doi.org/10.3390/coatings14081032

Academic Editor: Marko Petric

Received: 15 July 2024
Revised: 1 August 2024
Accepted: 12 August 2024
Published: 14 August 2024

Copyright: © 2024 by the authors. Licensee MDPI, Basel, Switzerland. This article is an open access article distributed under the terms and conditions of the Creative Commons Attribution (CC BY) license (https://creativecommons.org/licenses/by/4.0/).

1. Introduction

In order to create a more comfortable life, the study of furniture is very important [1–5]. Wood is widely used in furniture because of its natural beauty and ease of processing [6–9], but it is easy to breed bacteria during daily use [10,11]. Methods such as coating and substrate pretreatment are often used to improve the substrate performance of the furniture [12–19]. A water-based coating is an environmentally friendly coating with water as a solvent, which reduces the use of organic solvents and environmental pollution compared with traditional paints. However, water-based coatings have problems such as low hardness and single functionality. The enhanced antibacterial function of coatings can effectively reduce bacterial residue on the surface of household products, but more in-depth research is required. The addition of microcapsules gives the coatings antibacterial properties, significantly reducing bacteria on furniture products and further optimizing people's living spaces, which has practical application value.

Microcapsule technology refers to the encapsulation of the capsule wall material on the surface of the core material to form tiny spherical particles with micro-pores [20–22]. Through microcapsule technology, natural substances with antibacterial effects extracted from animals and plants can be wrapped and prepared into microcapsules, making them more stable, easier to process and use, and improving their wide application in life. Melamine resin is a common microcapsule wall material with an easy production process and relatively low cost, which has significant advantages in the preparation of microcapsules [23–25]. Although it has been shown in other fields that formaldehyde-free resins prepared by modification of melamine resins can be used as effective retanning agents for the production of formaldehyde-free leathers [26,27], this has been less reported in the field of microcapsule preparation. Chitosan, which is a natural substance, has antibacterial properties. As the wall material of microcapsules, chitosan is environmentally friendly and also makes the whole test process more in line with the concept of sustainable development. Chitosan has become one of the most important candidates for microcapsule wall materials because of its natural nontoxicity, biocompatibility, and degradability. Chitosan is a modified natural hydrocarbon polymer extracted from crustaceans like crabs and shrimps, among others, and has attracted much attention in different fields [28–31]. Zhang et al. [32] prepared chitosan nano-capsules of a cumin essential oil using the ionic gelation method and characterized and evaluated the antibacterial properties of these nano-capsules. It was shown that the microcapsules with a slow release of the encapsulated core were effective in inhibiting *Escherichia coli* and *Listeria monocytogenes* over a relatively long period of time. Flavonoids are widely distributed in plants and also have antibacterial properties [33,34]. Most flavonoids combine with sugars to form glycosides in the form of sugar ligands, and in a pomelo peel, the main one is 5,7,4,-trihydroxyflavonoids, naringin at the 7-position carbon [35]. Because the pomelo peel contains many flavonoids, it is often used to extract flavonoids [36]. The flavonoids in pomelo peel have a wide range of antibacterial effects and have inhibitory effects on a variety of bacteria and fungi. The extraction steps of the flavonoids from pomelo peel mainly include crushing, extraction, concentration, drying, and so on, which have broad application prospects. The safe and efficient antioxidant, antibacterial, and anti-corrosion active substances extracted from the pomelo peel are in line with the concepts of green environmental protection and healthy living. Zhang et al. [37] found that a grapefruit essential oil could reduce the reactive oxygen species production induced by *Staphylococcus aureus* metabolites and concluded that the flavonoid extracts from the grapefruit peels had a significant inhibitory effect on *Staphylococcus aureus*. The flavonoids extracted from the pomelo peel are usually large crystals and cannot be directly applied to coatings. By using microcapsule technology, the pomelo peel flavonoids can be prepared into microcapsules and added to water-based coatings.

Wood furniture can be divided into solid wood substrates and artificial board substrates according to the type of substrate [38–40]. Medium-density fiberboard (MDF) is a commonly used artificial board made from wood or plant fibers. Its structure is more uniform than that of natural wood, with less expansion and contraction, which can avoid problems such as decay and insect infestation. The MDF surface is flat, and the production process is simple. Therefore, MDF is often used in furniture. But the moisture resistance of MDF is poor, moisture in the air is easily absorbed, and humidity makes it a breeding ground for bacteria [41–43]. These issues increase the importance of antibacterial effects on the MDF surface in daily use. Coatings with antibacterial properties can reduce the bacteria on the surface of wood products and have practical application value. According to the pre-test analysis, it was found that the combination properties were poor when the content of microcapsules in the coating exceeded 9.0%. Two types of microcapsules with different wall materials were added to the water-based coatings at 0%, 1.0%, 3.0%, 5.0%, 7.0%, and 9.0%. When microcapsules are added to the coating, the oil absorption rate of the microcapsules will have a certain effect on the viscosity of the coating. The various properties of fiberboard surfaces with different microcapsule contents were tested to ensure that the coatings had antibacterial properties along with good mechanical and optical properties. The test results of each sample were also analyzed to explore the effect of antibacterial microcapsules on the coating properties, providing a technical basis for antibacterial microcapsules in the coating. In this experiment, in response to the problem that it is difficult to add the pomelo peel flavonoids directly into the water-based coatings, the pomelo peel flavonoid microcapsules successfully retained their original antibacterial properties and improved the antibacterial performance of water-based coatings for wood products when directly added to the coating.

2. Materials and Methods
2.1. Materials

The pomelo peel used in the test was taken from Sha Tin pomelo from Yulin, China. The degree of chitosan deacetylation was 80.0%–95.0%. The fiberboards were MDF with specifications of 50 mm × 50 mm × 6 mm, purchased from Daya Wood-based Panel Group Co., Ltd., Zhenjiang, China. The primer and topcoat used in the test were water-based acrylic varnishes (the non-volatile content of the material was 30.0%–35.0%), provided by Jiangsu Haitian Technology Co., Ltd., Nanjing, China. The materials required are shown in Table 1. The instruments used in the experiment are shown in Table 2.

Table 1. Materials required for the test.

Material	Molecular Formula	CAS No.	Purity	Manufacturer
Melamine	$C_3H_6N_6$	108-78-1	AR	Tianjin Huasheng Chemical Reagent Co., Ltd., Tianjin, China
37% Formaldehyde solution	-	-	-	Xilong Scientific Co., Ltd., Shantou, China
Triethanolamine	$C_6H_{15}NO_3$	102-71-6	AR	Sinopharm Chemical Reagent Co., Ltd., Shanghai, China
Chitosan	$(C_6H_{11}NO_4)_n$	9012-76-4	BR	Tianjin Huasheng Chemical Reagent Co., Ltd., Tianjin, China
Acetic acid	CH_3COOH	64-19-7	AR	Sinopharm Chemical Reagent Co., Ltd., Shanghai, China
Tween-80	$C_{24}H_{44}O_6$	9005-65-6	AR	Shanghai Puyu Industrial Technology Co., Ltd., Shanghai, China
Sodium tripolyphosphate	$Na_5P_3O_{10}$	7758-29-4	AR	Shandong Xiya Chemical Co., Ltd., Linyi, China
Citric acid monohydrate	$C_6H_{10}O_8$	5949-29-1	AR	Suzhou Changjiu Chemical Technology Co., Ltd., Suzhou, China
Sodium dodecyl benzene sulfonate	$C_{18}H_{29}NaO_3S$	25155-30-0	AR	Anhui Jinyueguan New Material Technology Co., Ltd., Huaibei, China
Anhydrous ethanol	C_2H_6O	64-17-5	AR	Xilong Scientific Co., Ltd., Shantou, China
0.5 mol/L NaOH standard solution	-	-	-	Phygene Biotechnology Co., Ltd., Fuzhou, China
Escherichia coli	-	-	-	Beijing Pharma and Biotech Center, Beijing, China
Staphylococcus aureus	-	-	-	Beijing Pharma and Biotech Center, Beijing, China

Table 2. Experimental equipment.

Equipment	Model of Equipment	Manufacturer
Thermostatic water bath	DF-101Z	Zhengzhou Tengyue Communication Equipment Co., Ltd., Zhengzhou, China
Scanning electron microscope	Quanta-200	Thermo Fisher Scientific, Waltham, MA, USA
Fourier transform infrared spectrometer	VERTEX 80V	Bruker Corporation, Karlsruhe, Germany
Constant temperature and humidity chamber	HWS-50	Shanghai Shangyi Instrument Equipment Co., Ltd., Shanghai, China
Electrothermal constant temperature blast drying oven	DHG-9643BS-BS-III	Shanghai Xinmiao Medical Treatment Apparatus Manufacturing Co., Ltd., Shanghai, China
Rotary evaporator	SN-RE-201D	Shanghai Shangyi Instrument Equipment Co., Ltd., Shanghai, China
Freeze-dryer	YTLG-10A	Shanghai Yetuo Technology Co., Ltd., Shanghai, China
Bacterial colony counter	XK97-A	Hangzhou Qiwei Instrument Co., Ltd., Hangzhou, China
Portable color difference meter	SC-10	Shanghai Hechen Scientific Instrument Co., Ltd., Shanghai, China
Gloss meter	X-rite ci60	Shenzhen Latte laboratory equipment Co., Ltd., Shenzhen, China
Ultraviolet spectrophotometer	U-3900	Hitachi Scientific Instruments (Beijing) Co., Ltd., Beijing, China
Precision roughness tester	JB-4C	Shanghai Taiming Optical Instrument Co., Ltd., Shanghai, China
Portable paint film hardness tester	HT-6510P	Quzhou Aipu Measuring Instrument Co., Ltd., Quzhou, China
Paint Impactor	QCJ-40	Quzhou Aipu Measuring Instrument Co., Ltd., Quzhou, China
Grid adhesion tester	QFH-A	Quzhou Aipu Measuring Instrument Co., Ltd., Quzhou, China

2.2. Preparation of Microcapsules

(1) Preparation of melamine-resin-coated pomelo peel flavonoid microcapsules (MR-CPPFMs)

When the MRCPPFMs were prepared, 37% of the formaldehyde solution and melamine were used to prepare wall material prepolymers. Then the core material emulsion was prepared with pomelo peel flavonoid and Tween 80. Finally, the wall material prepolymer solution was added drop by drop into a beaker containing the core material emulsion. The beaker was put into a thermostatic water bath, and the solution in the beaker was continuously stirred. The specific preparation steps can be referred to in Ref. [44].

(2) Preparation of chitosan-coated pomelo peel flavonoid microcapsules (CCPPFMs)

The CCPPFMs were prepared with a 2:1 oil phase compared to the aqueous phase in the mixed system, and the microcapsules were prepared in four steps.

Aqueous phase solution preparation: When preparing the 1% acetic acid solution, a certain mass of acetic acid and deionized water were added to a beaker. A total of 0.80 g of chitosan was dissolved with a 1% acetic acid solution of 39.20 g in the beaker, and the beaker was placed in the thermostatic water bath, which was set to 60 °C at 600 rpm, continuously mixed, and stirred for 1 h to prepare a 2 wt% chitosan solution, which was used as the microcapsule wall material solution.

Oil phase solution preparation: The emulsifier solution was made from 0.72 g of Tween-80 and 71.28 g of deionized water. The core material solution with a concentration of 10% was obtained by dissolving 0.80 g of pomelo peel flavonoids in 7.20 g of anhydrous ethanol. The core material solution was added to the beaker with the emulsifier solution, and the beaker was placed in the thermostatic water bath. The core material lotion was obtained by stirring continuously for 40 min at 60 °C at 600 rpm.

Oil phase mixed with the aqueous phase: The core emulsion was slowly added dropwise to the beaker with the wall material solution using a dropper. The beaker was placed in the thermostatic water bath at 60 °C, and the mixing reaction was continued for 1 h at a speed of 600 rpm.

Crosslinking reaction of microcapsule wall material: The pH value of the solution was adjusted to 7.5, and a 0.5 mol/L NaOH solution was used. A total of 0.80 g of STPP was added to the beaker and subjected to a crosslinking reaction with the wall material. The beaker was placed in the 60 °C thermostatic water bath and mixed continuously for 3 h at a speed of 600 rpm. The solution after the mixed reaction was precipitated, filtered, and then placed in the freeze-dryer. After 48 h of being freeze-dried, it was ground to obtain the CCPPFMs.

2.3. Coating Preparation Method

When the MDFs were coated using the hand finishing method, a total of 4–6 layers were usually applied, and the coating amount of each layer on the surface of the material was usually 60 g/m^2–80 g/m^2. In the actual application, there will be problems such as loss error, so the real coating amount will be greater than the theoretical coating amount. Therefore, the coating amount was set at 80 g/m^2 per layer, with two coats of the primer and two coats of the topcoat, and the total amount of coating used was 320 g/m^2. A coating preparation device was used to control the coating thickness to about 80 μm. The actual total amount of coating on each fiberboard was set at 1.8 times the theoretical amount. On average, the actual coating mass per single layer of fiberboard was 0.36 g, and the total mass was 1.44 g. The microcapsules were added to the topcoat during testing because bacteria typically adhere to the surface first. The addition of microcapsules to the coating increased the non-volatile content of the coating. The specific mass of coatings and microcapsules is shown in Table 3.

Table 3. The specific mass of the coatings and microcapsules.

Content of Microcapsules (%)	Mass of Primers (g)	Mass of Microcapsules (g)	Mass of Topcoats (g)	Non-Volatile Content of the Coating after Adding Microcapsules (%)
0	0.720	0	0.720	30.0
1.0	0.720	0.007	0.713	30.3
3.0	0.720	0.022	0.698	31.1
5.0	0.720	0.036	0.684	31.8
7.0	0.720	0.050	0.670	32.4
9.0	0.720	0.065	0.655	33.2

The total mass of the water-based coatings was controlled to be constant, and the MRCPPFMs and CCPPFMs were added to the topcoat at 0%, 1.0%, 3.0%, 5.0%, 7.0%, and 9.0%, respectively. First of all, 500-mesh sandpaper was used to remove burrs on the fiberboard surface so that the substrate surface was clean, smooth, and flat. Then a brush was used to evenly apply the primer to the fiberboard surface. After painting the first time, the coated MDF was placed at a room temperature of 25 °C for 30 min, then transferred to a 50 °C oven for 30 min, then polished with an 800-grit sandpaper sanding treatment. Then a second primer coat was added, and the same steps were followed again for the coated MDF. The topcoat with microcapsules was applied twice, according to the above steps. After the final drying of the topcoat, the MDF was removed from the oven and dried at room temperature for 12 h.

2.4. Testing and Characterization

2.4.1. Chemical Composition and Morphology Characterization

A VERTEX 80V fourier transform infrared spectrometer (FTIR) was used to test the chemical composition. Before the coating samples were tested, background spectra were collected. The samples to be tested were prepared to the appropriate size and placed on an ATR attachment, after which the spectra of the samples were taken. A KBr pressing method was used for testing the microcapsule samples. A total of 1 mg of the microcapsule sample was placed in a mortar along with 100 mg of KBr powder for thorough grinding. Subsequently, the powder was poured into a mold for tableting. Finally, an insert plate with a mold cover and a sample ingot piece were inserted directly into the sample holder of the instrument for measurement.

Quanta-200 scanning electron microscopy (SEM) was used to characterize the microstructure of the coatings. When the SEM was used to characterize the sample morphology, the sample was fixed to the sample stage with a conductive adhesive. The SEM sample chamber was opened, and the sample tray was mounted on the sample holder. The sample chamber was then closed, and a vacuum was applied to the sample chamber. After the vacuum was drawn, the samples were observed, and images were taken through the software interface.

2.4.2. Antibacterial Performance Testing of the Coatings

Escherichia coli (ATCC25922) and *Staphylococcus aureus* (ACTT6538) were used to test the antibacterial properties of water-based coatings. A strain activation was performed before testing. Then, according to GB/T 4789.2-2022 [45], a bacterial suspension was prepared. Finally, a sample test was carried out. Specific experimental steps were referred to in GB/T 21866-2008 [46]. A bacterial count in the sample was taken as an average of three parallel experiments. The average number of bacterial colonies recovered after 48 h of blank coating is recorded as B, and the average number of bacterial colonies recovered after 48 h of antibacterial coating is recorded as C, measured in CFU/piece. The calculation formula for the antibacterial rate R of the coating is shown in Formula (1).

$$R = (B - C)/B \times 100\% \tag{1}$$

2.4.3. Optical Performance Testing of the Coating

A SC-10 portable color difference meter was used to test the chromaticity value and calculate a color difference. GB/T 11186.3-1989 [47] was used as a standard. After calibration of the color difference meter, a certain point on the coating was selected to test and record the chromaticity values of L, a, and b. Three points were selected for each coating for testing, and the average value was calculated and recorded as the chromaticity value of the coating. The brightness of the measured sample was denoted as L, the red–green color change was recorded as a, and the yellow–blue color change was recorded as b. The values of the blank coating were denoted as L_1, a_1, b_1, and the values of the coating containing microcapsules were denoted as L_2, a_2, b_2. The color difference ΔE of the coating was calculated using Formula (2).

$$\Delta E = \left[(\Delta L)^2 + (\Delta a)^2 + (\Delta b)^2 \right]^{1/2} \tag{2}$$

A U-3900 ultraviolet spectrophotometer was used to test the visible light transmittance of the coatings, with a wavelength range of 380 nm to 780 nm. Transmittance was the ratio of a residual light intensity to an incident light intensity when the light beam passed through the coating.

According to GB/T 4893.6-2013 [48], a X-rite ci60 glossmeter was used to test the glossiness of the coating at three incidence angles (20°, 60°, and 85°). G_0 was the glossiness of the coating when it did not contain the microcapsules. G_1 was the glossiness of the coating when it contained the microcapsules. The gloss loss rate (G_L) of the coating at a 60° incident angle was calculated using Formula (3).

$$G_L = (G_0 - G_1)/G_0 \times 100\% \qquad (3)$$

2.4.4. Mechanical Performance Testing of the Coating

A HT-6510P portable coating hardness tester was used to test the hardness of the coating, and GB/T 6739-2022 [49] was used as the standard. The pencil was inserted diagonally at a 45° angle into the pencil hardness tester for testing. When the sample was scratched by the pencil and permanent indentation occurred, the surface hardness was recorded.

A QFH-A grid adhesion tester was used to test the adhesion of the substrate surface, and GB/T 4893.4-2013 [50] was used as the standard. The adhesive tape was applied to the cut area and peeled off. The adhesion level was evaluated based on the detachment of the coating.

A QCJ-40 coating impactor was used to test the impact resistance of the surface coating on the substrate, and then the impact resistance level was evaluated. GB/T 4893.9-2013 [51] was used as the standard. The sample was placed on a horizontal base, and the steel ball fell from a position 50 mm away from the sample. A magnifying glass was used to observe the cracking of the coating.

A JB-4C roughness tester was used to test the roughness. When the roughness of a sample was tested, the sample to be measured was placed on the table of the measuring instrument. A probe was moved parallel to the surface of the sample. The sampling was started when the cursor was at the 0 scale. The measurement results were recorded.

3. Results and Discussion

3.1. Surface Morphology Analysis

The microscopic morphology of two microcapsules is shown in Figure 1. The MRCPPFMs appeared to be relatively smooth and independent spheres. The CCPPFMs showed an obvious adhesion between them, and the agglomeration phenomenon was more obvious. The morphology of the CCPPFMs was poorer than that of the MRCPPFMs.

Figure 1. SEM images of microcapsules: (**A**) MRCPPFM and (**B**) CCPPFM.

The macroscopic morphology of the surface coating on MDFs with different contents of MRCPPFMs and CCPPFMs is shown in Figures 2 and 3, respectively. The color of the fiberboard itself is brown, and the water-based coating is transparent and colorless, so the water-based coating has little effect on the surface color of the fiberboard. This is because of the lighter color of the MRCPPFM compared to fiberboard and water-based coatings. From Figure 2, it can be observed that when the microcapsule content is 7.0% or 9.0%, there are significantly aggregated light-colored microcapsule particles on the surface of the fiberboard, and the dispersion of microcapsules using melamine resin as the wall material is poor. When the microcapsule content in the coating is high, it is easy to cause microcapsule aggregation and uneven distribution, and more obvious microcapsule aggregation particles appear on the fiberboard surface. This indicates that when the content of the MRCPPFM in the coating exceeds 7.0%, it will affect its aesthetic appearance during actual use. From Figure 3, it can be seen that the CCPPFMs do not exhibit significant aggregation on the surface of the MDF, but they are uniformly dispersed in small particles, increasing the roughness of the MDF surface. Because of the darker color of CCPPFMs, which are similar to the color of the MDF itself and are evenly dispersed, the impact on the surface appearance of the MDF is relatively small.

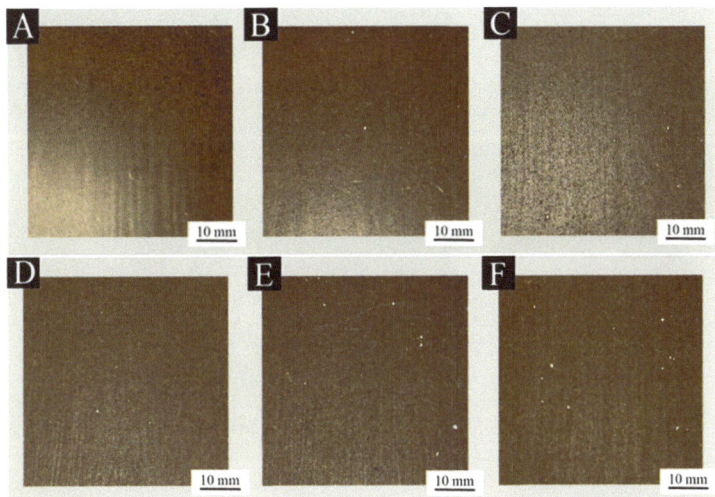

Figure 2. Macroscopic morphology of the fiberboard surface with different contents of MRCPPFMs: (**A**) without microcapsules, (**B**) with 1.0% MRCPPFMs, (**C**) with 3.0% MRCPPFMs, (**D**) with 5.0% MRCPPFMs, (**E**) with 7.0% MRCPPFMs, and (**F**) with 9.0% MRCPPFMs.

Figure 3. Macroscopic morphology of the fiberboard surface with different contents of CCPPFMs: (**A**) without microcapsules, (**B**) with 1.0% CCPPFMs, (**C**) with 3.0% CCPPFMs, (**D**) with 5.0% CCPPFMs, (**E**) with 7.0% CCPPFMs, and (**F**) with 9.0% CCPPFMs.

Figure 4 shows the microscopic morphology of the MDF surface without microcapsules, as well as those added with 7.0% MRCPPFM and CCPPFM. From Figure 4, it can be seen that the surface of the MDF without microcapsules is the smoothest because the MDF itself is relatively flat and smooth, and the influence of coatings on the surface morphology of MDF is relatively small. Although there are raised wrinkles on the surface of the MDF with MRCPPFMs and CCPPFMs added separately, the coating on the MDF with MRCPPFMs is more uneven than that with CCPPFMs, indicating that the aggregation and adhesion problem of the MRCPPFMs on the surface of the fiberboard is more serious than that of CCPPFMs, and the dispersion performance of MRCPPFMs is poor.

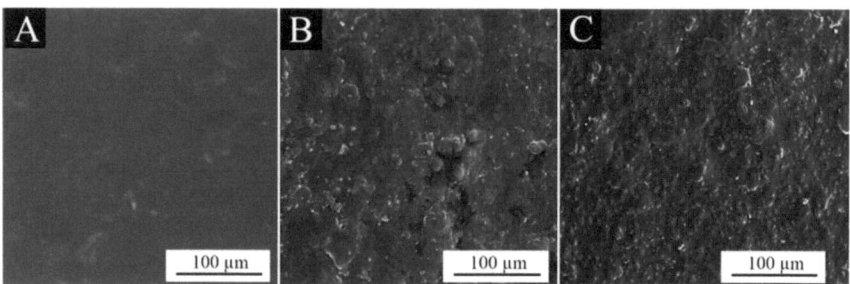

Figure 4. SEM images of the MDF surface: (**A**) without microcapsules, (**B**) with 7.0% MRCPPFMs, and (**C**) with 7.0% CCPPFMs.

The microstructure of the cross-section between the MDF, primer, and topcoat is shown in Figure 5. It can be seen that the primer on the MDF surface has a penetration effect on the fiberboard, and the primer has a sealing and isolation effect on the wood surface without a large number of obvious microcapsules penetrating into the wood. The microcapsules added to the topcoat can exert antibacterial effects and have an inhibitory effect on bacteria adhering to the surface of the MDF.

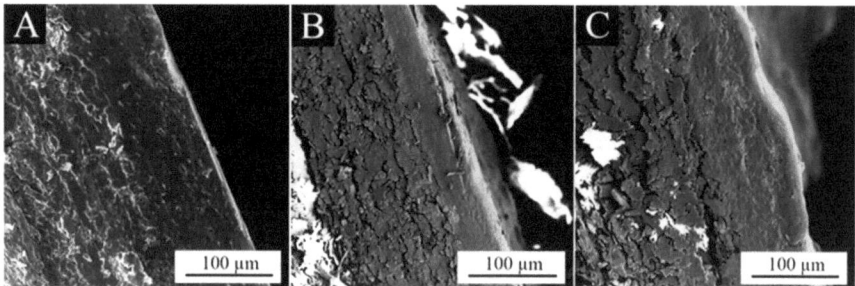

Figure 5. SEM images of the cross-section of the interface between the MDF, primer, and topcoat: (**A**) without microcapsules, (**B**) 7.0% MRCPPFMs in the topcoat, and (**C**) 7.0% CCPPFMs in the topcoat.

3.2. Chemical Composition Analysis

The infrared spectra in Figure 6 show the surface coating on the MDF without microcapsules, coating added with MRCPPFMs, and coating added with CCPPFMs. In all three infrared spectra, a characteristic peak of C=O at 1726 cm^{-1} and a vibrational peak of C-O at around 1144 cm^{-1}, representing the water-based acrylic resin in water-based coatings, were observed. This proves that adding microcapsules to water-based acrylic coatings does not alter their main chemical composition. A bending vibration absorption peak of the triazine ring in melamine resin was observed at 813 cm^{-1}, indicating that the chemical composition of the MRCPPFM in the surface coating on the MDF was not significantly altered. An absorption peak of C-O-C in the chitosan structure appeared at around 1088 cm^{-1}, indicating that the chemical composition of the CCPPFM in the surface coating on the MDF was not significantly altered. The absorption peak of CH_2- was at 2932 cm^{-1}, and the absorption peak formed by hydroxyl group association was around 3390 cm^{-1}. The appearance of these peaks proves that the addition of microcapsules did not cause a significant change in the chemical composition of the surface coating on the MDF.

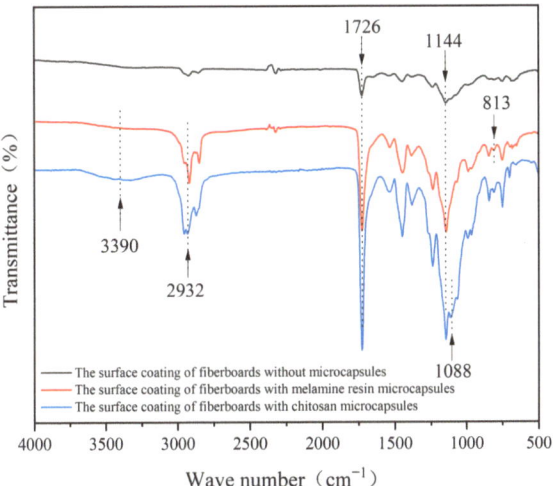

Figure 6. FTIR image of the coatings on the MDF surface.

3.3. Antibacterial Performance Analysis

Figure 7 shows the antibacterial test results of the surface coatings on the MDF with two different microcapsules added, namely the antibacterial rates against *Escherichia coli* and *Staphylococcus aureus*. From Figure 7, it can be seen that the antibacterial rates of both coatings against *Staphylococcus aureus* are slightly higher than those against *Escherichia coli*. The more microcapsules added to the surface coating on the MDF, the higher the antibacterial rate of the coating against two types of bacteria. When the content of the two microcapsules was between 3.0% and 6.0%, the growth rate of the antibacterial rate against the two bacteria was relatively slow. The coating with CCPPFMs has better antibacterial effects on two types of bacteria than the coating with MRCPPFMs. When coated with 9.0% MRCPPFM, the antibacterial rate was 67.7% against *Escherichia coli* and 71.1% against *Staphylococcus aureus*. When the content of CCPPFMs in the coating was 9.0%, the antibacterial rate was 73.7% against *Escherichia coli* and 77.4% against *Staphylococcus aureus*. This indicated that the MRCPPFMs and CCPPFMs also have antibacterial activity in the surface coating on the fiberboard, which better enhanced the antibacterial effect on the MDF. The CCPPFMs have a better antibacterial effect on the surface coating of the MDF.

The antibacterial rate of applying MRCPPFMs in the water-based coating on MDF was decreased compared to the application of MRCPPFMs on a glass board [44]. This is because the surface of a glass board is smoother, while the surface of a MDF contains more pores and fiber structure, and the higher roughness of the surface provides more space for bacteria to attach and grow [52], which decreases the antibacterial property of the surface coating on the MDF. However, compared with the *Toddalia asiatica* (L.) Lam. extract microcapsules [53], which also used plant-based antibacterial agents as the core material, the microcapsules with pomelo peel flavonoids as the core material have a slightly lower antibacterial rate. This is because *Toddalia asiatica* (L.) Lam. contains antibacterial active substances such as alkaloids, coumarins, triterpenoids, and flavonoids [54], while the effective antibacterial substances in pomelo peel extract are relatively few.

Although the addition of CCPPFMs gives the fiberboard surface coating better antibacterial properties, this proves that the application of antibacterial microcapsules in coating has achieved initial success. At present, when 9% microcapsules are added to the coating, the comprehensive performance of the coating is the best. At this time, the antibacterial rate of the coating against *Escherichia coli* and *Staphylococcus aureus* has reached 73.7% and 77.4%, respectively, but it still fails to meet the requirements of Level 2 antibacterial rate [46]. Therefore, in future work, the preparation process of CCPPFMs should be further optimized so as to improve the antibacterial rate of CCPPFMs. The exploration of antibacterial activity data for coatings over time is an important part of the follow-up work.

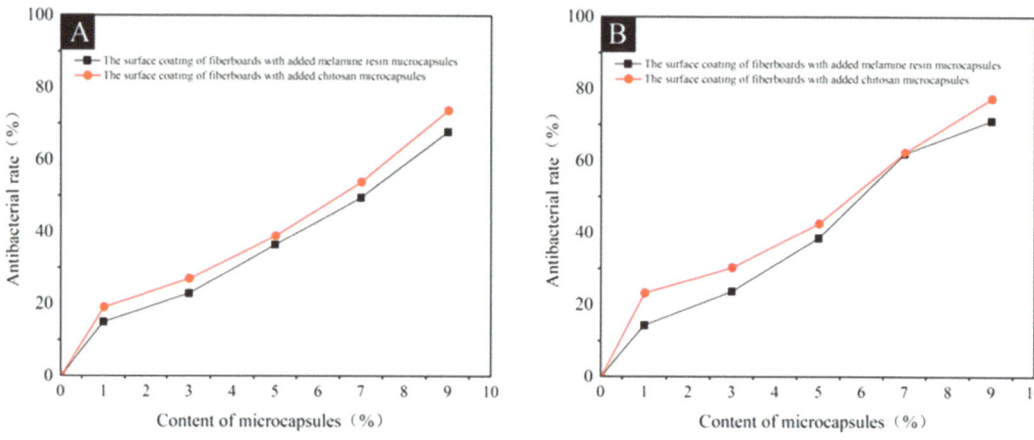

Figure 7. Antibacterial rate on the MDF surface: (**A**) against *Escherichia coli* and (**B**) against *Staphylococcus aureus*.

3.4. Optical Performance Analysis

The influence of microcapsules on the chromaticity value and color difference of the coating is shown in Table 4. The value of L, which represents brightness, increased with the microcapsule content in both groups of coated fiberboards. This is because the color of microcapsules is lighter than that of fiberboards. The addition of microcapsules neutralizes the dark color of fiberboards, resulting in an increase in their brightness. The a value representing the red–green value and the b value representing the yellow–blue value in the two groups of fiberboards are both positive, indicating that the fiberboard with microcapsules added has a reddish or yellowish color, but the overall values are relatively similar with small changes in values. This indicates that the two types of microcapsules have a relatively small impact on the red–green and yellow–blue values of the MDF. The color difference between the two groups of MDF increases with the increase in microcapsule content, but the overall value is small. When the microcapsule content reaches 9.0%, the color difference between the two groups of fiberboards is 9.76 and 3.85, respectively. This indicates that both types of microcapsules have a relatively small impact on the color difference of MDF. The color difference of the coated MDF added with CCPPFMs was generally smaller than that of the coatings added with MRCPPFMs. This is because the fiberboard itself has a darker color, while the color of the CCPPFM is darker than that of the MRCPPFM, which is closer to the color of the fiberboard itself, so the impact on the color difference of the MDF is smaller.

Table 5 shows the effect of the microcapsules on glossiness and gloss loss rate. The glossiness of the MDF surface decreased with the increase in microcapsule content at different incidence angles. This is attributed to the fact that the increase in microcapsule particles reduces the smoothness of the MDF surface, making it rough and exacerbating the phenomenon of diffuse reflection. The gloss loss rate of the fiberboard at a 60° incidence angle gradually increased with the increase in the microcapsule content. The gloss loss rate of the surface coating on the MDF with MRCPPFMs was usually lower than that with CCPPFMs. This is because CCPPFMs have good dispersibility, and the CCPPFMs are dispersed more uniformly on the surface of the MDF. However, the MRCPPFMs have poor dispersibility, which can cause local aggregation and non-dispersion on the surface of the fiberboard. Therefore, the part of the MDF surface without the microcapsules is relatively flat, and the gloss loss rate is low. This phenomenon is basically consistent with the macroscopic morphology of the MDF.

Table 4. Chromaticity and color difference.

Type	Content of Microcapsules (%)	L	a	b	ΔE
MRCPPFM	0	41.30	10.20	20.23	-
	1.0	42.70	9.67	18.90	2.00
	3.0	46.07	9.07	17.33	5.70
	5.0	47.17	8.67	15.80	7.51
	7.0	47.43	8.03	15.67	7.94
	9.0	49.67	7.77	15.83	9.76
CCPPFM	0	41.30	10.20	20.23	-
	1.0	41.40	9.90	19.70	0.62
	3.0	42.20	9.56	18.50	2.05
	5.0	43.33	10.67	18.00	3.05
	7.0	43.50	10.00	18.07	3.09
	9.0	44.93	9.60	19.10	3.85

Table 5. Glossiness and gloss loss rate.

Type	Content of Microcapsules (%)	20°	60°	85°	Gloss Loss Rate (%)
MRCPPFM	0	1.40	12.00	43.90	-
	1.0	1.17	9.70	19.23	19.2
	3.0	0.83	6.67	7.57	44.4
	5.0	0.53	4.07	2.23	66.1
	7.0	0.43	3.43	1.83	71.4
	9.0	0.40	2.80	1.03	76.7
CCPPFM	0	1.40	12.00	43.90	-
	1.0	1.07	9.00	18.50	25.0
	3.0	0.73	6.47	12.93	46.1
	5.0	0.63	5.70	10.03	52.5
	7.0	0.50	1.27	3.83	89.4
	9.0	0.50	1.20	6.20	90.0

When light shines on wooden substrates, the coating with high reflectivity on the surface of the MDF will reflect the light, which can prevent heat accumulation, slow down the thermal aging of the MDF, and extend its service life. Table 6 shows the reflectivity of the coating, while Figure 8 shows the variation trend of surface reflectance curves for two groups of coatings. With the increase in microcapsule content, the reflectivity of visible light increased slightly for both groups on the MDF. The visible light reflectance of the water-based coating with MRCPPFMs was slightly higher than that of the fiberboard with CCPPFMs. The water-based coating with two types of microcapsules has the highest reflectivity of 22.52% and 20.19%, respectively. The difference in numerical values between the two groups of data is small, and the trend of the two groups of curves is basically the same. This indicates that the addition of microcapsules has no significant effect on the reflectivity of the coating.

Table 6. The reflectivity of the water-based coating.

Type	Content of Microcapsules (%)	Reflectivity (%)
MRCPPFM	0	17.66
	1.0	19.68
	3.0	21.23
	5.0	22.15
	7.0	22.28
	9.0	22.52
CCPPFM	0	17.66
	1.0	18.48
	3.0	18.56
	5.0	19.84
	7.0	19.15
	9.0	20.19

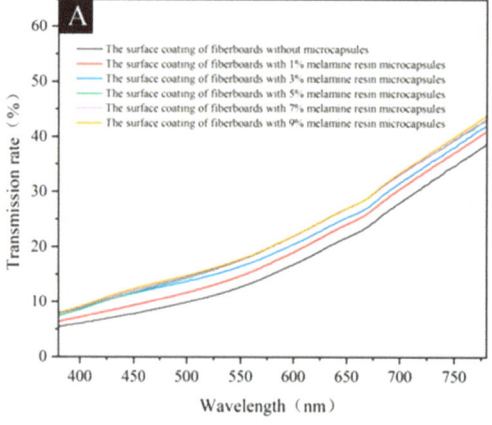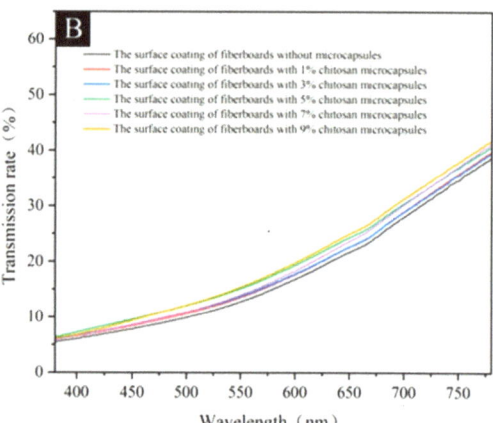

Figure 8. Reflectivity of the visible light band of the water-based coating: (**A**) with MRCPPFMs and (**B**) with CCPPFMs.

The optical properties of the water-based coating with different contents of MR-CPPFMs and CCPPFMs were compared. As the microcapsule content increased, the color difference on the surface of the two groups on the MDF gradually increased, the glossiness decreased, and the gloss loss rate and reflectivity both showed an upward trend. The trend of optical performance change on the MDF surface of the two groups is relatively similar, indicating that the change in microcapsule wall material does not have a significant impact on the optical performance of the surface coating on the MDF.

3.5. Mechanical Performance Analysis

Table 7 shows the influence of different microcapsule contents on the mechanical properties of the MDF surface. The adhesion level of the coating with the two microcapsules with different contents was 1, which was the same as that of the surface coating without microcapsules, and there was no change. This shows that when the content of the two microcapsules is 9.0% or less, it will not affect the adhesion of the surface coating on the MDF and can be applied in practice.

Table 7. The mechanical performance.

Type	Content of Microcapsules (%)	Adhesion Level	Hardness	Impact Resistance Level	Roughness (μm)
Surface coating with MRCPPFMs	0	1	HB	3	0.30
	1.0	1	H	3	0.89
	3.0	1	H	3	1.74
	5.0	1	3H	3	1.96
	7.0	1	3H	3	2.23
	9.0	1	4H	2	3.20
Surface coating with CCPPFMs	0	1	HB	3	0.30
	1.0	1	H	3	0.37
	3.0	1	H	3	0.67
	5.0	1	H	3	1.00
	7.0	1	HB	3	1.67
	9.0	1	HB	3	1.94

The hardness of the surface coating on the MDF without microcapsules was HB, which was poor because the water-based coating used was prepared from a one-component water-based acrylic resin. The surface hardness of the MDF gradually increased as the MRCPPFM content increased. This is because the MRCPPFMs use melamine resin as the wall material, which has good compactness and film-forming properties, and the dispersion in the water-based coating can improve the density of the coating, thereby improving the hardness of the surface coating on the MDF. The content of the CCPPFMs at 1.0%–5.0% on the MDF surface was slightly higher than that without adding microcapsules, which was H. The hardness dropped to HB when the CCPPFM content exceeded 5.0%, which is because the CCPPFMs use chitosan as the wall material and have a good dispersion in the water-based coatings. But the chitosan itself has a certain viscosity, and blending it with a water-based coating can thicken the coatings. If the content of microcapsules in the coating is too high, the viscosity of water-based coatings will be adjusted to a certain extent. This leads to a decrease in the hardness of the surface coating on the MDF. However, the hardness of the coating without microcapsules was not lower when the microcapsule content was 1.0%–9.0%. This indicates that CCPPFMs with a content of 9.0% or less are suitable for practical application.

When the content of MRCPPFMs was not higher than 7.0%, there were 1–2 circles of mild cracks on the coating surface, and the impact resistance level was 3. When the MRCPPFM content was 9.0%, the coating surface had impact marks without cracks, and the impact resistance level was 2. The results indicated that increasing the MRCPPFM content in the surface coating slightly improved the impact resistance on the MDF surface because the addition of MRCPPFMs increases the surface hardness of the MDF, resulting in an improvement in the impact resistance. The impact resistance level of the fiberboard surface with the CCPPFMs was the same as that without microcapsules, which was level 3, indicating that the CCPPFMs had little effect on the impact resistance.

The surface roughness on the MDF with the MRCPPFMs and the CCPPFMs gradually increased with the microcapsule content increasing. The roughness value of the surface coating on the MDF with the MRCPPFMs was higher than that with the CCPPFMs because the dispersion of the CCPPFMs in water-based coatings was better than that of the MRCPPFMs, and the MRCPPFMs were more likely to form a large agglomeration phenomenon, which makes the roughness of the MDF surface increase.

The mechanical properties of the coating with different contents of MRCPPFMs and CCPPFMs were compared, and both microcapsules had an impact on the mechanical properties of the MDF surface. With microcapsule content increasing, there was no significant change in the adhesion of the surface coatings of the two groups on the MDF, and the roughness gradually increased. The surface hardness of the MDF with the MRCPPFMs gradually increased, and the impact resistance level slightly improved. The surface hardness of the MDF with the CCPPFMs changed less, which was overall smaller than the surface on the MDF with the MRCPPFMs, but both were not lower than the coating without microcapsules, and the impact resistance level did not change significantly. Among them, the surface hardness on the MDF with the MRCPPFMs was generally higher, while the surface roughness on the MDF with the CCPPFMs was generally lower.

4. Conclusions

Two types of water-based coatings were subjected to antibacterial performance testing. The antibacterial effect of coating with CCPPFMs was better than that with MRCPPFMs. The optical performance trends of the two groups of coatings were similar. The mechanical performance of the surface coating on the MDF with MRCPPFMs was better. With 5.0% MRCPPFMs, the comprehensive performance on the MDF was better, with the antibacterial rates of 36.4% and 38.5% for *Escherichia coli* and *Staphylococcus aureus*, respectively, the color difference of 7.51, the loss of light rate of 66.1%, the reflectivity of 22.15%, the adhesion level of 1, the impact resistance level of 3, the hardness of 3H, and the roughness of 1.96 µm. When the coating had 9.0% CCPPFMs, the comprehensive performance was better. The antibacterial rate was 73.7% against *Escherichia coli* and 77.4% against *Staphylococcus aureus*, the color difference was 3.85, the loss of light was 90.0%, the reflectivity was 20.19%, the adhesion was level 1, the impact resistance level was 3, the hardness was HB, and the roughness was 1.94 µm. The results showed that the CCPPFMs were more suitable for application in the coating of MDFs. By exploring the effects of the addition of two types of microcapsules on the coating performance, a foundation for the application of antibacterial microcapsules on MDF surfaces was laid.

Author Contributions: Conceptualization, methodology, validation, resources, data management, supervision, J.D.; writing—review and editing, T.D.; formal analysis, investigation X.Y. All authors have read and agreed to the published version of the manuscript.

Funding: This project was partly supported by the Postgraduate Research and Practice Innovation Program of Jiangsu Province (SJCX24_0399) and the Natural Science Foundation of Jiangsu Province (BK20201386).

Institutional Review Board Statement: Not applicable.

Informed Consent Statement: Not applicable.

Data Availability Statement: Data are contained within the article.

Conflicts of Interest: The authors declare that there are no conflicts of interest.

References

1. Zhang, Z.Y.; Zhu, J.G.; Qi, Q. Research on the recyclable design of wooden furniture based on the recyclability evaluation. *Sustainability* **2023**, *15*, 16758. [CrossRef]
2. Wang, C.; Yu, J.H.; Jiang, M.H.; Li, J.Y. Effect of slicing parameters on the light transmittance of 3D-printed polyethylene terephthalate glycol products. *BioResources* **2024**, *19*, 500–509. [CrossRef]
3. Hu, W.G.; Yu, R. Study on the strength mechanism of the wooden round-end mortise-and-tenon joint using the digital image correlation method. *Holzforschung* **2024**. [CrossRef]
4. Wang, C.; Zhang, C.Y.; Ding, K.Q.; Jiang, M.H. Immersion polishing post-treatment of PLA 3D printed formed parts on its surface and mechanical performance. *BioResources* **2023**, *18*, 7995–8006. [CrossRef]
5. Zhang, X.Y.; Xu, W.; Li, R.R.; Zhou, J.C.; Luo, Z.Y. Study on sustainable lightweight design of airport waiting chair frame structure based on ANSYS workbench. *Sustainability* **2024**, *16*, 5350. [CrossRef]
6. Hu, J.; Liu, Y.; Wang, J.X.; Xu, W. Study of selective modification effect of constructed structural color layers on European beech wood surfaces. *Forests* **2024**, *15*, 261. [CrossRef]
7. Liu, Y.; Hu, W.; Kasal, A.; Erdil, Y.Z. The state of the art of biomechanics applied in ergonomic furniture design. *Appl. Sci.* **2023**, *13*, 12120. [CrossRef]
8. Hu, W.G.; Fu, W.J.; Zhao, Y. Optimal design of the traditional Chinese wood furniture joint based on experimental and numerical method. *Wood Res.-Slovakia* **2024**, *69*, 50–59. [CrossRef]

9. Hu, W.G.; Yu, R.Z. Mechanical and acoustic characteristics of four wood species subjected to bending load. *Maderas-Cienc. Tecnol.* **2023**, *25*, 39. [CrossRef]
10. Wang, C.F.; Abidin, S.Z.; Toyong, N.M.P.; Zhu, W.K.; Zhang, Y.C. Mildew resistance and antibacterial activity of plywood decorated with ZnO/TiO$_2$. *J. Saudi Chem. Soc.* **2024**, *28*, 101877. [CrossRef]
11. Chen, S.X.; Wei, B.C.; Fu, Y.L. A study of the chemical composition and biological activity of michelia macclurei dandy heartwood: New sources of natural antioxidants, enzyme inhibitors and bacterial inhibitors. *Int. J. Mol. Sci.* **2024**, *24*, 792. [CrossRef] [PubMed]
12. Zhang, N.; Xu, W.; Tan, Y. Multi-attribute hierarchical clustering for product family division of customized wooden doors. *Bioresources* **2023**, *18*, 7889–7904. [CrossRef]
13. Wang, C.; Zhang, C.Y.; Zhu, Y. Reverse design and additive manufacturing of furniture protective foot covers. *BioResources* **2024**, *19*, 4670–4678. [CrossRef]
14. Hu, J.; Liu, Y.; Xu, W. Influence of cell characteristics on the construction of structural color layers on wood surfaces. *Forests* **2024**, *15*, 676. [CrossRef]
15. Hu, J.; Liu, Y.; Xu, W. Impact of cellular structure on the thickness and light reflection properties of structural color layers on diverse wood surfaces. *Wood Mater. Sci. Eng.* **2024**. [CrossRef]
16. Wu, S.S.; Zhou, J.C.; Xu, W. A convenient approach to manufacturing lightweight and high-sound-insulation plywood using furfuryl alcohol/multilayer graphene oxide as a shielding layer. *Wood Mater. Sci. Eng.* **2024**. [CrossRef]
17. Wang, X.Y.; Liu, X.; Wu, S.S.; Xu, W. The influence of different impregnation factors on mechanical properties of silica sol-modified Populus tomentosa. *Wood Fiber Sci.* **2024**, *56*, 65–71.
18. Zhou, J.C.; Xu, W. A fast method to prepare highly isotropic and optically adjustable transparent wood-based composites based on interface optimization. *Ind. Crops Prod.* **2024**, *218*, 118898. [CrossRef]
19. Wang, C.; Zhou, Z.Y. Optical properties and lampshade design applications of PLA 3D printing materials. *Bioresources* **2023**, *18*, 1545–1553. [CrossRef]
20. Zhang, C.; Wang, H.R.; Zhou, Q.X. Preparation and characterization of microcapsules based self-healing coatings containing epoxy ester as healing agent. *Prog. Org. Coat.* **2018**, *125*, 403–410. [CrossRef]
21. Thakur, T.; Gaur, B.; Singha, A.S. Bio-based epoxy/imidoamine encapsulated microcapsules and their application for high performance self-healing coatings. *Prog. Org. Coat.* **2021**, *159*, 106436. [CrossRef]
22. Zotiadis, C.; Patrikalos, I.; Loukaidou, V.; Korres, D.M.; Karantonis, A.; Vouyiouka, S. Self-healing coatings based on poly(urea-formaldehyde) microcapsules: In situ polymerization, capsule properties and application. *Prog. Org. Coat.* **2021**, *161*, 106475. [CrossRef]
23. Ding, T.T.; Huang, N.; Yan, X.X. Effect of different emulsifiers on the preparation process of aloe-emodin microcapsules and waterborne coating properties. *Coatings* **2023**, *13*, 1355. [CrossRef]
24. Han, S.J.; Li, J.P.; Lu, Y.L.; Zang, J.; Ding, Q.Y.; Su, J.Y.; Wang, X.Y.; Song, J.A.; Lu, Y. Synthesis and characterization of microencapsulated paraffin with melamine-urea-formaldehyde shell modified with lignin. *Int. J. Biol. Macromol.* **2024**, *261*, 129640. [CrossRef] [PubMed]
25. Liu, H.S.; Deng, Y.H.; Ye, Y.; Liu, X.Q. Reversible thermochromic microcapsules and their applications in anticounterfeiting. *Material* **2023**, *16*, 5150. [CrossRef] [PubMed]
26. Ashraf, M.N.; Ali, A.; Shakoor, M.B.; Ahmad, S.R.; Hussain, F.; Oh, S.E. Development of novel formaldehyde-free melamine resin for retanning of leather and reduced effluent discharge in water. *Separations* **2022**, *9*, 368. [CrossRef]
27. Ashraf, M.N.; Khan, S.M.; Munir, S.; Saleem, R. Comparative study of retanning properties of melamine-glyoxal resins produced by different sulfonating agents. *J. Soc. Leath. Leather Technol. Chem.* **2023**, *106*, 205–214.
28. Meng, Q.Y.; Zhong, S.L.; Wang, J.; Gao, Y.; Cui, X.J. Advances in chitosan-based microcapsules and their applications. *Carbohyd. Polym.* **2023**, *300*, 120265. [CrossRef]
29. Wang, R.; Hu, H.; He, X.; Liu, W.B.; Li, H.Y.; Guo, Q.; Yuan, L.Y. Synthesis and characterization of chitosan/urea-formaldehyde shell microcapsules containing dicyclopentadiene. *J. Appl. Polym. Sci.* **2011**, *121*, 2202–2212. [CrossRef]
30. Phan, C.; Nguyen, T.T.C.; Do, T.V.T.; Tang, G.P. Preparation and characterization of sorafenib-loading microcapsules by complex coacervation of gum Arabic with chitosan or modified chitosan. *Materia* **2023**, *28*, e20230015. [CrossRef]
31. Wang, S.; Ren, Z.H.; Li, H.L.; Xue, Y.; Zhang, M.Y.; Li, R.; Liu, P.F. Preparation and sustained-release of chitosan-alginate bilayer microcapsules containing aromatic compounds with different functional groups. *Int. J. Biol. Macromol.* **2024**, *271 Pt 2*, 132663. [CrossRef] [PubMed]
32. Zhang, M.C.; Li, M.Y.; Zhang, D.Y.; Yu, Y.; Zhu, K.X.; Zang, X.D.; Liu, D.Y. Preparation and investigation of sustained-release nanocapsules containing cumin essential oil for their bacteriostatic properties. *Food* **2024**, *13*, 945. [CrossRef] [PubMed]
33. Tang, N.A.; Zhang, X.; Li, J.Y.; Lu, R.H.; Luo, H.W.; Li, Y.H.; Liu, H.T.; Liu, S.C. Hyperbranched porous boronate affinity imprinted hydrogels for specific separation of flavonoids under physiological pH: A emulsion interfacial assembly imprinted strategy. *Chem. Eng. J.* **2024**, *493*, 152769. [CrossRef]
34. Hasnat, H.; Shompa, S.A.; Islam, M.M.; Alam, S.; Richi, F.T.; Emon, N.U.; Ashrafi, S.; Ahmed, N.U.; Chowdhury, M.N.R.; Fatema, N.; et al. Flavonoids: A treasure house of prospective pharmacological potentials. *Heliyon* **2024**, *10*, e27533. [CrossRef]
35. Prasad, A.; Kumar, R.; Kumari, S. Recent advances in synthetic aspects of naringenin flavonoid and its bioprotective effect (A Review). *Russ. J. Bioorg. Chem.* **2024**, *49*, 1177–1197. [CrossRef]

36. Addi, M.; Elbouzidi, A.; Abid, M.; Tungmunnithum, D.; Elamrani, A.; Hano, C. An overview of bioactive flavonoids from citrus fruits. *Appl. Sci.* **2022**, *12*, 29. [CrossRef]
37. Zhang, X.N.; Xu, H.R.; Hua, J.L.; Zhu, Z.Y.; Wang, M. Protective effects of grapefruit essential oil against staphylococcus aureus-Induced inflammation and cell damage in human epidermal keratinocytes. *Chem. Biodivers.* **2022**, *19*, e202200205. [CrossRef] [PubMed]
38. Hu, W.G.; Luo, M.Y.; Liu, Y.Q.; Xu, W.; Konukcu, A.C. Experimental and numerical studies on the mechanical properties and behaviors of a novel wood dowel reinforced dovetail joint. *Eng. Fail. Anal.* **2023**, *152*, 107440. [CrossRef]
39. Hu, W.G.; Luo, M.Y.; Hao, M.M.; Tang, B.; Wan, C. Study on the effects of selected factors on the diagonal tensile strength of oblique corner furniture joints constructed by wood dowel. *Forests* **2023**, *14*, 1149. [CrossRef]
40. Hu, W.G.; Liu, Y.; Konukcu, A.C. Study on withdrawal load resistance of screw in wood-based materials: Experimental and numerical. *Wood Mater. Sci. Eng.* **2023**, *18*, 334–343. [CrossRef]
41. Yontar, A.K.; Çevik, S.; Akbay, S. Production of environmentally friendly and antibacterial MDF (Medium-density fiberboard) surfaces with green synthesized nano silvers. *Inorg. Chem. Commun.* **2024**, *159*, 111865. [CrossRef]
42. Costa, D.; Serra, J.; Quinteiro, P.; Dias, A.C. Life cycle assessment of wood-based panels: A review. *J. Clean. Prod.* **2024**, *444*, 140955. [CrossRef]
43. Cao, Y.H.; Yang, Z.Y.; Ou, J.H.; Jiang, L.; Chu, G.C.; Wang, Y.F.; Chen, S.G. Ultra-transparent, hard and antibacterial coating with pendent quaternary pyridine salt. *Prog. Org. Coat.* **2023**, *175*, 107369. [CrossRef]
44. Ding, T.T.; Yan, X.X. Preparation Process Optimization for Melamine Resin-Covered Pomelo Peel Flavonoid Antibacterial Microcapsules and Their Effect on Waterborne Paint Film Performance. *Coatings* **2024**, *14*, 654. [CrossRef]
45. *GB/T 4789.2-2022*; National Food Safety Standard Food Microbiological Examination: Aerobic Plate Count. Standardization Administration of the People's Republic of China: Beijing, China, 2022.
46. *GB/T 21866-2008*; Test Method and Effect for Antibacterial Capability of Paints Film. Standardization Administration of the People's Republic of China: Beijing, China, 2008.
47. *GB/T 11186.3-1989*; Methods for Measuring the Colour of Paint Films. Part III: Calculation of Colour Differences. Standardization Administration of the People's Republic of China: Beijing, China, 1990.
48. *GB/T 4893.6-2013*; Test of Surface Coatings of Furniture-Part 6: Determination of Gloss Value. Standardization Administration of the People's Republic of China: Beijing, China, 2013.
49. *GB/T 6739-2022*; Paints and Varnishes—Determination of Film Hardness by Pencil Test. Standardization Administration of the People's Republic of China: Beijing, China, 2022.
50. *GB/T 4893.4-2013*; Test of Surface Coatings of Furniture—Part 4: Determination of Adhesion—Cross Cut. Standardization Administration of the People's Republic of China: Beijing, China, 2013.
51. *GB/T 4893.9-2013*; Test of Surface Coatings of Furniture—Part 9: Determination of Resistance to Impact. Standardization Administration of the People's Republic of China: Beijing, China, 2013.
52. Sekar, H.; Tirumkudulu, M.; Gundabala, V. Film Formation of Iodinated Latex Dispersions and Its Role in Their Antimicrobial Activity. *Langmuir* **2024**, *40*, 9197–9204. [CrossRef] [PubMed]
53. Wang, Y.; Yan, X.X. Preparation of *Toddalia asiatica* (L.) Lam. Extract Microcapsules and Their Effect on Optical, Mechanical and Antibacterial Performance of Waterborne Topcoat Paint Films. *Coatings* **2024**, *14*, 655. [CrossRef]
54. Zeng, Z.; Tian, R.; Feng, J.; Yan, N.A.; Yuan, L. A systematic review on traditional medicine *Toddalia asiatica* (L.) Lam.: Chemistry and medicinal potentia. *Saudi Pharm. J.* **2021**, *29*, 781–798. [CrossRef]

Disclaimer/Publisher's Note: The statements, opinions and data contained in all publications are solely those of the individual author(s) and contributor(s) and not of MDPI and/or the editor(s). MDPI and/or the editor(s) disclaim responsibility for any injury to people or property resulting from any ideas, methods, instructions or products referred to in the content.

Article

The Effects of Urea–Formaldehyde Resin-Coated *Toddalia asiatica* (L.) Lam Extract Microcapsules on the Properties of Surface Coatings for Poplar Wood

Ye Zhu [1,2], Ying Wang [1,2] and Xiaoxing Yan [1,2,*]

[1] Co-Innovation Center of Efficient Processing and Utilization of Forest Resources, Nanjing Forestry University, Nanjing 210037, China; zhuye@njfu.edu.cn (Y.Z.); wangying1214@njfu.edu.cn (Y.W.)
[2] College of Furnishings and Industrial Design, Nanjing Forestry University, Nanjing 210037, China
* Correspondence: yanxiaoxing@njfu.edu.cn

Abstract: Urea–formaldehyde resin was used as a wall material and *Toddalia asiatica* (L.) Lam extract was used as a core material to prepare urea–formaldehyde resin-coated *Toddalia asiatica* (L.) Lam extract microcapsules (UFRCTEMs). The effects of UFRCTEM content and the mass ratio of core-to-wall material ($M_{core}:M_{wall}$) on the performance of waterborne coatings on poplar surfaces were investigated by adding microcapsules to the waterborne topcoat. Under different $M_{core}:M_{wall}$ of microcapsules, as the content of microcapsules increased, the glossiness and adhesion of the coatings gradually decreased, and the color difference value of the coatings gradually increased. The cold liquid resistance, hardness, and impact resistance of the coatings were all improved, and the roughness of the coatings increased. The antibacterial rates of the coatings against *Escherichia coli* and *Staphylococcus aureus* were both on the rise, and the antibacterial rate against *Staphylococcus aureus* was slightly higher than that against *Escherichia coli*. When the microcapsule content was 7.0% and the $M_{core}:M_{wall}$ was 0.8:1, the surface coating performance on poplar wood was excellent. The glossiness was 3.43 GU, light loss was 75.55%, color difference ΔE was 3.23, hardness was 2H, impact resistance level was 3, adhesion level was 1, and roughness was 3.759 μm. The cold liquid resistance was excellent, and resistance grades to citric acid, ethanol, and cleaning agents were all 1. The antibacterial rates against *Escherichia coli* and *Staphylococcus aureus* were 68.59% and 75.27%, respectively.

Keywords: microcapsule; *Toddalia asiatica* (L.) Lam extract; waterborne coatings; antibacterial coating

Citation: Zhu, Y.; Wang, Y.; Yan, X. The Effects of Urea–Formaldehyde Resin-Coated *Toddalia asiatica* (L.) Lam Extract Microcapsules on the Properties of Surface Coatings for Poplar Wood. *Coatings* **2024**, *14*, 1011. https://doi.org/10.3390/coatings14081011

Academic Editor: Marko Petric

Received: 20 June 2024
Revised: 31 July 2024
Accepted: 2 August 2024
Published: 9 August 2024

Copyright: © 2024 by the authors. Licensee MDPI, Basel, Switzerland. This article is an open access article distributed under the terms and conditions of the Creative Commons Attribution (CC BY) license (https://creativecommons.org/licenses/by/4.0/).

1. Introduction

Wood, as a renewable biomass material, plays an extremely important role in ecological civilization, national economic construction, and people's daily lives [1–3]. Wood is mainly composed of three components: lignin, cellulose, and hemicellulose, which make it susceptible to pests and diseases, as well as bacterial and fungal erosion [4,5]. Hard broad-leaved trees have a long growth cycle and are scarce in resources [6–8]. Poplar has the advantages of fast production speed, moderate material, tough wood, corrosion resistance, and strong stability [9,10]. Poplar is a fast-growing wood that can alleviate the current shortage of wood resources and has a wide range of application prospects in wooden furniture [11]. However, poplar has problems such as loose fiber structure, low density, easy moisture absorption and deformation, and difficulty in drying, which greatly limits its application range [12,13]. These issues have resulted in higher requirements for the antibacterial effect on the surface of poplar wood. Therefore, the antibacterial treatment of wood is an important means to extend its service life, improve its utilization level, and save wood resources [14]. As an environmentally friendly coating, waterborne coatings can have special properties that affect wood, such as antibacterial, anti-corrosion, heat-resistant, waterproof, and fireproof characteristics [15–17]. It has been found that excessive use of chemical antibacterial agents can easily cause environmental pollution, carcinogenic effects on humans and other organisms, and corrosion of metal objects [18–20]. Plant-derived antibacterial agents obtained from natural plants are extracted through physical or chemical separation and have advantages such as safety, efficiency, wide range of sources, wide variety, and minimal toxic side effects [21–23]. Therefore, a large number of plant extracts have been developed and applied in fields such as cosmetics, natural fungicides, and feed [24]. However, when plant extracts are directly used as wood antibacterial agents, there are various shortcomings such as a narrow range of insect-resistant bacteria, easy degradation, easy loss, short efficacy period, and sensitivity to external environmental factors (such as temperature, air humidity, light, rain, etc.), which result in overall poor antibacterial effect and limit its application range. Therefore, further research and expansion are needed [25].

The extract from the roots, stems, and leaves of *Toddalia asiatica* (L.) Lam contains a certain degree of bioactive substances such as antibacterial, antioxidant, and insecticidal properties [26]. The chemical components of *Toddalia asiatica* (L.) Lam anhydrous ethanol extract contain alkaloids, coumarins, triterpenes, and flavonoids with antibacterial activity [27]. Coumarin compounds can reduce the pathogenicity and drug resistance of bacteria by inhibiting the bacterial quorum sensing system, reducing the expression of related virulence factors, and forming biofilms [28]. Flavonoids act on the cell membrane of microorganisms, altering the permeability, and achieving the goal of inhibiting or killing bacteria [29]. Alkaloids can induce the release of membrane-bound cell wall autolytic enzymes, ultimately leading to lysis [30]. The antibacterial effect of terpenes is mainly attributed to their ability to interact with microbial membranes and destroy them, as well as the increased antibacterial and antifungal activity of terpenes by hydroxyl, ketone, and aldehyde groups [31]. However, it was not easily dispersed when directly applied to coatings, and antibacterial properties were not guaranteed. It is even more difficult to apply them in surface coatings for wooden products, and further research is needed to expand their application functions. The use of microcapsule technology can encapsulate some natural antibacterial agents or antibacterial substances to make antibacterial microcapsules, improve the processing performance of antibacterial agents, enhance the stability of natural antibacterial agents, and improve their usability, expanding their application scope. Therefore, *Toddalia asiatica* (L.) Lam extract was prepared into microcapsules and applied in wood antibacterial coatings. Raj et al. used different materials to sequentially extract the active ingredients from the leaves of *Toddalia asiatica* (L.) Lam. The results showed that the extract exhibited antibacterial activity against selected bacteria (*Staphylococcus epidermidis, Enterobacter aerogenes, Shigella flexneri, Klebsiella pneumoniae,* and *Escherichia coli*) and fungi (*Aspergillus flavus, Candida krusei,* and *Botrytis cinereal*) [32]. Roshan et al. prepared chitosan-based nanocapsules of *Toddalia asiatica* (L.) Lam essential oil (neTAEO) and the results confirmed that neTAEO exhibited stronger antifungal and aflatoxin B1 inhibitory activity than *Toddalia asiatica* (L.) Lam essential oil, and had greater development prospects [33].

Three types of urea–formaldehyde resin-coated *Toddalia asiatica* (L.) Lam extract microcapsules (UFRCTEMs) with the mass ratio of core-to-wall material ($M_{core}:M_{wall}$) of 0.6:1, $M_{core}:M_{wall}$ of 0.8:1, and $M_{core}:M_{wall}$ of 1.2:1 were prepared. Three types of UFRCTEMs were added to waterborne coatings at concentrations of 1.0%, 3.0%, 5.0%, 7.0%, and 9.0%, respectively. Then, the waterborne coatings were applied on the surface of poplar wood. By testing and analyzing the microstructure, chemical composition, optical properties, mechanical properties, cold liquid resistance, and antibacterial properties of the surface coating of poplar wood, the influence of different UFRCTEMs and addition contents on the comprehensive performance of the surface coating on poplar wood was explored, providing a reference for the coating process of antibacterial coatings.

2. Materials and Methods

2.1. Materials and Instruments

The leaves of *Toddalia asiatica* (L.) Lam were obtained from Lingshan County, Qinzhou, China. The leaves were placed in a 40 °C oven to dry to a constant weight. The leaves were pulverized into powder by a crusher. The size of the poplar wood was 50 mm × 50 mm × 8 mm, which was smoothed using #800 and #1000 sandpaper. The coatings used were a waterborne acrylic topcoat and a primer, both from Jiangsu Haitian Technology Co., Ltd., Nanjing, China.

2.2. Preparation Method of Microcapsules

2.2.1. Preparation Method of *Toddalia asiatica* (L.) Lam Extract

The leaf powder was mixed with anhydrous ethanol, heated in a water bath, and centrifuged. A vacuum pump combined with a Buchner funnel was used to filter the solution. The specific preparation method used in this study is the same as in Reference [34].

2.2.2. Preparation Method of Microcapsules

According to Reference [35], three types of UFRCTEMs with 0.6:1, 0.8:1, and 1.2:1 $M_{core}:M_{wall}$ were prepared. Table 1 shows the proportion of the UFRCTEM raw materials.

Table 1. The proportion of the microcapsule raw materials.

Sample (#)	$M_{core}:M_{wall}$	Urea (g)	37% Formaldehyde (g)	Wall Material (g)	*Toddalia asiatica* (L.) Lam Extract (g)	Ethanol (g)	Core Material (g)	Emulsifier (g)	Deionized Water (g)
1	0.6:1	10.00	16.22	16.00	0.19	9.41	9.60	1.04	50.96
2	0.8:1	10.00	16.22	16.00	0.26	12.54	12.80	1.43	70.07
3	1.2:1	10.00	16.22	16.00	0.38	18.82	19.20	2.09	102.41

2.3. Painting Method for Poplar Board

According to the technical specifications for the application of waterborne wood coatings on furniture surfaces, a uniform layer of coating was manually applied to the poplar board, with a one-time application amount of 60 g/m² to 80 g/m². The coating was applied with a brush-painting technique. A coating method of two layers of primer and two layers of topcoat was adopted. The selected coating amount for each layer was 80 g/m², and the total coating amount for the primer and topcoat was 320 g/m². The thickness of each coating on the surface of the wood was about 80 μm, and the total thickness was about 320 μm. Taking into account the loss error during the coating process, the actual consumption of the coating was 1.8 times the theoretical application amount. The total mass of the coating applied to the surface of the poplar wood was 1.44 g. Table 2 shows the materials used for the waterborne coatings.

Table 2. List of materials used for the waterborne coatings.

Microcapsule Content (%)	Waterborne Primer (g)	Microcapsules (g)	Waterborne Topcoat (g)
0	0.720	0	0.720
1.0	0.720	0.007	0.713
3.0	0.720	0.022	0.698
5.0	0.720	0.036	0.684
7.0	0.720	0.051	0.669
9.0	0.720	0.065	0.655

The specific coating process for the surface coating of poplar wood was as follows: A #360 sandpaper was used to treat the rough edges of the poplar board, and a #800 sandpaper was used to remove the surface of the poplar board and polish it smooth. Then, the first layer of primer was applied with a brush. After the first application of primer, the coating was leveled at room temperature for 20 min before being transferred to an oven to dry. After the coating was completely cured, it was taken out and polished with #800 sandpaper. The above-described procedures of brushing, leveling, and curing were repeated. The total mass of the topcoat was kept constant, and the three UFRCTEM samples, #1, #2, and #3, prepared in the early stage were added to the topcoat in a mass ratio of 1.0%, 3.0%, 5.0%, 7.0%, and 9.0%, and they were mixed evenly. The first layer of the topcoat was applied using a brush. The above-described procedures of brushing, leveling, and curing the topcoat were repeated. Then, #1000 sandpaper was used to polish the coating before the second layer of topcoat was applied. In addition, a pure primer and a pure topcoat were applied to the poplar board as control group specimens for future use.

2.4. Testing and Characterization

2.4.1. Performance Characterization of Microcapsules

(1) Coverage rate (C): UFRCTEMs with a mass of M_1 were weighed. M_2 was the weight of the weighing filter paper. The UFRCTEMs were soaked in ethanol, filtered, and dried after 24 h. The total mass of the dried filter paper and wall material was M_3. The calculation of the coverage rate is shown in Formula (1).

$$C = \frac{(M_1 + M_2) - M_3}{M_1} \tag{1}$$

(2) Yield rate (Y): The total mass of materials used for preparing the UFRCTEM samples was denoted as M_1. The mass of the UFRCTEMs after drying was recorded as M_2. The calculation of the yield rate is shown in Formula (2).

$$Y = \frac{M_2}{M_1} \tag{2}$$

(3) Analysis of microstructure and chemical composition: a Zeiss optical microscope (OM, Carl Zeiss AG, Oberkochen, Germany) was used to observe the morphology of the UFRCTEMs. Scanning electron microscopy (SEM, Tescan, Brno, the Czech Republic) was used to analyze the microstructure of the UFRCTEMs and coatings. Fourier-transform infrared spectroscopy (FTIR, Brucker AG, Karlsruhe, Germany) was used to analyze the chemical composition of the UFRCTEMs and coatings.

2.4.2. Color Difference Testing of Coatings

In the light of GB/T 11186.3-1989 [36], the chromaticity value of the coatings was measured and recorded using a SEGT-J colorimeter (Zhuhai Tianchuang Instrument Co., Ltd., Zhuhai, China) and a color difference was calculated. The color difference ΔE between the coating with UFRCTEMs added and the pure coating was calculated using the color difference shown in Formula (3), where $\Delta L = L_1 - L_2$, $\Delta a = a_1 - a_2$, and $\Delta b = b_1 - b_2$; ΔL represent the difference in brightness of the coating; Δa represents the red–green difference in the coating; and Δb represents the yellow–blue difference in the coating.

$$\Delta E = \left[(\Delta L)^2 + (\Delta a)^2 + (\Delta b)^2 \right]^{\frac{1}{2}} \tag{3}$$

2.4.3. Test for Glossiness and Reflectivity of Coatings

The sample was treated according to the requirements of GB/T 4893.6-2013 [37]. The glossiness values of the coating at three incidence angles of $20°$, $60°$, and $85°$ were tested and recorded using a glossmeter (Shenzhen Linshang Technology Co., Ltd., Shenzhen, China), with the unit being GU.

The reflection curve of the coating in the visible light wavelength range (380–780 nm) was tested and recorded using a Hitachi UV spectrophotometer (Zhuhai Tianchuang Instrument Co., Ltd., Zhuhai, China). The reflectance R value was calculated using Formula (4).

$$R = \frac{\int_{380}^{780} r(\lambda)i(\lambda)d(\lambda)}{\int_{380}^{780} i(\lambda)d(\lambda)} \tag{4}$$

where $i(\lambda)$ is the standard radiation intensity for sunlight, and the unit is $W \cdot m^{-2} \cdot nm^{-1}$. $r(\lambda)$ is the reflectance value obtained through testing.

2.4.4. Roughness Testing of Coatings

The roughness value was tested and recorded using a JB-4C roughness tester (Cangzhou Oupu Testing Instrument Co., Ltd., Cangzhou, China). The macro knob was rotated to fine tune the probe position until the red display point was at the zero-scale line. The test button was activated and data were recorded. The unit of roughness value is μm.

2.4.5. Cold Liquid Resistance Test of Coatings

According to GB/T 4893.1-2021 [38], 10% citric acid solution, undescended ethanol with a volume fraction of 96%, and a cleaning agent (Guangdong Baiyun Cleaning Group Co., Ltd., Guangzhou, China) were chosen as the experimental liquids. The damage to the surface was inspected under specified lighting conditions. The test results were evaluated using numerical levels.

2.4.6. Antibacterial Performance Testing of Coatings

According to GB/T 21866-2008 [39], test operations were carried out. Firstly, *Escherichia coli* (ATCC25922, Shanghai Shifeng Biotechnology Co., Ltd., Shanghai, China) and *Staphylococcus aureus* (ACTT6538, Shanghai Shifeng Biotechnology Co., Ltd., Shanghai, China) were subjected to live bacterial manipulation. An amount of 24 g of agar medium (Huankai Microbial Technology Co., Ltd., Guangdong, China) and 1000 mL of distilled water were weighed to prepare an agar plate medium, and a sterilization treatment was performed. Slanted preserved bacterial strains were inoculated onto agar plates and placed in a constant temperature and humidity incubator (Shanghai Zhetu Scientific Instrument Co., Ltd., Shanghai, China) with a relative temperature of 38 °C for cultivation for 18–20 h. Next, the required bacterial suspension was prepared. Finally, sample testing was conducted according to References [35,40]. The formula for calculating the antibacterial rate is shown in Formula (5), where R represents the antibacterial rate and the unit is %. B represents the average number of recovered colonies of pure coating samples after 48 h, in CFU/piece. C represents the average number of recovered bacteria in the antibacterial coating sample after 48 h, in CFU/piece.

$$R = \frac{B - C}{B} \times 100\% \tag{5}$$

2.4.7. Hardness, Impact Resistance, and Adhesion Testing of Coatings

(1) Hardness: according to GB/T 6739-2022 [41], a pencil with a hardness of 9B-9H was used, which was determined by a QHQ-A portable pencil hardness tester (Shenzhen Weichuangjie Testing Instrument Co., Ltd., Shenzhen, China). The pencil was inserted diagonally at a 45° angle into the pencil hardness tester with a load of 750 g for hardness testing.

(2) Impact resistance: according to the content of GB/T 4893.9-2013 "Physical and chemical properties testing of furniture surface coating—Part 9: determination of impact resistance" [42], the impact resistance of the wood surface coatings was tested with a coating impactor (Dongguan Jiaxin Measuring Instrument Co., Ltd., Dongguan, China). A magnifying glass was used to observe the number of cycles of surface rupture of the coating to evaluate the impact resistance level. Each sample was subjected to 5 impacts. The nearest integer to the arithmetic mean of the evaluation level was taken as the result of

the level evaluation. The impact resistance level increased sequentially from level 5 to level 1, and the evaluation table for the coating impact site level is shown in Table 3.

Table 3. Evaluation table of coating impact position grade.

Level	Changes in Coating on Wood Surface
1	No visible changes (no damage).
2	No cracks on the surface of the coating, but visible impact marks.
3	There are mild cracks on the surface of the coating, usually 1–2 circular or arc cracks.
4	There are moderate to severe cracks on the surface of the coating, usually 3–4 circular or arc cracks.
5	The surface of the coating is severely damaged, with usually more than 5 cycles of ring cracks, arc cracks, or coating detachment.

(3) Adhesion: According to GB/T 4893.4-2013 [43], the adhesion of the coating was tested using a coating adhesion tester (Quzhou Aipu Measuring Instrument Co., Ltd., Quzhou, China). The coating was cross-cut with the blade at a vertical angle of 90 degrees. A 3M adhesive tape was applied on the grid surface and peeled off at an angle close to 60°, quickly and smoothly. The adhesion level decreased from level 0 to level 5.

3. Results and Discussion

3.1. Morphology and Chemical Composition Analysis of Microcapsules

3.1.1. Microscopic Morphology Analysis of Microcapsules

SEM images of the UFRCTEMs are shown in Figure 1. Figure 1A–C show the morphology of UFRCTEM samples #1–#3 under low magnification, while Figure 1D–F show the morphology of the UFRCTEMs under high magnification. The UFRCTEMs with $M_{core}:M_{wall}$ of 0.6:1 had more spherical and rounded shapes, with a small difference in particle size. The UFRCTEMs with $M_{core}:M_{wall}$ of 0.8:1 were spherical, plump, and had a relatively uniform particle size distribution, but they aggregated more severely. The UFRCTEMs with $M_{core}:M_{wall}$ of 1.2:1 had large adherent and irregularly shaped substances, and the difference in UFRCTEM particle size was relatively small. As the $M_{core}:M_{wall}$ increased, the agglomeration phenomenon of the microcapsules continued to strengthen.

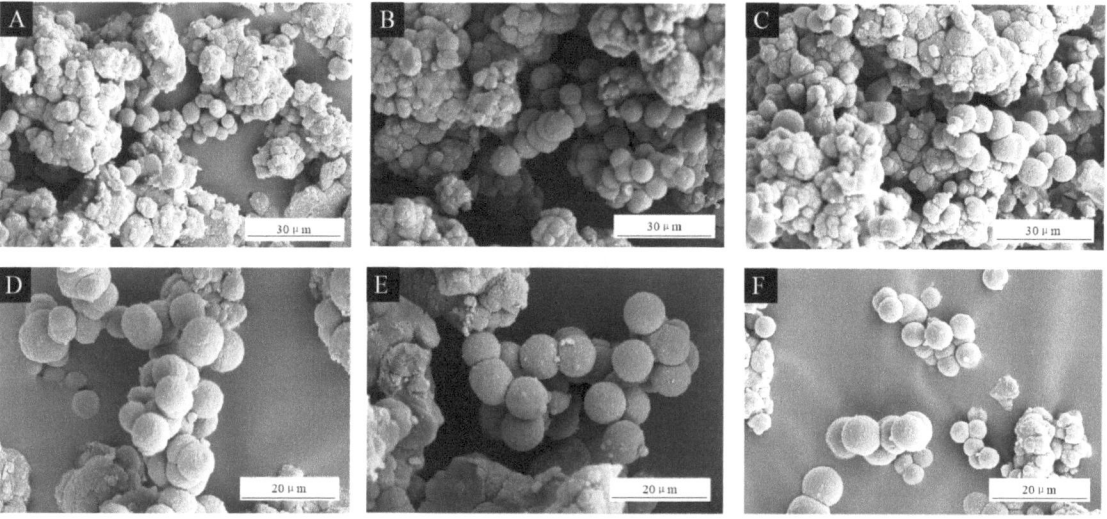

Figure 1. SEM images of UFRCTEMs with different $M_{core}:M_{wall}$. Under low magnification: (**A**) 0.6:1, (**B**) 0.8:1%, and (**C**) 1.2:1. Under high magnification: (**D**) 0.6:1, (**E**) 0.8:1, and (**F**) 1.2:1.

3.1.2. Chemical Composition Analysis of Microcapsules

As shown in Figure 2, the absorption peak at 3350 cm^{-1} was the stretching vibration peak of C-O in the core material and -NH and -OH in the wall material [44]. The characteristic peaks of C=O and C=N in the urea–formaldehyde resin appeared at 1639 cm^{-1} and 1550 cm^{-1}, respectively. The absorption peak at 1247 cm^{-1} was caused by the stretching vibration of C-N and the deformation vibration of N-H in the urea–formaldehyde resin, indicating the presence of the chemical composition of urea–formaldehyde resin in the UFRCTEMs [45]. The characteristic peaks of C=N and C-O in coumarin compounds in the core material were located at 1600 cm^{-1} and 1110 cm^{-1}, which existed on the absorption curve of the microcapsules, proving the presence of *Toddalia asiatica* (L.) Lam extract in the UFRCTEMs [46].

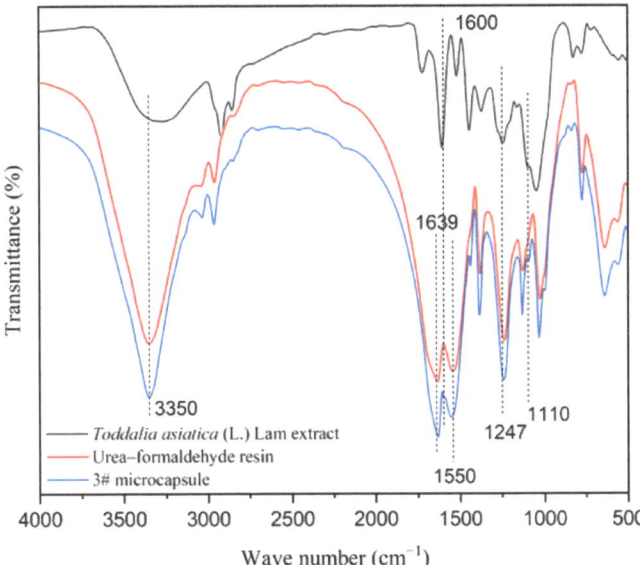

Figure 2. Infrared spectra of core materials, wall materials, and UFRCTEMs.

The above results indicate that the UFRCTEMs contain characteristic peaks of the core and wall materials.

3.2. Optical Performance Analysis of Coatings

The results of the coating glossiness and light loss rate for the UFRCTEMs with different $M_{core}:M_{wall}$ and different contents are shown in Table 4. The data at a 60° incidence angle are plotted in Figure 3. Compared with the surface coating of poplar wood without UFRCTEMs added, the glossiness of the three coatings prepared with UFRCTEMs with different $M_{core}:M_{wall}$ decreased with the increase in UFRCTEM content, and the difference in glossiness data between them was relatively small. When the $M_{core}:M_{wall}$ of the added UFRCTEMs was 0.8:1, the overall glossiness of the coating was slightly higher, and the optical performance was relatively excellent. When the content of the UFRCTEMs in the coating was between 1.0% and 5.0%, the glossiness on the surface of poplar wood decreased rapidly. When the content of the UFRCTEMs was greater than 5.0%, the trend of glossiness changes tended to be gentle. When the $M_{core}:M_{wall}$ of the UFRCTEMs was 0.8:1, the optical performance of the coating was better, and the maximum glossiness of the coating was 7.00 GU.

Table 4. Glossiness and light loss rate of poplar surface coatings with different $M_{core}:M_{wall}$ and different content.

Sample (#)	$M_{core}:M_{wall}$	Microcapsule Content (%)	Glossiness (GU)			Light Loss Rate (%)
			20°	60°	85°	
-	-	0	2.20	14.03	33.23	-
Surface coating on poplar wood with #1 microcapsules added	0.6:1	1.0	1.87	7.17	10.57	48.90
		3.0	1.33	5.20	2.33	62.94
		5.0	1.10	4.40	1.53	68.64
		7.0	1.17	4.30	1.10	69.35
		9.0	0.90	2.53	0.87	81.97
Surface coating on poplar wood with #2 microcapsules added	0.8:1	1.0	1.60	7.00	9.93	50.11
		3.0	1.40	6.23	2.77	55.60
		5.0	1.23	4.60	1.13	67.21
		7.0	1.13	3.43	0.93	75.55
		9.0	0.97	2.83	0.77	79.83
Surface coating on poplar wood with #3 microcapsules added	1.2:1	1.0	1.63	8.50	14.30	39.42
		3.0	1.40	6.17	3.43	56.02
		5.0	1.27	4.30	1.57	69.35
		7.0	1.13	3.80	1.00	72.92
		9.0	0.87	2.80	0.93	80.04

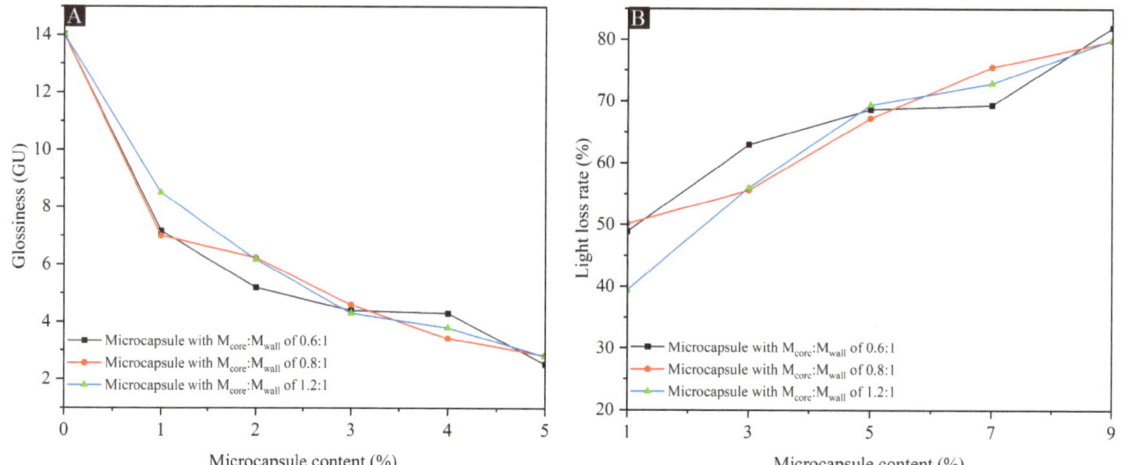

Figure 3. The effects of UFRCTEMs with different $M_{core}:M_{wall}$ on glossiness and light loss rate of poplar surface coatings: (A) glossiness and (B) light loss rate.

The lower the light loss of the wood surface coating, the better the optical performance of the coating [47,48]. Table 4 shows that when the UFRCTEMs with the same $M_{core}:M_{wall}$ were added to the coating, the light loss was positively correlated with the content of UFRCTEMs in the coating.

Table 5 and Figure 4, respectively, show the results and trends of coating chromaticity values for UFRCTEMs with different $M_{core}:M_{wall}$ and contents. Compared with the coating without UFRCTEMs in Figure 4A, the brightness value decreased with the continuous increase in UFRCTEM content. The addition of UFRCTEMs causes protrusions on the coating, weakening the coating's ability to reflect light and reducing the brightness value. Figure 4B,C show a positive correlation between the red and green values and the content of UFRCTEMs. The yellow and blue values generally show a trend of first increasing, then decreasing, and then increasing with the increase in UFRCTEM content. In Figure 4D, the color difference value shows an overall upward trend with the increase in UFRCTEM content. For the wood surface coating containing the UFRCTEMs with $M_{core}:M_{wall}$ of 0.6:1, the minimum color difference of the coating was 0.98 when the microcapsule content was 1.0%. The minimum color difference value with 0.8:1 $M_{core}:M_{wall}$ of UFRCTEMs added to the coating was 1.35 when the UFRCTEM content was 3.0%. When $M_{core}:M_{wall}$ of 1.2:1 UFRCTEMs was added to the coating, the minimum color difference of the coating was 1.22 when the UFRCTEM content was 3.0%.

Table 5. Chromaticity and color difference of poplar surface coatings with different $M_{core}:M_{wall}$ and different content.

Sample (#)	$M_{core}:M_{wall}$	Microcapsule Content (%)	Chromaticity Parameter						ΔE
			L_1	a_1	b_1	L_2	a_2	b_2	
-	-	0	82.35	5.30	29.95	82.25	5.55	29.85	-
Surface coating on poplar wood with #1 UFRCTEM added	0.6:1	1	81.90	6.25	28.85	81.50	6.10	28.70	0.98
		3	78.35	8.75	27.00	78.20	8.20	26.60	2.41
		5	76.25	10.20	27.15	76.05	9.80	27.10	3.39
		7	77.50	8.20	30.45	77.20	8.05	29.80	2.01
		9	69.25	12.35	29.50	69.00	11.95	29.45	6.53
Surface coating on poplar wood with #2 UFRCTEM added	0.8:1	1	81.55	4.25	26.65	81.50	4.20	26.70	1.83
		3	79.65	7.15	28.05	79.25	6.85	27.75	1.35
		5	75.95	10.40	27.35	75.65	10.15	26.80	3.60
		7	76.35	8.15	25.75	76.15	7.90	25.45	3.23
		9	75.60	11.95	28.55	75.10	11.70	28.05	4.13
Surface coating on poplar wood with #3 UFRCTEM added	1.2:1	1	78.90	6.85	27.70	78.70	6.45	27.60	1.49
		3	78.40	6.30	31.00	78.05	6.20	30.70	1.22
		5	75.55	7.25	30.30	75.50	6.85	30.40	2.54
		7	76.10	7.65	30.00	76.05	7.35	29.60	2.36
		9	76.65	9.15	27.65	76.10	8.95	27.25	2.94

Comparing the color difference values corresponding to the coatings prepared by three types of UFRCTEMs, the overall color difference with $M_{core}:M_{wall}$ of 1.2:1 added was relatively small. Because the UFRCTEMs with $M_{core}:M_{wall}$ of 1.2:1 have a higher content of core material, the prepared UFRCTEMs have a darker color compared to the other two samples, which can balance the color difference caused by the wood grain on the surface of the poplar.

Figure 5 shows the effects of UFRCTEMs with different $M_{core}:M_{wall}$ on the surface coating reflectivity of poplar wood. The reflectivity curves of wood surface coatings prepared using three types of UFRCTEMs were highly similar. As the content of UFRCTEMs increased, the overall reflectivity showed an upward trend. Table 6 shows the effect of UFRCTEM content on the coating reflectivity R value of UFRCTEMs with different $M_{core}:M_{wall}$. The content of UFRCTEMs had a relatively small impact on the reflectivity R value of the coating, and it was consistent with the change in the reflectivity curve. When adding UFRCTEMs with $M_{core}:M_{wall}$ of 0.6:1, the coating reflectivity showed a trend of first increasing, decreasing, and then increasing again. When the UFRCTEM content was 1.0%, the largest reflectivity value was 0.6472. When the $M_{core}:M_{wall}$ of UFRCTEMs was 0.8:1, the reflectivity fluctuated and generally showed an upward trend. When the content was 7.0%, the largest reflectivity value was 0.6699. When the $M_{core}:M_{wall}$ of UFRCTEMs was 1.2:1, the reflectivity showed a trend of first increasing and then decreasing. When the UFRCTEM content was 7.0%, the largest reflectivity value was 0.6310.

The higher the reflectivity of the coating, the weaker the light absorption ability, and thus the corresponding heat absorption ability was also weaker. The absorption capacity of coatings for solar infrared and ultraviolet rays decreases, weakening the accumulated heat on the coating surface, thereby extending the lifespan of wooden substrates and their surface coatings [49].

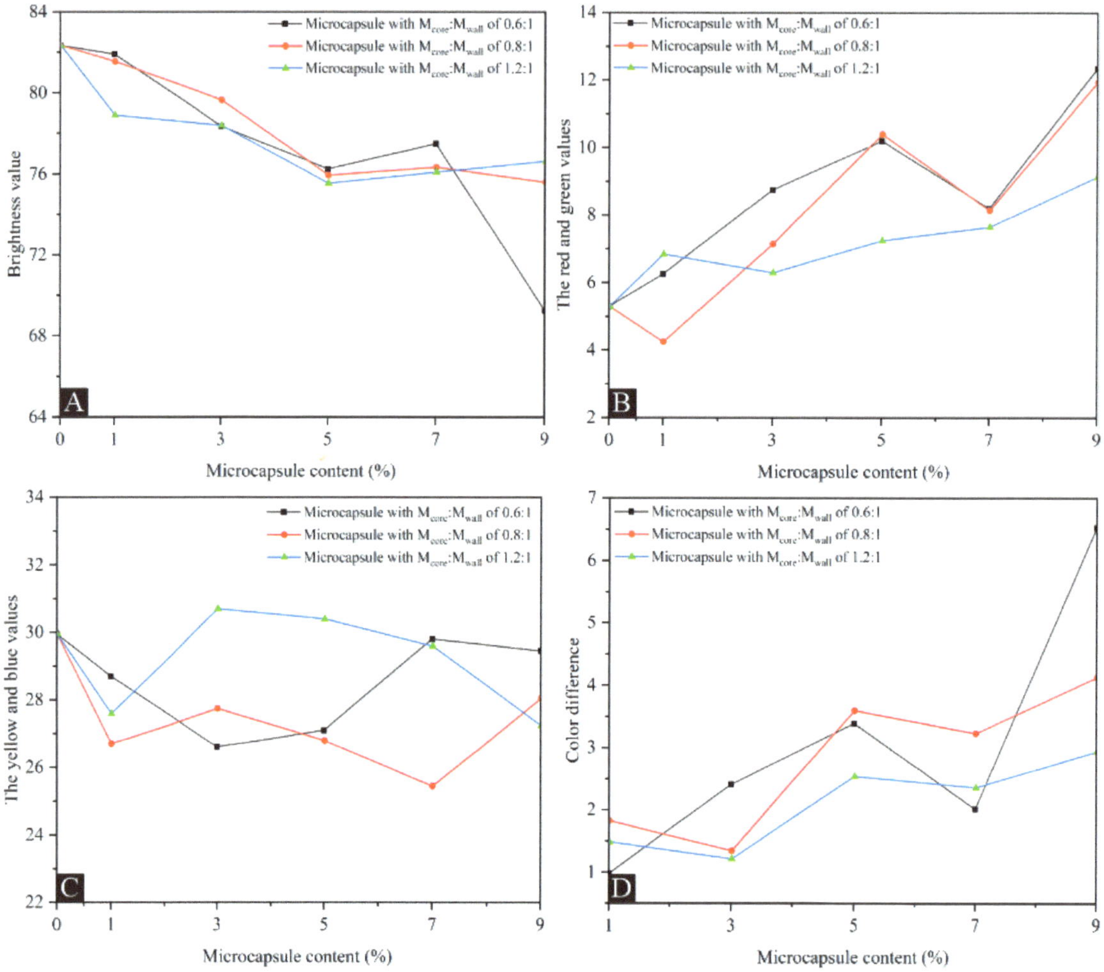

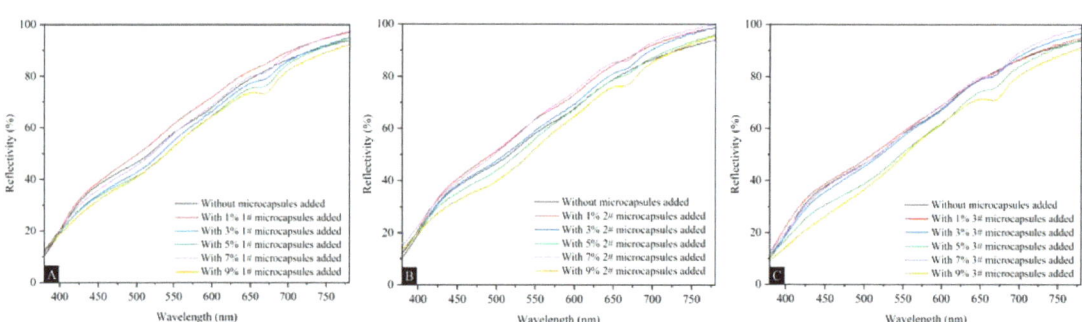

Figure 4. The effects of UFRCTEMs with different $M_{core}:M_{wall}$ on chromaticity and color difference of the poplar surface coatings: (**A**) L value, (**B**) a value, (**C**) b value, and (**D**) ΔE.

Figure 5. The effects of UFRCTEMs with different $M_{core}:M_{wall}$ on the reflectivity of the poplar surface coating: (**A**) 0.6:1, (**B**) 0.8:1, and (**C**) 1.2:1.

Table 6. The effects of UFRCTEMs with different $M_{core}:M_{wall}$ on the poplar surface coating reflectance R value.

Microcapsule Content (%)	Reflectance R Value of Surface Coating on Poplar Wood		
	0.6:1	0.8:1	1.2:1
0	0.6193	0.6193	0.6193
1.0	0.6472	0.5652	0.6298
3.0	0.6014	0.6399	0.6176
5.0	0.5890	0.6144	0.5715
7.0	0.6250	0.6699	0.6310
9.0	0.5770	0.5899	0.5471

3.3. Cold Liquid Resistance of Coatings

Table 7 shows the cold liquid resistance level of surface coatings on poplar wood with different $M_{core}:M_{wall}$ and different amounts. The citric acid resistance level of the surface coatings on poplar prepared by using three types of UFRCTEMs increased to level 1. The ethanol and cleaning agent resistance levels upgraded from level 3 to level 1 or 2. The addition of UFRCTEMs had a positive protective effect on the liquid resistance of the surface coating on poplar wood. As the $M_{core}:M_{wall}$ of UFRCTEMs increased, the liquid resistance of the coating became increasingly excellent. When the $M_{core}:M_{wall}$ of UFRCTEMs was 0.6:1 and the UFRCTEM content was 9.0%, the liquid resistance performance of the coating was level 1. The liquid resistance of the coating was level 1 when the $M_{core}:M_{wall}$ of UFRCTEMs was 0.8:1 and the UFRCTEM content was greater than or equal to 7.0%. When the $M_{core}:M_{wall}$ of UFRCTEMs was 1.2:1 and the UFRCTEM content was greater than or equal to 5.0%, the liquid resistance performance of the coating was level 1.

Table 7. Cold liquid resistance grades of the poplar surface coatings with different $M_{core}:M_{wall}$ and different content.

Sample (#)	$M_{core}:M_{wall}$	Microcapsule Content (%)	Cold Liquid Resistance Level of Surface Coating on Poplar Wood (Level)		
			Citric Acid	Ethanol	Cleaning Agents
-	-	0	2	3	3
Surface coating on poplar wood with #1 UFRCTEMs added	0.6:1	1	1	2	3
		3	1	2	2
		5	1	2	2
		7	1	1	2
		9	1	1	1
Surface coating on poplar wood with #2 UFRCTEMs added	0.8:1	1	1	2	2
		3	1	2	2
		5	1	2	2
		7	1	1	1
		9	1	1	1
Surface coating on poplar wood with #3 UFRCTEMs added	1.2:1	1	1	2	2
		3	1	2	2
		5	1	1	1
		7	1	1	1
		9	1	1	1

The addition of UFRCTEMs to waterborne coatings can protect the coating from erosion by the test liquid. Poplar is affected by various stains, watermarks, and other factors during use. Improving the cold liquid resistance of the coating can provide protection for the substrate and surface coating on poplar wood, thereby extending the service life of the coating and the poplar wood [50].

3.4. Mechanical Properties of Surface Coatings

3.4.1. Hardness of Surface Coating on Poplar Wood

The hardness of wood surface coating refers to the resistance of the coating to a series of mechanical forces such as scratches, impacts, and squeezing, and is an important indicator of the mechanical strength of the coating. As shown in Table 8, the hardness of the surface coating on poplar wood changed. The hardness of the surface coating on wood without the addition of UFRCTEMs was B. For UFRCTEMs with $M_{core}:M_{wall}$ of 0.6:1 and 0.8:1, the hardness of the coating increased from B to 2H with the increase in microcapsule content. For UFRCTEMs with $M_{core}:M_{wall}$ of 0.8:1, the coating hardness increased from B to 3H with the increase in microcapsule content.

Table 8. The hardness of the poplar surface coatings with different $M_{core}:M_{wall}$ of UFRCTEMs and different content.

Microcapsule Content (%)	Hardness		
	0.6:1	0.8:1	1.2:1
0	B	B	B
1.0	B	B	B
3.0	HB	HB	HB
5.0	HB	HB	2H
7.0	HB	2H	2H
9.0	2H	2H	3H

The larger the $M_{core}:M_{wall}$ of microcapsules, the higher the hardness of the surface coating of the poplar wood. This is because UFRCTEMs are small particles with a certain volume; adding UFRCTEMs to waterborne coatings can fill the pores in the coating matrix. The addition of microcapsules increases the solid content of waterborne coatings, thereby increasing the density of the coating and gradually increasing the hardness [51].

3.4.2. Impact Resistance of Surface Coating on Poplar Wood

The impact resistance of a layer, also known as impact strength, is the ability of a surface coating to withstand heavy impact without cracking. Table 9 shows the changes in the impact resistance level of surface coatings on poplar wood. Figures 6–8 show the surface condition of the poplar coatings after impact resistance testing. The impact resistance level of the wood surface coating without UFRCTEMs was level 5, and the performance was poor and did not meet the qualified standards. When the $M_{core}:M_{wall}$ of UFRCTEMs in the coating was 0.6:1, the impact resistance level of the coating was increased from level 4 to level 3. When the content of UFRCTEMs exceeded 7.0%, the surface coating of the poplar wood met the qualified standard for surface coating of wooden furniture (GB/T 3324-2017) [52]. When the $M_{core}:M_{wall}$ was 0.8:1, the impact resistance level of the coating was increased from level 4 to level 3. When the content of UFRCTEMs exceeded 5.0%, the surface coating of the poplar wood met the qualified standard. When the $M_{core}:M_{wall}$ was 1.2:1 and the microcapsule content was greater than 3.0%, the impact resistance level of the coating reached the qualified standard.

Table 9. Impact resistance grades of poplar surface coatings with different $M_{core}:M_{wall}$ of UFRCTEMs and different content.

Microcapsule Content (%)	Impact Resistance Level (Level)		
	0.6:1	0.8:1	1.2:1
0	5	5	5
1.0	4	4	4
3.0	4	4	3
5.0	4	3	3
7.0	3	3	3
9.0	3	3	3

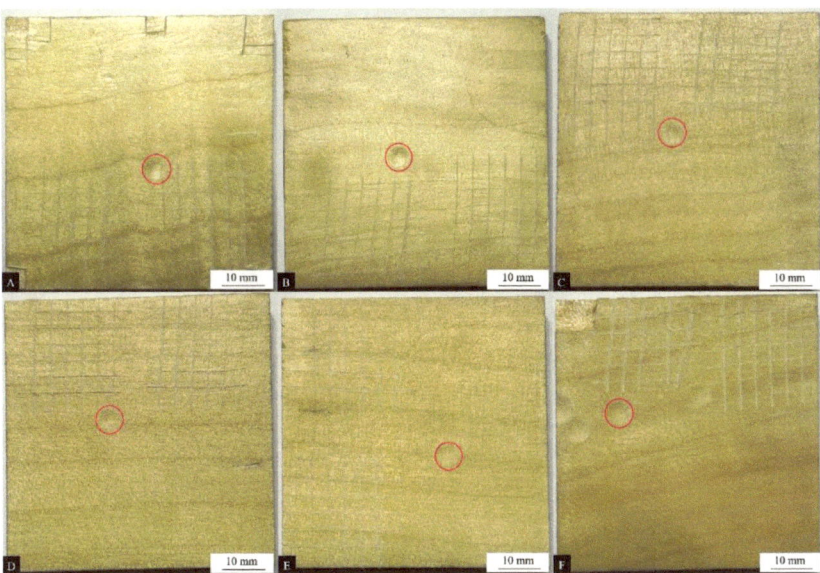

Figure 6. Impact resistance test results of coatings (the red circle) with different contents of the #1 microcapsule type: (**A**) 0%, (**B**) 1%, (**C**) 3%, (**D**) 5%, (**E**) 7%, and (**F**) 9%.

Figure 7. Impact resistance test results of coatings (the red circle) with different contents of the #2 microcapsule type: (**A**) 1%, (**B**) 3%, (**C**) 5%, (**D**) 7%, and (**E**) 9%.

Figure 8. Impact resistance test results of coatings (the red circle) with different contents of the #3 microcapsule type: (**A**) 1%, (**B**) 3%, (**C**) 5%, (**D**) 7%, and (**E**) 9%.

The $M_{core}:M_{wall}$ of UFRCTEMs had a significant impact on the impact resistance of the coating. As the $M_{core}:M_{wall}$ of UFRCTEMs increased, the impact resistance gradually improved. This is because the structure of UFRCTEMs is dense, and when they are added as solid powder fillers to waterborne coatings, UFRCTEMs can enhance the mechanical strength and suppress the cracking of the coating due to stress [53].

3.4.3. Adhesion of Surface Coating on Poplar Wood

Table 10 reflects the changes in the adhesion level of the surface coating of poplar wood. For UFRCTEMs with $M_{core}:M_{wall}$ of 0.6:1 and 0.8:1, when the microcapsule content was between 1.0% and 7.0%, the adhesion level of the coating remained stable at level 1. When the content of microcapsules exceeded 7.0%, the adhesion of the coating decreased to level 2. For UFRCTEMs with $M_{core}:M_{wall}$ of 1.2:1, when the microcapsule content was between 1.0% and 5.0%, the coating adhesion was level 1. When the content of microcapsules exceeded 5.0%, the adhesion of the coating decreased to level 2. Overall, the adhesion performance of UFRCTEMs in the coating was superior when the $M_{core}:M_{wall}$ was 0.6:1 and 0.8:1, while the surface adhesion level of UFRCTEM coatings with $M_{core}:M_{wall}$ of 1.2:1 was poor.

Table 10. Adhesion grades of the poplar surface coatings with different $M_{core}:M_{wall}$ and different content.

Microcapsule Content (%)	Adhesion Level (Level)		
	0.6:1	0.8:1	1.2:1
0	0	0	0
1.0	1	1	1
3.0	1	1	1
5.0	1	1	1
7.0	1	1	2
9.0	2	2	2

The addition of UFRCTEMs disrupts the uniformity of waterborne coatings, causing an agglomeration and stress concentration in the coating, and thus reducing the coating adhesion [54].

3.4.4. Roughness of Surface Coating on Poplar Wood

The roughness values of coatings on poplar wood are shown in Table 11. The roughness value of the surface coating on poplar without microcapsules was 0.260 μm. When adding UFRCTEMs with $M_{core}:M_{wall}$ of 0.6:1, the coating roughness increased from 1.030 μm to 4.571 μm as the microcapsule content gradually increased. When the $M_{core}:M_{wall}$ of UFRCTEMs was 0.8:1, the coating roughness increased from 0.951 μm to 4.961 μm with the increase in microcapsule content. When the $M_{core}:M_{wall}$ of the added UFRCTEMs was 1.2:1, the coating roughness increased from 1.387 μm to 4.226 μm as the microcapsule content gradually increased.

Table 11. Surface coating roughness of the poplar with different $M_{core}:M_{wall}$ and different content.

Microcapsule Content (%)	Roughness (μm)		
	0.6:1	0.8:1	1.2:1
0	0.260	0.260	0.260
1.0	1.030	0.951	1.387
3.0	2.806	2.457	2.194
5.0	2.161	3.407	2.508
7.0	3.522	3.759	3.947
9.0	4.571	4.961	4.226

This is because the UFRCTEMs contain solid powder and cannot be uniformly dispersed when mixed with waterborne coatings, resulting in an uneven coating surface. In addition, the manual brushing process used also affects the uniformity and smoothness of the coating, thus increasing the roughness of the coating to a certain extent. Overall, the $M_{core}:M_{wall}$ of three UFRCTEM samples had a certain degree of influence on the coating roughness on the wood surface. The coating with 0.8:1 of $M_{core}:M_{wall}$ had a higher roughness, followed by the coating with 0.6:1 of $M_{core}:M_{wall}$, and the coating with 1.2:1 of $M_{core}:M_{wall}$ had a lower roughness. This is because the morphology of UFRCTEMs can affect the surface smoothness of the coating. The UFRCTEMs with $M_{core}:M_{wall}$ of 0.8:1 tend to agglomerate more severely, resulting in an uneven dispersion of UFRCTEMs in waterborne topcoats and an increase in coating roughness values.

3.5. Antibacterial Properties of Surface Coatings on Poplar Wood

Table 12 shows the antibacterial rates of surface coatings on poplar against *Escherichia coli* and *Staphylococcus aureus*, and the average number of recovered colonies. The overall antibacterial rate of the three types of microcapsule coatings against *Staphylococcus aureus* was slightly higher than that of *Escherichia coli*, and the coating with $M_{core}:M_{wall}$ of 0.8:1 had the best comprehensive antibacterial rate. As the content of UFRCTEMs in the coating increased, the antibacterial rate of the coating also increased. Figure 9 shows the trend of antibacterial rate changes. Figures 10 and 11 show the bacterial colonies of *Escherichia coli* in the culture dish after antibacterial testing with microcapsules #2 and #3. For *Escherichia coli*, when the content of UFRCTEMs in the coating was between 1.0% and 7.0%, the antibacterial rate increased significantly. When the content of UFRCTEMs was greater than 7.0%, the increase in antibacterial rate was gradual. For *Staphylococcus aureus*, when the content of UFRCTEMs in the coating was less than 5.0%, the antibacterial rate curve increased significantly. When the content of UFRCTEMs was greater than 7.0%, the antibacterial rate curve gradually stabilized. When adding UFRCTEMs with $M_{core}:M_{wall}$ of 0.6:1, the maximum antibacterial rate against *Escherichia coli* was 72.62%, and the maximum antibacterial rate against *Staphylococcus aureus* was 79.08%. When the $M_{core}:M_{wall}$ of UFRCTEMs was 0.8:1, the maximum antibacterial rate against *Escherichia coli* was 71.16%, and the maximum antibacterial rate against *Staphylococcus aureus* was 80.16%. When the $M_{core}:M_{wall}$ of UFRCTEMs was 1.2:1, the maximum antibacterial rate against *Escherichia coli* was 72.33%, and the maximum antibacterial rate against *Staphylococcus aureus* was 76.90%.

The addition of antibacterial microcapsules enhanced the antibacterial performance of waterborne coatings on the surface of poplar wood, proving that UFRCTEMs do indeed exert antibacterial effects.

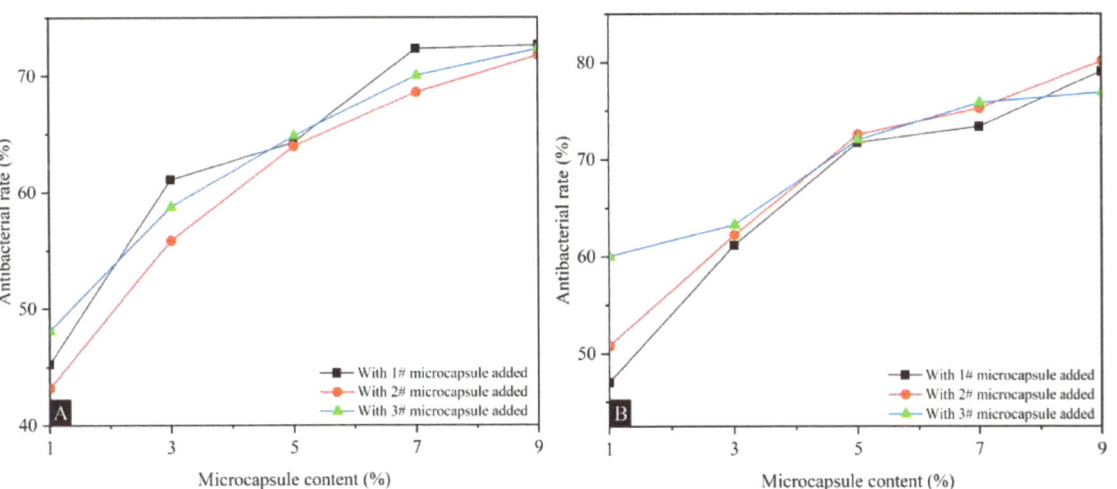

Figure 9. The effects of UFRCTEMs with different $M_{core}:M_{wall}$ on the antibacterial rate of coatings: (**A**) antibacterial rate of *Escherichia coli*, (**B**) antibacterial rate of *Staphylococcus aureus*.

Table 12. Average number of recovered colonies and antibacterial rate of the poplar surface coatings.

$M_{core}:M_{wall}$	Microcapsule Content (%)	Average Number of Recovered *Escherichia coli* (CFU·Piece^{-1})	Antibacterial Rate of *Escherichia coli* (%)	Average Number of Recovered *Staphylococcus aureus* (CFU·Piece^{-1})	Antibacterial Rate of *Staphylococcus aureus* (%)
-	0	347	-	368	-
0.6:1	1	190	45.25	195	47.01
	3	135	61.10	143	61.14
	5	124	64.27	104	71.74
	7	96	72.33	98	73.37
	9	95	72.62	77	79.08
0.8:1	1	197	43.22	181	50.82
	3	153	55.91	139	62.23
	5	125	63.98	101	72.55
	7	109	68.59	91	75.27
	9	98	71.76	73	80.16
1.2:1	1	180	48.13	147	60.05
	3	143	58.79	135	63.32
	5	122	64.84	103	72.01
	7	104	70.03	89	75.82
	9	96	72.33	85	76.90

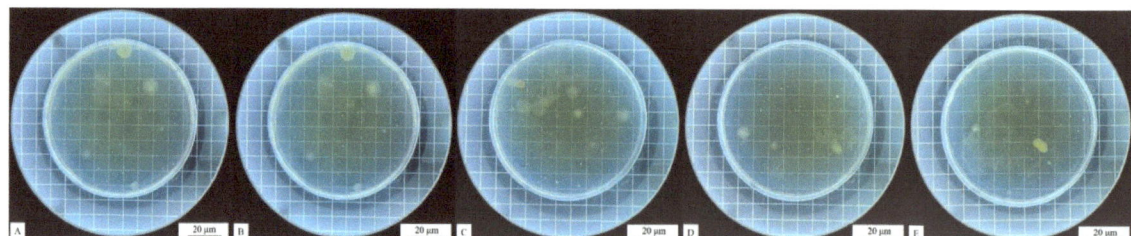

Figure 10. Colony recovery of coatings with different contents of the #2 microcapsule type after the antibacterial test against *Escherichia coli*: (**A**) 1%, (**B**) 3%, (**C**) 5%, (**D**) 7%, and (**E**) 9%.

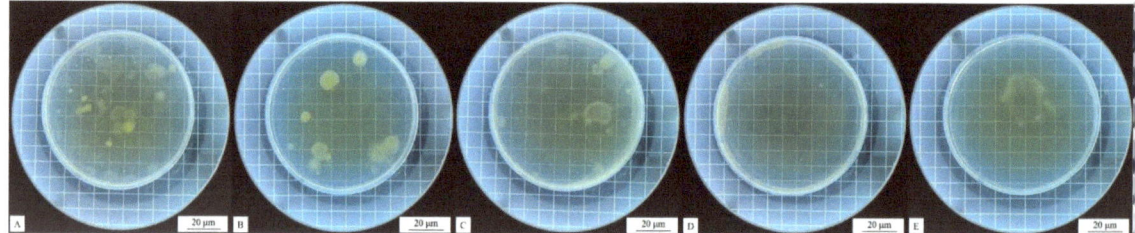

Figure 11. Colony recovery of coatings with different contents of the #3 microcapsule type after the antibacterial test against *Escherichia coli*: (**A**) 1%, (**B**) 3%, (**C**) 5%, (**D**) 7%, and (**E**) 9%.

3.6. Microscopic Morphology and Chemical Composition of Surface Coating on Poplar Wood

The coating with a 7.0% addition of UFRCTEMs had a better overall performance. Therefore, SEM analysis was performed on the coatings prepared using microcapsules with different $M_{core}:M_{wall}$ at 7.0% addition (Figure 12). The surface of the coating without UFRCTEMs was relatively flat, while the coating with $M_{core}:M_{wall}$ of 0.8:1 and 1.2:1 was relatively rough. Among them, the coating with $M_{core}:M_{wall}$ of 1.2:1 had significant protrusions on the surface. After adding microcapsules with $M_{core}:M_{wall}$ of 0.6:1, the coating showed less agglomeration and a smoother surface. At 7.0% additive content, the coating roughness values for $M_{core}:M_{wall}$ of 0.6:1, 0.8:1, and 1.2:1 were 3.522 μm, 3.759 μm, and 3.947 μm, respectively. The observed microstructure of the surface coating on the poplar wood was consistent with the roughness value analysis results.

   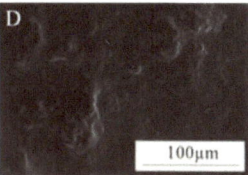

Figure 12. SEM images of wood surface coatings prepared by adding 7.0% UFRCTEMs with different $M_{core}:M_{wall}$: (**A**) without UFRCTEMs, (**B**) 0.6:1, (**C**) 0.8:1, and (**D**) 1.2:1.

Figure 13 shows the infrared spectra of the waterborne coating on the surface of poplar with UFRCTEMs added and without UFRCTEMs added. The characteristic peak of around 1727 cm^{-1} belonged to the stretching vibration peak of C=O in waterborne coatings. The stretching vibration peaks of -CH$_3$ and C=O in the urea–formaldehyde resin wall material of UFRCTEMs were located at 2924 cm^{-1} and 1639 cm^{-1}, respectively [55]. The characteristic peak at 1114 cm^{-1} was the absorption peak of C-O in the UFRCTEM core material.

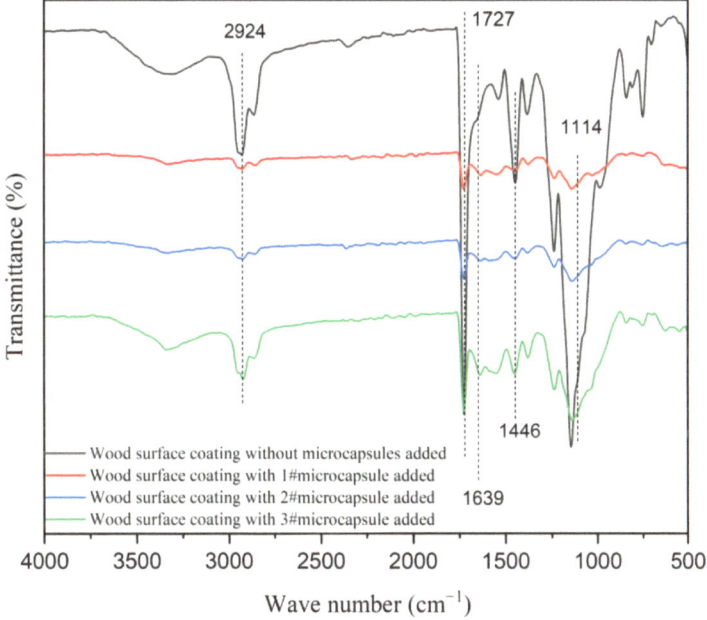

Figure 13. Infrared spectra of UFRCTEM coatings with different $M_{core}:M_{wall}$ on a wood surface.

The above findings indicate that the chemical composition of the waterborne coating did not change after mixing with UFRCTEMs, and these results are consistent with the infrared analysis results of the UFRCTEMs.

3.7. Sectional Views Coatings with Different $M_{core}:M_{wall}$ on Wood Surface

Figure 14 shows views of poplar wood coated with a UFRCTEM coating and a waterborne coating. Figure 14A shows the poplar layer, waterborne primer layer, and waterborne topcoat layer with UFRCTEMs added. The yellow circle indicates a small amount of waterborne primer penetrating and filling the conduit holes of poplar wood, while the red circle represents the UFRCTEMs distributed in the waterborne topcoat. In Figure 14A, the waterborne topcoat layer containing UFRCTEMs had poor flatness, while in Figure 14B, the section of the waterborne coating was relatively smooth and even. The results in Figure 9 are consistent with the optical and mechanical analysis results of the coating.

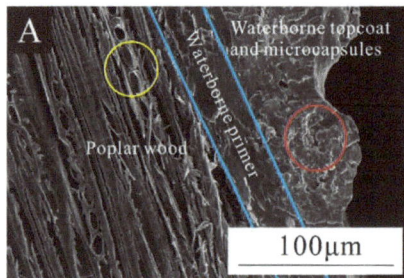 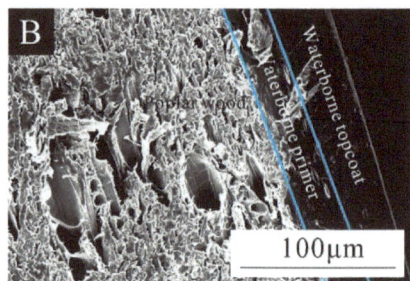

Figure 14. SEM of the interface between the wood and the coating: (**A**) with UFRCTEM; (**B**) without UFRCTEM.

4. Conclusions

UFRCTEMs were added to waterborne paint and applied to poplar wood to explore the effects of different contents and $M_{core}:M_{wall}$ on the performance of waterborne coatings on the surface of poplar wood. Under different $M_{core}:M_{wall}$ of UFRCTEMs, as the content of UFRCTEMs increased, the glossiness and the visible light reflectance gradually decreased, and the light loss and the color difference gradually increased. Adding microcapsules to the coating improved the cold liquid resistance, enhanced the hardness, increased the surface roughness, and enhanced the impact resistance. However, to some extent, it weakened the adhesion. As the content of UFRCTEMs increased, the antibacterial rate against both bacteria showed an upward trend, and the overall antibacterial rate against *Staphylococcus aureus* was slightly higher than that against *Escherichia coli*. The UFRCTEMs in the coating did not react chemically with waterborne coatings, ensuring excellent performance of the UFRCTEMs. The coating on a poplar surface with $M_{core}:M_{wall}$ of 0.8:1 and UFRCTEM content of 7.0% had excellent comprehensive performance, ensuring good optical, cold liquid resistance, mechanical, and antibacterial properties of the coating. The glossiness was 3.43 GU, light loss was 75.55%, reflectance R value was 0.6699, color difference ΔE was 3.23, hardness was 2H, impact resistance level was level 3 at an impact height of 50 mm, adhesion level was level 1, and roughness value was 3.759 μm. The cold liquid resistance was excellent, with resistance to citric acid, ethanol, and cleaning agents of grade 1. The antibacterial rates against *Escherichia coli* and *Staphylococcus aureus* were 68.59% and 75.27%, respectively. These findings provide technical references for the application of UFRCTEMs in antibacterial coatings on poplar surfaces and expand their potential application prospects in industries such as coatings. The obtained microcapsule powder was relatively rough and adhesive, with a low encapsulation rate. Therefore, it was necessary to separate and purify the main antibacterial components of *Toddalia asiatica* (L.) Lam extract to prepare antibacterial microcapsules with better performance. The antibacterial performance of waterborne coatings on poplar wood surfaces was investigated after 48 h, without testing and analyzing the durability of microcapsules and the maximum antibacterial time of waterborne coatings containing microcapsules. Only microcapsules were added to waterborne topcoats, without exploring the effects of different methods of adding microcapsules and different types of wood on the surface coating. In the future, more in-depth exploration will be conducted on the above limitations.

Author Contributions: Conceptualization, methodology, writing—review and editing, Y.Z.; validation, resources, and data management, Y.W.; formal analysis, investigation, and supervision, X.Y. All authors have read and agreed to the published version of the manuscript.

Funding: This project was partly supported by Qing Lan Project and the Natural Science Foundation of Jiangsu Province (BK20201386).

Institutional Review Board Statement: Not applicable.

Informed Consent Statement: Not applicable.

Data Availability Statement: Data are contained within the article.

Conflicts of Interest: The authors declare no conflicts of interest.

References

1. Brezina, D.; Michal, J.; Hlaváčková, P. The Impact of Natural Disturbances on the Central European Timber Market—An Analytical Study. *Forests* **2024**, *15*, 592. [CrossRef]
2. Wang, C.; Yu, J.H.; Jiang, M.H.; Li, J.Y. Effect of selective enhancement on the bending performance of fused deposition methods 3D-printed PLA models. *BioResources* **2024**, *19*, 2660–2669. [CrossRef]
3. Wu, W.; Xu, W.; Wu, S. Mechanical performance analysis of double-dovetail joint applied to furniture T-shaped components. *BioResources* **2024**, *19*, 5862–5879. [CrossRef]
4. Zhang, N.; Xu, W.; Tan, Y. Multi-attribute hierarchical clustering for product family division of customized wooden doors. *Bioresources* **2023**, *18*, 7889–7904. [CrossRef]
5. Singh, A.P.; Kim, Y.S.; Chavan, R.R. Relationship of wood cell wall ultrastructure to bacterial degradation of wood. *IAWA J.* **2019**, *40*, 845–870. [CrossRef]
6. Hu, J.; Liu, Y.; Wang, J.X.; Xu, W. Study of selective modification effect of constructed structural color layers on European beech wood surfaces. *Forests* **2024**, *15*, 261. [CrossRef]
7. Hu, W.; Yu, R. Study on the strength mechanism of the wooden round-end mortise-and-tenon joint using the digital image correlation method. *Holzforschung* **2024**. [CrossRef]
8. Hu, W.G.; Li, S.; Liu, Y. Vibrational characteristics of four wood species commonly used in wood products. *BioResources* **2021**, *16*, 7100–7110. [CrossRef]
9. Wu, S.S.; Tao, X.; Xu, W. Thermal conductivity of poplar wood veneer impregnated with graphene/polyvinyl alcohol. *Forests* **2021**, *12*, 777. [CrossRef]
10. Wu, S.S.; Zhou, J.C.; Xu, W. A convenient approach to manufacturing lightweight and high-sound-insulation plywood using furfuryl alcohol/multilayer graphene oxide as a shielding layer. *Wood Mater. Sci. Eng.* **2024**, 1–8. [CrossRef]
11. Hu, W.G.; Liu, N.; Guan, H.Y. Experimental and numerical study on methods of testing withdrawal resistance of mortise-and-tenon joint for Wood products. *Forests* **2020**, *11*, 280. [CrossRef]
12. Qi, Y.Q.; Zhang, Z.Q.; Sun, Y.; Shen, L.M.; Han, J.L. Study on the process optimization of peanut coat pigment staining of poplar wood. *Forests* **2024**, *15*, 504. [CrossRef]
13. Qi, Y.Q.; Sun, Y.; Zhou, Z.W.; Huang, Y.; Li, J.X.; Liu, G.Y. Response surface optimization based on freeze-thaw cycle pretreatment of poplar wood dyeing effect. *Wood Res.* **2023**, *68*, 293–305. [CrossRef]
14. Wu, X.Y.; Yang, F.; Gan, J.; Kong, Z.Q.; Wu, Y. A Superhydrophobic, Antibacterial, and Durable Surface of Poplar Wood. *Nanomaterials* **2021**, *11*, 1885. [CrossRef]
15. Wan, B.Y.; Tian, L.X.; Fu, M.; Zhang, G.Y. Green development growth momentum under carbon neutrality scenario. *J. Clean.* **2021**, *316*, 128327. [CrossRef]
16. Wang, C.; Zhang, C.Y.; Zhu, Y. Reverse design and additive manufacturing of furniture protective foot covers. *BioResources* **2024**, *19*, 4670–4678. [CrossRef]
17. Zhao, Z.N.; Niu, Y.T.; Chen, F.Y. Development and Finishing Technology of Waterborne UV Lacquer-Coated Wooden Flooring. *BioResources* **2021**, *16*, 1101–1114. [CrossRef]
18. Tompros, A.; Wilber, M.Q.; Fenton, A.; Carter, E.D.; Gray, M.J. Efficacy of Plant-Derived Fungicides at Inhibiting Batrachochytrium salamandrivorans Growth. *J. Fungi* **2022**, *8*, 1025. [CrossRef]
19. Sang, R.J.; Yang, F. Effect of TiO_2@$CaCO_3$ waterborne primer on the coloring performance of inkjet-printed wood product coatings. *Coatings* **2023**, *13*, 2071. [CrossRef]
20. Sang, R.J.; Yang, F.; Fan, Z.X. The effect of water-based primer pretreatment on the performance of water-based inkjet coatings on wood surfaces. *Coatings* **2023**, *13*, 1649. [CrossRef]
21. Martins, C.H.G. Antibacterial Agents from Natural Sources. *Molecules* **2024**, *29*, 644. [CrossRef]
22. Qi, Y.Q.; Zhou, Z.W.; Xu, R.; Dong, Y.T.; Liu, M.J.; Shen, L.M.; Han, J.L. Research on the dyeing properties of Chinese fir using ultrasonic-assisted mulberry pigment dyeing. *Forests* **2023**, *14*, 1832. [CrossRef]
23. Qi, Y.Q.; Liu, G.Y.; Zhang, Z.; Zhou, Z.W. Optimization of green extraction process of Cinnamomum camphora fruit dye and its performance by response surface methodology. *BioResources* **2023**, *18*, 4916–4934. [CrossRef]
24. Faccio, G. Plant Complexity and Cosmetic Innovation. *iScience* **2020**, *23*, 101358. [CrossRef]
25. Aziz, I.M.; Alshalan, R.M.; Rizwana, H.; Alkhelaiwi, F.; Almuqrin, A.M.; Aljowaie, R.M.; Alkubaisi, N.A. Chemical Composition, Antioxidant, Anticancer, and Antibacterial Activities of Roots and Seeds of *Ammi visnaga* L. Methanol Extract. *Pharmaceuticals* **2024**, *17*, 121. [CrossRef] [PubMed]
26. Li, P.; Lei, K.; Ji, L.S. Characterization of the complete chloroplast genome of *Toddalia asiatica* (L.) Lam. *Mitochondrial DNA B* **2021**, *16*, 1650–1651. [CrossRef] [PubMed]

27. Zeng, Z.; Tian, R.; Feng, J.; Yang, N.A.; Yuan, L. A systematic review on traditional medicine *Toddalia asiatica* (L.) Lam.: Chemistry and medicinal potential. *Saudi Pharm. J.* **2021**, *29*, 781–798. [CrossRef]
28. Kowalczyk, P.; Koszelewski, D.; Brodzka, A.; Kramkowski, K.; Ostaszewski, R. Evaluation of Antibacterial Activity against Nosocomial Pathogens of an Enzymatically Derived α-Aminophosphonates Possessing Coumarin Scaffold. *Int. J. Mol. Sci.* **2023**, *24*, 14886. [CrossRef]
29. Li, Y.D.; Guan, J.P.; Tang, R.C.; Qiao, Y.F. Application of Natural Flavonoids to Impart Antioxidant and Antibacterial Activities to Polyamide Fiber for Health Care Applications. *Antioxidants* **2019**, *8*, 301. [CrossRef]
30. Yan, Y.M.; Li, X.; Zhang, C.H.; Lv, L.J.; Gao, B.; Li, M.H. Research Progress on Antibacterial Activities and Mechanisms of Natural Alkaloids: A Review. *Antibiotics* **2021**, *10*, 318. [CrossRef]
31. Masyita, A.; Sari, R.M.; Astuti, A.D.; Yasir, B.; Rumata, N.R.; Bin Emran, T.; Nainu, F.; Simal-Gandara, J. Terpenes and terpenoids as main bioactive compounds of essential oils, their roles in human health and potential application as natural food preservatives. *Food Chem. X* **2022**, *13*, 100217. [CrossRef] [PubMed]
32. Raj, M.K.; Balachandran, C.; Duraipandiyan, V.; Agastian, P.; Ignacimuthu, S. Antimicrobial activity of Ulopterol isolated from *Toddalia asiatica* (L.) Lam.: A traditional medicinal plant. *J. Ethnopharmacol.* **2012**, *140*, 161–165.
33. Roshan, A.B.; Venkatesh, H.N.; Dubery, N.K.; Mohana, D.C. Chitosan-based nanoencapsulation of Toddalia asiatica (L.) Lam. essential oil to enhance antifungal and aflatoxin B_1 inhibitory activities for safe storage of maize. *Int. J. Biol. Macromol.* **2022**, *204*, 476–484. [CrossRef] [PubMed]
34. Zhu, Y.; Wang, Y.; Yan, X.X. Preparation of chitosan-coated *Toddalia asiatica* (L.) Lam extract microcapsules and its effect on coating antibacterial properties. *Coatings* **2024**, *14*, 942. [CrossRef]
35. Wang, Y.; Yan, X.X. Preparation of *Toddalia asiatica* (L.) Lam. Extract Microcapsules and Their Effect on Optical, Mechanical and Antibacterial Performance of Waterborne Topcoat Paint Films. *Coatings* **2024**, *14*, 655. [CrossRef]
36. *GB/T 11186.3-1989*; Methods for Measuring the Color of Coatings—Part 3: Calculation of Color Difference. Standardization Administration of the People's Republic of China: Beijing, China, 1989.
37. *GB/T 4893.6-2013*; Testing of Physical and Chemical Properties of Furniture Surface Paint Films—Part 6: Gloss Determination Method. Standardization Administration of the People's Republic of China: Beijing, China, 2013.
38. *GB/T 4893.1-2021*; Test of Surface Coatings of Furniture—Part 1: Determination of Surface Resistance to Cold Liquids. Stanardization Administration of the People's Republic of China: Beijing, China, 2021.
39. *GB/T 21866-2008*; Test Method and Effect for Antibacterial Capability of Paints Film. Standardization Administration of the People's Republic of China: Beijing, China, 2008.
40. *GB/T 4789.2-2022*; Microbiological Examination of Food Hygiene—Aerobic Plate Count. Standardization Administration of the People's Republic of China: Beijing, China, 2022.
41. *GB/T 6739-2022*; Paints and Varnishes—Determination of Film Hardness by Pencil Test. Standardization Administration of the People's Republic of China: Beijing, China, 2022.
42. *GB/T 4893.9-2013*; Test of Surface Coatings of Furniture—Part 9: Determination of Resistance to Impact. Standardization Administration of the People's Republic of China: Beijing, China, 2013.
43. *GB/T 4893.4-2013*; Test of Surface Coating of Furniture—Part 4: Determination of Adhesion by Cross-Cut. Standardization Administration of the People's Republic of China: Beijing, China, 2013.
44. Ma, W.Y.; Ali, I.; Li, Y.L.; Hussain, H.; Zhao, H.Z.; Sun, X.; Xie, L.; Cui, L.; Wang, D.J. A Simple and Efficient Two-Dimensional High-Speed Counter-Current Chromatography Linear Gradient and Isocratic Elution Modes for the Preparative Separation of Coumarins from Roots of *Toddalia asiatica* (Linn.) Lam. *Molecules* **2021**, *26*, 5986. [CrossRef] [PubMed]
45. Yan, X.X.; Tao, Y.; Qian, X.Y. Preparation and Optimization of Waterborne Acrylic Core Microcapsules for Waterborne Wood Coatings and Comparison with Epoxy Resin Core. *Polymers* **2020**, *12*, 2366. [CrossRef] [PubMed]
46. Gangopadhyay, A. Plant-derived natural coumarins with anticancer potentials: Future and challenges. *J. Herb. Med.* **2023**, *42*, 100797. [CrossRef]
47. Zhou, J.C.; Xu, W. A fast method to prepare highly isotropic and optically adjustable transparent wood-based composites based on interface optimization. *Ind. Crops Prod.* **2024**, *218*, 118898. [CrossRef]
48. Hu, J.; Liu, Y.; Xu, W. Impact of cellular structure on the thickness and light reflection properties of structural color layers on diverse wood surfaces. *Wood Mater. Sci. Eng.* **2024**, 1–11. [CrossRef]
49. Lim, T.; Bae, S.H.; Yu, S.H.; Baek, K.Y.; Cho, S. Near-Infrared Reflective Dark-Tone Bilayer System for LiDAR-Based Autonomous Vehicles. *Macromol. Res.* **2022**, *30*, 342–347. [CrossRef]
50. Hu, W.G.; Wan, H. Comparative study on weathering durability properties of phenol formaldehyde resin modified sweetgum and southern pine specimens. *Maderas Cienc. Tecnol.* **2022**, *24*, 100417. [CrossRef]
51. Hazir, E.; Koc, K.H. Evaluation of wood surface coating performance using water based, solvent based and powder coating. *Maderas Cienc. Tecnol.* **2019**, *21*, 467–480. [CrossRef]
52. *GB/T 3324-2017*; General Technical Requirements for Wooden Furniture. Standardization Administration of the People's Republic of China: Beijing, China, 2017.
53. Jia, X.; Zeng, H.H.; Gao, Q.; Huang, Z.X.; Bai, X.J.; Zhao, Y.; Zhao, H.F. Impact Resistance and Structural Optimization of POZD Coated Composite Plates. *Int. J. Crashworthiness* **2023**, *28*, 601–615. [CrossRef]

54. Tao, X.; Tian, D.X.; Liang, S.Q.; Li, S.M.; Peng, L.M.; Fu, F. Enhanced Paint Adhesion of Puffed Wood-Based Metal Composites via Surface Treatment with Silane Coupling Agent. *Wood Mater. Sci. Eng.* **2023**, *19*, 573–579. [CrossRef]
55. Yan, X.X.; Zhao, W.T.; Wang, L.; Qian, X.Y. Effect of Microcapsule Concentration with Different Core-Shell Ratios on Waterborne Topcoat Film Properties for Tilia europaea. *Coatings* **2021**, *11*, 1013. [CrossRef]

Disclaimer/Publisher's Note: The statements, opinions and data contained in all publications are solely those of the individual author(s) and contributor(s) and not of MDPI and/or the editor(s). MDPI and/or the editor(s) disclaim responsibility for any injury to people or property resulting from any ideas, methods, instructions or products referred to in the content.

Article

Preparation and Characterisation of UV-Curable Flame Retardant Wood Coating Containing a Phosphorus Acrylate Monomer

Solène Pellerin [1,2], Fabienne Samyn [2], Sophie Duquesne [2] and Véronic Landry [1,*]

[1] NSERC–Canlak Industrial Research Chair in Interior Wood-Product Finishes, Department of Wood and Forest Sciences, Université Laval, Québec City, QC G1V 0A6, Canada
[2] CNRS, INRAE, Centrale Lille, UMR 8207-UMET-Unité Matériaux et Transformations, Université de Lille, F-59000 Lille, France
* Correspondence: veronic.landry@sbf.ulaval.ca; Tel.: +1-(418)-656-2131

Abstract: The application of a flame retardant coating is an effective solution to enhance the fire retardancy of wood flooring. However, finding the right balance between reducing the flame propagation and good overall coating properties while conserving wood appearance is complex. In order to answer this complex problem, transparent ultraviolet (UV)-curable flame retardant wood coatings were prepared from an acrylate oligomer, an acrylate monomer, and the addition of the tri(acryloyloxyethyl) phosphate (TAEP), a phosphorus-based monomer, at different concentrations in the formulation. The coatings' photopolymerisation, optical transparency, hardness, water sorption and thermal stability were assessed. The fire behaviour and the adhesion of the coatings applied on the yellow birch panels were evaluated, respectively, using the cone calorimeter and pull-off tests. Scanning electron microscopy (SEM) and energy-dispersive spectroscopy (EDS) analyses were performed on the collected burnt residues to obtain a better understanding of the flame retardancy mechanism. Our study reveals that phosphorus monomer addition improved the coating adhesion and the fire performance of the coated wood without impacting the photopolymerisation. The conversion percentage remained close to 70% with the TAEP addition. The pull-off strength reached 1.12 MPa for the coating with the highest P-monomer content, a value significantly different from the non-flame retarded coating. For the same coating formulation, the peak of heat release rate decreased by 13% and the mass percentage of the residues increased by 37% compared to the reference. However, the flame-retarded coatings displayed a higher hygroscopy. The action in the condensed phase of the phosphorus flame retardant is highlighted in this study.

Keywords: flame retardant; wood coating; phosphorus acrylate; UV-curable coating

Citation: Pellerin, S.; Samyn, F.; Duquesne, S.; Landry, V. Preparation and Characterisation of UV-Curable Flame Retardant Wood Coating Containing a Phosphorus Acrylate Monomer. *Coatings* **2022**, *12*, 1850. https://doi.org/10.3390/coatings12121850

Academic Editor: Marko Petric

Received: 12 October 2022
Accepted: 25 November 2022
Published: 29 November 2022

Publisher's Note: MDPI stays neutral with regard to jurisdictional claims in published maps and institutional affiliations.

Copyright: © 2022 by the authors. Licensee MDPI, Basel, Switzerland. This article is an open access article distributed under the terms and conditions of the Creative Commons Attribution (CC BY) license (https://creativecommons.org/licenses/by/4.0/).

1. Introduction

With sustainability questions arising in recent decades, wood is more than ever a material of choice for construction. When this renewable resource is supplied from sustainable forests, wood transformation and use are key in the reduction of the energy embodied in building materials. Along with the use of biobased materials, wood constructions provide further energy and greenhouse gas emission savings. In a perspective of limiting global warming, industries are encouraged to replace high-carbon materials with lower-carbon alternatives and when possible, with products based on renewable materials [1]. Hence, using wood instead of steel or cement in the construction sector is one example of emission reduction options. Used since a long time ago for its mechanical properties and visual aspect, wood can be used for both structural and appearance products such as floors, wall panels, ceiling tiles and mouldings [2]. However, the construction industry has to face one limitation of wood: its flammability. To promote their use in non-residential buildings, increasing flame retardancy is thus essential. Different strategies can be employed to improve the fire performance of wood, including the impregnation of flame retardants and the application of flame retardant coatings [3]. In the flooring industry, the second option is usually preferred as it is easier to implement.

Achieving flame retardancy by the application of a flame retardant (FR) coating is mainly dependent on its formulation and the incorporation of FR components. There are different types of flame retardants, the main ones being halogenated compounds, mineral fillers as well as phosphorus- and nitrogen-based compounds. The phosphorus FR have gained interest in recent decades [4]. Phosphorus-based FR may combine a flame retardancy action in the gaseous phase and in the solid phase [5]. Upon phosphorus FR decomposition, various radicals are released including PO• or HPO$_2$• [6]. These are able to react by radical recombination with the highly reactive hydrogen H• and hydroxy OH• radicals present in the flame. Consequently, oxidation reactions in the flame are slowed down. The production of heat being reduced, flame propagation is limited. The action of the phosphorus FR in the solid phase mainly results in its contribution to the char formation by esterification and dehydration [6–8]. The generation of acid species upon temperature elevation leads to further reactions with the surrounding polymer matrix and its decomposition by-products. The efficacy of the phosphorus FR is strongly related to its decomposition products' interaction with the chemical environment in the pyrolysis zone [9].

High-solid-content ultraviolet (UV)-curable coating is the most common technology used in the manufacturing of pre-finished flooring. The finishing system is typically composed of a stain, a tie coat, sealer layers and a topcoat. These coatings are most of the time acrylate based, as this technology offers an overall good price/property balance. The curing is performed within seconds and no drying step is required, offering a time gain compared to waterborne or solvent-borne coatings. The obtained coatings achieve good hardness and excellent chemical resistance [10–12].

In this context, the development of P-containing UV-curable coatings to improve the fire performance of wood may represent an interesting approach to widen the use of wood in the construction industry. Few studies deal with transparent UV-curable coatings offering a good flame retardancy of coated wood while still preserving its appearance [13–18]. The addition of FR monomers, reactive diluents, is generally the investigated approach. Chambhare, Lokhande et al. studied phosphorus- and nitrogen-based monomer addition in UV-curable wood coating formulations by adding, respectively, the tris-diethanolamine spirocyclic pentaerythritol bisphosphorate and a difunctional (hydroxypropyl methacrylate) piperazine modified with cyclic phosphates [19,20]. They reported an improved fire and water resistance, higher crosslinking density, but a loss of adhesion on wood. However, the flame retardancy has only been assessed on the coating and not on the coating/wood assembly. Naik et al. also reported the preparation of FR wood coating based on phosphorus and silicon monomers [21]. The coating showing a good adhesion and hardness as well as the best UL-94 rating was a blend with both monomers. The authors attributed the enhancement in flame retardancy to the synergistic effect of phosphorous and silicon. A similar study was conducted with a P,N-containing monomer [22]. In both studies, the intrinsic flame retardant properties of the coatings were assessed by the UL-94 test and the limiting oxygen index (LOI) assessment. The synergy of three monomers containing, respectively, P, S and N elements, the triallyl orthophosphate, the pentaerythritol tetrakis (3-mercaptopropionate), and, the N-dimethylacrylamide, was also investigated in another publication [14]. UV-curable coating formulations were applied on Chinese fir wood samples and UL-94 tests were performed. After fire exposure, the authors reported the formation of an intumescent char layer for some coatings thanks to the release of NH$_3$ and CO$_2$, non-flammable gases during the degradation process. In those studies, the mechanical properties of the coatings were not always thoroughly studied. Furthermore, it appears that the flame retardancy characteristic of the coatings is not systematically assessed in the final condition of use, particularly on the wood/coating assembly.

By enlarging the literature review of acrylic-based FR coatings to non-wood applications, a monomer seems particularly interesting, namely the tri(acryloyloxyethyl) phosphate (TAEP). The fire retardancy of the monomer has already been highlighted in several studies [23–25]. The synthesis of the trifunctional phosphate acrylate monomer was firstly reported by Liang and Shi [26]. They further studied the thermal degradation of different blends of the TAEP monomer with an epoxy acrylate oligomer or with a polyurethane one [23]. The LOI improved with an increased TAEP content. The highest loading allowed to reach a V0 rating at the UL-94 test for both oligomer mixes. With the thermal degradation of the TAEP phosphate groups, polyphosphoric acids were formed, contributing to the char development. The authors highlighted a particularly effective degradation catalysation action on the epoxy acrylate oligomer. The TAEP addition to an epoxy acrylate oligomer led to a viscosity decrease while promoting the photopolymerisation rates and the thermal stability of the UV-cured resins [24]. The authors reported a V0 rating during the UL-94 test with a TAEP content of 50% in the formulation. TAEP was also used to prepare a silicon-containing hyperbranched polyphosphonate acrylate (HPA), which was then blended with an epoxy acrylate oligomer [25]. In this study, Wang et al. also highlighted the effectiveness of the phosphorus derivative to catalyse the degradation of the oligomer, helping in the formation of a char layer. The transparent HPA-based coatings displayed an improved LOI value and a decreased pHRR and THR. However, the coatings failed to reach a V0 rating for the UL-94 classification with the FR addition. TAEP was also used to develop a FR coating for polyester fabrics [27]. The textiles were plasma-treated before being dipped in the FR formulations and then exposed to UV light. The authors reported an excellent flame retardant property obtained from dipping the fabrics in a blend of TAEP with another FR compound, a 9,10-dihydro-9-oxa-10-phosphaphenanthrene 10-oxide (DOPO) derivative. The improved washing resistance was attributed to the surface plasma treatment.

Using flame-retarded UV-curable finishing layers could be a strategy to achieve the flame retardant performances of flooring. However, there is no solution thoroughly described in the literature that enables to fulfil all the specifications targeted. The aim of this present work is thus to evaluate the use of TAEP in a high-solid-content UV-curable wood sealer coat formulations not only with regard to the fire properties but also the other key performances expected from a coating. High-solid-content UV-curable coatings were thus prepared with different phosphorus monomer percentages. Two layers were applied on wood samples, with a low thickness of 25 µm per coat. The photopolymerisation kinetics were monitored by photo-differential scanning calorimetry (photo-DSC). The coating optical transparency was measured as a wood flooring coating is usually expected to preserve or improve wood appearance. Other coating properties such as hardness, water sorption and thermal stability were also studied. The wood coating adhesion was evaluated by pull-off tests. This study also highlights the flame retardance behaviour of the wood/coating assembly, studied by cone calorimetry. Scanning electron microscopy (SEM) and energy-dispersive spectroscopy (EDS) analyses were performed on the collected residues after the cone calorimetry test to complete the characterisation of the FR coating property.

2. Experimental Section

2.1. Materials

The coatings were prepared with different phosphorus monomer concentrations. Formulations are presented in Table 1. They are based on typical formulations used as sealers for prefinished wood flooring. Both the difunctional epoxy acrylate oligomer (referred as oligomer) and the acrylate monomer tripropylene glycol diacrylate (TPGDA) were supplied by Arkema (Burlington, ON, Canada). The photoinitiator, 2-hydroxy-2-methylpropiophenone (Sigma-Aldrich, Oakville, ON, Canada), was added at 3 wt%. A polybutadiene additive (Cray Valley, Stratford, CT, USA) was used for its defoaming properties at 0.01 wt% in the formulation. All products were graciously provided by Canlak (Daveluyville, QC, Canada).

Table 1. Composition of the UV-curable wood coating formulations.

Raw Material	F0	F1	F2
Oligomer	47.0	47.0	47.0
TPGDA	50.0	30.9	11.7
TAEP	0.0	19.1	38.3
Ricon 130	0.01	0.01	0.01
Photoinitiator	3.0	3.0	3.0

The phosphorus monomer, TAEP, was prepared by the addition of phosphorus oxychloride to hydroxylethyl acrylate (Sigma-Aldrich, Oakville, ON, Canada) following the synthesis described by Liang and Shi [26]. All the products were used as received without further purification. Proton nuclear magnetic resonance (^{1}H NMR) and Fourier-transform infrared spectroscopy (FTIR) analyses, respectively, displayed in Figures S1 and S2 in the supplementary information, showed the expected structure of TAEP. Its phosphorus concentration was also confirmed by inductively coupled plasma optical emission spectrometry (ICP-OES) analysis.

Yellow birch (*Betula alleghaniensis* Britt.) 4.5 mm thick lamellas were provided by Mirage (Saint-Georges, Canada). Raw sawn planks were placed in a conditioning room until mass equilibrium. Relative humidity was set at 42% and temperature at 23 °C, allowing wood samples to reach a moisture content of approximately 8%. Wood samples free from knots and defects were used in this study.

2.2. Sample Preparation

Depending on the analysis performed, coatings were either applied on wood samples, steel plates (Type R Q-panels, Q-LAB, Westlake, OH, USA) or on glass panels cleaned with ethanol.

Sample preparation occurred in different stages as shown on Figure 1, with the example of the steps followed prior to the cone calorimeter test. Before coating applications, the wood sample surface was prepared using a 3-step gradual sanding. A first pass was performed at a P-120-grit, then P-150 and the sanding was finalised at P-180. To do so, an automatic sander 36 CCK 1150 (Costa levigatrici S.p.A., Schio, Italy) equipped with paper sanding belts (aluminium oxide type, Abrasifs JJS Inc., Newmarket, ON, Canada) was used.

Prepared formulations were homogenised using a spatula and placed in an ultrasonic bath during 15 min for degassing before use. Formulation viscosity was measured with a piston-type viscometer (ViscoLab 4100, Cambridge Applied Systems Inc., Boston, MA, USA) at 23 °C. Coatings were applied on wood samples, glass or steel plates with a 25 µm wire-wound rod (BYK Gardner, Columbia, SC, USA). To be closer to what is performed in the industry, it was applied in two layers, each of 25 µm. Before applying the second layer, the coating surface was lightly sanded with a P-400-grit paper to promote interlayer adhesion. A 4-sided applicator (BYK Gardner, Columbia, SC, USA) was used for the application when a higher film thickness was necessary for characterisation tests. When the free films were required, coatings were applied on glass panels, cured, and then carefully pealed of the substrate with a razor blade.

Curing was performed under a medium-pressure mercury vapour lamp on a conveyor at a 5 m·min^{-1} speed under atmospheric conditions. Lamp intensity was set at 500 mW·cm^{-2}, corresponding to an energy dose of 430–450 mJ·cm^{-2} in the UVA range. The intensity and energy doses were measured before and after curing with a Power Puck II radiometer (EIT, Leesburg, VA, USA).

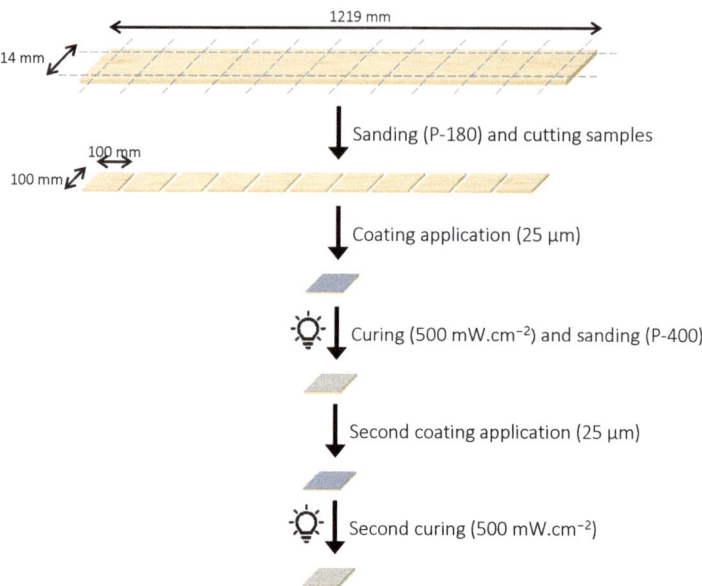

Figure 1. Wood sample preparation for the cone calorimeter test.

2.3. Study of the Coating Photopolymerisation

The impact of trifunctional phosphorus monomer addition on the curing behaviour was monitored by photo-DSC. A DSC822e differential scanning calorimeter (Mettler Toledo, Columbia, USA) coupled with a Lightningcure LC5 mercury-xenon lamp (Hamamatsu Photonics K.K., Iwata, Japan) was used in order to measure the enthalpies of reaction during photopolymerisation. To ensure repeatability, lamp intensity was verified with a radiometer (C6080-02, Hamamatsu Photonics K.K., Iwata, Japan). Experiments were performed under air atmosphere at 30 °C. A coating drop of 2.5 ± 0.1 mg was centred in an open aluminium pan. The sample was then irradiated for 60 s at 50 mW·cm^{-2}. Each formulation was tested four times. All DSC exotherms were normalised to sample weight. The photopolymerisation exotherms resulting from photo-DSC allow the calculation of Rp, the rate of polymerisation, the unsaturation conversion and t_{ind}, the induction time. The induction time corresponds to the necessary time to achieve 1% of conversion. The polymerisation rate Rp and the conversion percentage were calculated according to Equations (1) and (2), respectively. Both depend on ΔH_{theo}, the theoretical heat flow for the complete conversion of the acrylate system expressed in J·g^{-1}. For calculations, a value of 86 kJ·mol^{-1} was taken for the heat of the polymerisation per acrylic unsaturation [28]. dH/dt corresponds to the measured heat flow at time t and ΔH_t to the heat flow generated up to time t.

$$Rp = \frac{(dH/dt)}{\Delta H_{theo}} \qquad (1)$$

$$\%_{Conversion} = \left(\frac{\Delta H_t}{\Delta H_{theo}}\right) \times 100 \qquad (2)$$

2.4. Characterisation of the Coating

The coating optical transparency was evaluated by transmittance on a Cary 60 ultraviolet–visible (UV–Vis) spectrophotometer (Agilent Technologies, Santa Clara, CA, USA). Analyses were carried out on 100 µm thick free films with a scanning wavelength ranging from 400 to 800 nm and using a medium scanning speed (600 nm·min^{-1}). An approximation of the glass transition temperature (T_g) and the crosslinking density of the cured films were calculated from dynamic mechanical analysis (DMA) measurements, performed on a DMA Q800 supplied by TA Instruments (New Castle, USA). DMA apparatus was operated in a tensile mode with the frequency set at 1 Hz and the amplitude at 10 µm. A force track of 125% was applied. Measurements were performed at a constant heating rate of 3 °C·min^{-1} in a temperature range from 30 °C to 150 °C. Coating-free films were laser cut in a rectangular shape of 25 mm per 5 mm, with a thickness of approximately 100 µm. Prior to the test, films were dried at 100 °C during 24 h to avoid any influence of entrapped water. Five repetitions were run per coating formulation. The storage modulus (E') and loss modulus (E'') resulting from the dynamic mechanical analysis allow the calculation of the loss factor (tan δ), defined as the ratio of E'' over E'. The storage modulus (E') represents the elastic portion, while the loss modulus (E'') characterises the viscous behaviour of the material. In this study, the maximum of the loss factor was used to approximate the glass transition temperature. The crosslinking density was calculated according to Equation (3). E'_{min} corresponds to the minimum of storage modulus expressed in Pa, T is the temperature at the minimum of storage modulus and R refers to the ideal gas constant.

$$\nu_{LD} = \frac{E'_{min}}{3RT} \qquad (3)$$

Pendulum hardness tests were performed according to the standardised test method ASTM D4366-16 using a König pendulum tester (BYK Gardner, Columbia, USA). The pendulum was placed on the coating applied on a glass plate and set at a 6° angle to the coating surface normal. The number of oscillations were recorded during pendulum damping from a 6° to 3° angle. Five replicas per formulations were tested for the calculations of the mean values and standard deviations. Dynamic instrumented indentations were carried out on a Micro-Combi tester (CSM Instruments, Peseux, Switzerland) with a diamond Berkovich indenter calibrated on fumed silica beforehand. The indenter tip was led normal to a coating surface applying a maximum load of 300 mN. The force was gradually applied in 5 s, held during 2 s before being unloaded on a 3 s period. Indentation tests were performed 5 times for each sample, on at least 3 repetitions per formulation. For this test, coatings were applied on steel panels with a thickness of 125 µm. Higher thickness than the attended application is mandatory in order to have an adequate indenter penetration depth. Steel panels were preferred over wood to ensure substrate hardness homogeneity. Water content was determined with an 890 Titrando titrator for volumetric Karl Fischer titrations equipped with the software Tiamo^{TM} Light (Metrohm AG, Herisau, Switzerland). Coating flakes with a thickness of 100 µm were dispersed in N,N-dimethylformamide and then placed in an ultrasonic bath for an hour prior to the Karl Fischer titrations. The analyses were performed twice per formulation. To assess the potential influence of adding phosphorus derivatives on the water uptake of the coating films, sorption and desorption isotherms were performed in triplicate on 10 mg film samples with a thickness of 100 µm. Before the experiments, the samples were dried overnight in an oven at 100 °C. Tests were performed on a DVS Adventure water vapour sorption analyser (Surface Measurement Systems, Allentown, USA) at a constant temperature of 25 °C. First, a step at 0% relative humidity (RH) was performed to allow mass stabilisation. Then, RH was increased from 0% to 95% by means of 5% steps and scaled back down using the same increment. The relative humidity of the analysis chamber was maintained until the sample mass percentage varies less than 0.0001% over a 5 min period or the step time exceeds 24 h. Thermogravimetric analyses (TGAs) were performed on a TGA 851e analyser (Mettler Toledo, Greifense, Switzerland). Experiments were carried out under a nitrogen flow (50 mL·min^{-1}) from 25 °C to 800 °C at a heating rate of 10 °C·min^{-1}. To avoid any interaction with the phosphorus derivatives, a sapphire crucible (70 µL) was used. All experiments were performed in triplicate on free 50 µm thick coating films. Characteristic data such as $T_{5\%}$, the temperature corresponding to 5% of degradation, T_{max}, the temperatures of the maximum degradation rates and the percentages of residues at 800 °C were determined.

2.5. Characterisation of the Coating Adhesion on Wood

Adhesion was investigated using pull-off tests based on the test method described in the ASTM D4541-17 standard. Aluminium dollies with a diameter of 20 mm were glued on the coating using a two-component epoxy adhesive (Lepage Marine Epoxy, Henkel, Mississauga, ON, Canada). Prior to the glue application, the dolly surface was sanded with a P-150-grit paper and cleaned with a wipe. The same procedure was performed on the coating surface with a finer sanding paper (P-1500-grit). Steel hex nuts (79 g) were placed on dollies during the adhesive curing to ensure an even pressure applied on each sample. After 48 h, the test area around the dolly was scored by means of a dolly cutter, until the wood substrate was reached. This ensures testing the same surface area. Pull-off tests were carried out on a QTest/5 Elite electromechanical test system with a 5 kN load cell (MTS, Eden Prairie, MN, USA) at a 2 mm·min^{-1} pull rate. The coating tensile strength was calculated according to Equation (4), where F_{max} corresponds to the load at fracture (N) and d, to the dolly diameter (mm). Fracture type was visually assessed based on the described method in the ISO 4624:2016 standard. Twenty repetitions per formulation were performed.

$$\sigma = \frac{4F_{max}}{\pi d^2} \quad (4)$$

The statistical analysis of data was achieved on the software R: a language and environment for statistical (R Foundation for Statistical Computing, Vienna, Austria). Prior to analysis, the normal distribution of residuals was examined as well as the homogeneity of variances to ensure the validity of the analysis. One-way ANOVA was executed to compare the impact of phosphorus monomer addition in the coating formulations on the pull-off strength. The Tukey's HSD test allowed a pairwise comparison between the group to identify significant differences at a 0.05 level.

2.6. Fire Performance of Coated Wood Samples

The fire behaviour of the treated wood samples was assessed on a dual cone calorimeter (Fire Testing Technology, East Grinstead, UK). Cone calorimeter is a bench-scale test method widely used to assess a material's fire behaviour. It gives different parameters as the heat release rate, the time to ignition or the mass loss of the combusting material. However, it offers a limited interpretation of a fire scenario as it does not consider flame spreading.

Experiments were carried out on 100 × 100 mm samples placed in the horizontal position. The distance between the sample and the cone heater was set at 25 mm. Heat flux was set at 35 kW·m^{-2} for the cone calorimeter tests, corresponding to an early developing fire scenario. Such a heat flux is recommended for explanatory testing [29]. It is particularly appropriate to assess flame-retarded polymers as it matches the fire stage in which their action is crucial. Ignition was forced and the test was stopped 120 s after flame extinction. Wood sample backsides were wrapped in aluminium foil before being put in the sample holder. All experiments were performed in quadruplicate. In order to characterise the fire behaviour of the sample, the following data were gathered: the average time to ignition (TTI), the average peak of the heat release rate (pHRR) and the average total heat released (THR).

Cone calorimeter residues were collected and characterised with an Inspect F50 SEM equipped with a field emission gun source (FEI, Hillsboro, OR, USA). Elemental analysis on the residues was performed by means of energy dispersive spectroscopy (Octane Super-A, EDAX-AMETEK, Mahwah, NJ, USA). The acceleration voltage was set at 15 kV. Prior to analysis, all samples were covered with a conductive Au/Pd deposit.

3. Results

3.1. Photopolymerisation

DSC analysis assesses the heat flow of a material as a function of time and temperature compared to a reference. When coupled with a UV lamp, it enables the measurement of different parameters specific to photo-curable systems. Photo-DSC results are presented in Figure 2 and Table 2.

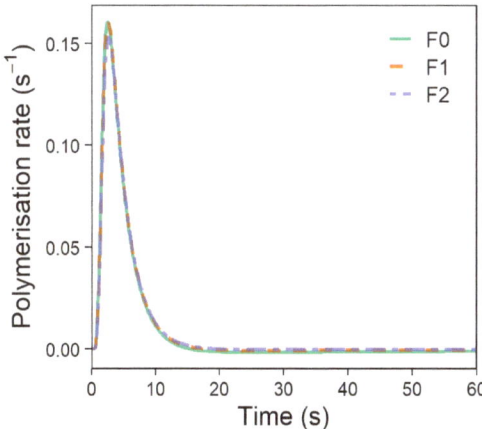

Figure 2. Polymerisation rate of the different coatings.

Table 2. Coating viscosity and photo-DSC results including the induction time, the maximum heat flow, the rate of polymerisation and the unsaturation conversion.

Formulation	Viscosity (cP, 23 °C)	t_{ind} (s)	H_{max} (W.g^{-1})	Rp^{max} (s^{-1})	%Conversion (%)
F0	630	1.06 ± 0.04	58 ± 2	16.1 ± 0.4	68 ± 3
F1	840	1.10 ± 0.02	60 ± 1	16.1 ± 0.4	69 ± 1
F2	1230	1.13 ± 0.01	60 ± 3	15.3 ± 0.7	67 ± 2

All the rates of polymerisation curves demonstrated a steep increase within a few seconds after the beginning of the irradiation. Then, a rapid decrease in the rate of polymerisation was observed. Such a heat flow or a Rp behaviour was already reported in the literature for the free radical polymerisation of acrylate systems [30,31]. The rate of polymerisation of free radical polymerisation can be divided into different steps. The induction period is followed by the autoacceleration step, also called the Trommsdorff gel effect or the Norrish–Smith effect. Rp drastically increases due to the increasing viscosity of the polymerising system that slows termination reactions [32]. After reaching its maximum, Rp diminishes during the phase of autodeceleration. The propagation is then controlled through diffusion [33]. In this study, Rp^{max} were in the same range, with a value of 15.3 s^{-1} for F2 and 16.1 s^{-1} for the other systems. The conversion percentages were close to 70% for all coating formulations. The conversion percentages being kept in the same range when replacing the TPGDA by the phosphorus monomer demonstrated its active participation in the curing reaction. Thus, TAEP was successfully incorporated in the polymeric network created during the photopolymerisation of F1 and F2 coatings. No major difference in the induction time nor in the maximum heat flow, H_{max}, was noticed.

In the literature, a sharp peak in the rate of photopolymerisation was also observed for TAEP containing formulations [34,35]. However, Rp^{max} varied with the phosphorus monomer content. Tests were performed at 2.4 mW.cm^{-2} or at a lower irradiance, greatly inferior to the 50 mW.cm^{-2} used in this study. Such an irradiance was chosen so as to be more realistic to assess the photopolymerisation behaviour for an industrial application. Lower irradiance generates less energy, and hence, the polymerisation rate is slower. Moreover, the gelation effect induced by the trifunctional monomer addition in the system is greater, explaining the difference in the Rp^{max} value. The literature also reports that

R_p tends to increase when the formulation viscosity decreases [36]. The viscosity is an important parameter for UV-curable systems as it affects different properties such as the ease of application, levelling, polymerisation rate and adhesion. In the present study, the viscosity varied significantly between the formulations (Table 2). However, the addition of TAEP, a non-linear phosphorus monomer, even at a high percentage in the formulation, did not affect the UV-curing behaviour.

3.2. Coating Properties

3.2.1. Optical Transparency

Achieving transparency is an important parameter for wood coatings as they are often designed to protect the substrates while keeping their natural aspect. Film transparency was assessed by spectrophotometry on the visible spectrum, between 400 and 800 nm (Figure 3). F0 film was transparent with an 88% transmittance at 600 nm. With the addition of the FR monomer, the F1 and the F2 coatings also exhibited good optical transparency with a transmittance higher than 87% at 600 nm. All coatings preserved wood aspect as shown on Figure S3 in the supplementary information.

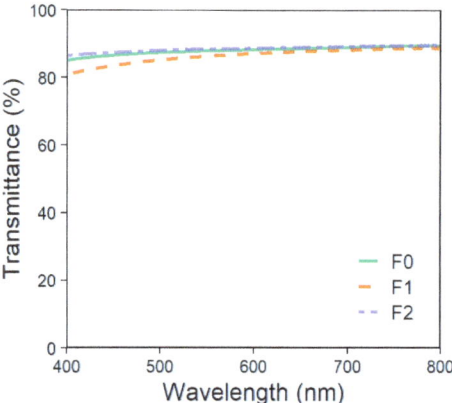

Figure 3. UV–Vis transmittance spectra of the coating film.

3.2.2. Coating Mechanical Properties

All the factors impacting the chain mobility, such as the steric hindrance, the polymer functionality, the nature of the moving segment or the free volume available for motion, will affect the glass transition temperature. Both monomers added in the formulations in this study, TPGDA and TAEP, differ in their backbone structures and their number of acrylate functionalities. Such differences in chemical composition may lead to polymers exhibiting different T_g and crosslinking densities [37].

DMA results are displayed in Figure 4; the storage modulus (E′) and the tan δ curves are shown in Figure S4 of the supplementary information. The glass transition temperature increased along with TAEP loading. T_g reached 75 °C for F2, ten degrees higher than the one of F0. The oligomer supplier indicates a T_g at 51 °C, measured by DSC. The T_g of the TAEP homopolymer is also known (105 °C) as it has already been measured and reported in the literature [34,35]. While the value of T_g increased with the phosphorus monomer percentage in the formulation, the crosslinking densities were in the same range. Given the standard deviation, the crosslinking densities were not considered to be different. The width of the tan δ curves remained constant, evidencing good miscibility between the phosphate monomer and the acrylate monomer, TPGDA, and the epoxy acrylate oligomer.

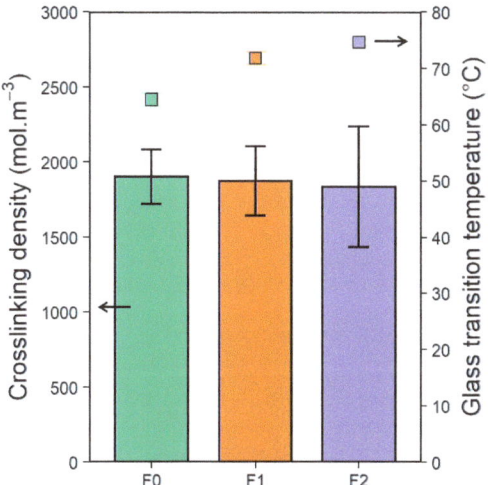

Figure 4. Crosslinking density and glass transition temperature of the coatings.

Hardness refers to an imprecise term since its definition varies in the literature [38]. A broad definition of the surface hardness corresponds to the coating resistance to deformation by means of the indentation or the penetration of a solid object [39,40]. This is the reason why two technics (pendulum and indentation) were used in this study to determine the hardness of the coatings. The results are reported in Figure 5.

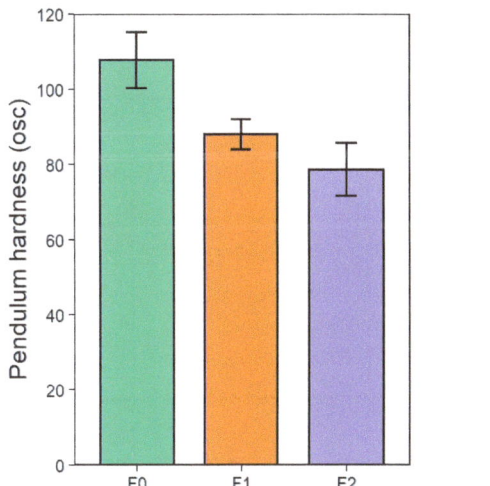
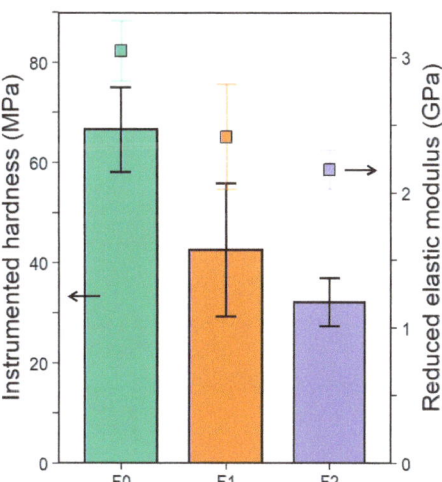

Figure 5. Pendulum hardness and indentation results of the coating films.

The pendulum hardness results exhibited a higher damping property for the coatings containing the phosphorus monomer. The F0 film reached 108 oscillations, while F1 and F2 achieved 89 and 78 oscillations, respectively. Replacing TPGDA by the TAEP monomer led to a change in the viscoelastic properties of the coating. The phosphorus-containing films were able to dissipate the pendulum energy of motion more easily.

Indentation hardness quantifies the coating film ability to resist to the penetration of an indenter. Preliminary tests confirmed that the maximum applied load ensured that the penetration was restricted to the upper part of the coating film. Having an optimal penetration depth guarantees a minimised influence of the substrate mechanical property on the results. The coating film containing only the acrylate monomer attained an instrumented hardness of approximately 65 MPa. With the addition of phosphorus monomer, the capacity to withstand penetration distinctly dropped under 45 MPa for both F1 and F2 formulations. The reduced elastic modulus E_r also followed the same trend, decreasing with a higher phosphorus content. Even though the conversion rate as well as the crosslinking density were not affected by the phosphorus monomer addition, a change in mechanical properties was observed. While TAEP had a higher functionality, the presence of the P-monomer brought flexibility to the coating films. The presence of water in the P-containing films was evidenced by further tests. The plasticising effect of the trapped water in the coating film is one possible explanation for their increased flexibility.

3.2.3. Coating Water Sorption

Using phosphorus-containing polymers in a coating may cause some concern. Phosphorus polymers could be sensitive to moisture and in particular, P-O-C linkages present hydrolytic instability [41]. In order to assess the coating water affinity, Karl Fischer titration and dynamic vapour sorption experiments (DVS) measurements (Figure 6) were performed.

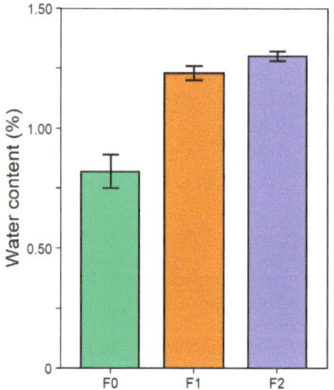
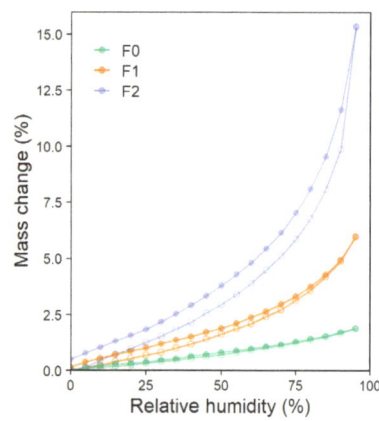

Figure 6. Karl Fischer titration results and, sorption and desorption curves of the coating films. The hollow circles correspond to the sorption step while the filled circles correspond to the desorption one.

Karl Fischer titration quantifies both free and bound water present in the coating film. The coating without P-monomer contained 0.82% of residual water. The water content rose with the phosphorus addition to achieve 1.30% for F2 coating. These results were consistent with the early stage TGA measurements (Figure 7) showing a higher mass decrease for the phosphorus-containing films. This gap in mass loss could be explained by their higher water content released below 150 °C.

DVS measures the quantity of water absorbed by the coating film in a given range of vapour concentrations. For the F0 film, water absorption was limited, and the mass gain reached 1.90% at the highest RH. The F2 sorption isotherm showed a steep increase in high-relative-humidity percentages. Such behaviour, corresponding to a low intake at low pressures followed by an increased in the high pressures, is typical of a Type III isotherm defined in the International Union of Pure and Applied Chemistry nomenclature [42]. F1 displayed a similar water uptake profile to a minor extent, as the mass change at the maximum RH reached 5.98% instead of 15.3% for F2. The F2 desorption isotherm did not return to a null mass change at a 0% RH. The hypothesis was made that the water is still present in the sample. Further energy seemed to be necessary to remove the bound water. This latter was removed prior to the DVS test by heating during the drying stage at 100 °C. F0 showed no hysteresis on the isotherms of sorption and desorption unlike the coating samples containing phosphorus. The mechanism of water uptake was changed with the presence of a P-monomer in the coating. The absence of hysteresis suggested a plausible surface adsorption mechanism with limited to no bulk absorption. Hence, the results displayed the influence of adding phosphorus derivatives on the water uptake of coating films, mainly in the high RH range.

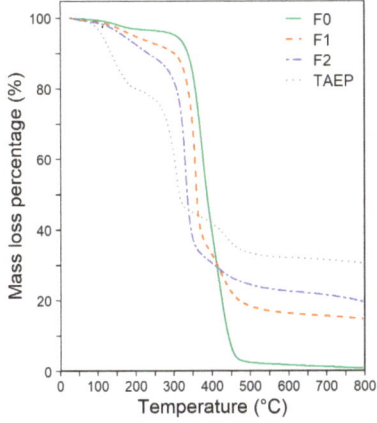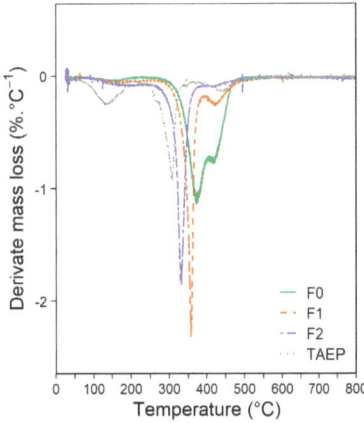

Figure 7. Curves of the thermogravimetric analysis performed under nitrogen and their derivatives for the TAEP monomer and the coating films with different phosphorus contents.

Although those results could appear as a limit for a coating application, the formulated coatings developed in this study are meant for a sealer use. In a complete wood finishing system, they would benefit from a topcoat applied above. This upper protective layer will limit water exchange and would be designed to resist abrasion and scratch. Once integrated in the finishing system, the sealer has a limited contribution to the surface hardness. Further investigations were thus performed on the coatings and the thermal stability was first studied.

3.2.4. Coating Thermogravimetric Analysis

The thermal stability of the coating films as well as the TAEP monomer was investigated by thermogravimetric analysis in an inert atmosphere. The mass loss curves and their derivatives are shown in Figure 7. The main results ($T_{5\%}$, T_{max} and the percentages of residues at 800 °C) are featured in Table 3.

Table 3. Results of thermogravimetric analysis. $T_{5\%}$ corresponds to the temperature at which 5% of the sample mass is lost, $T_{max\ 1}$, $T_{max\ 2}$ and $T_{max\ 3}$ are the temperatures of the degradation rate maxima.

Formulation	$T_{5\%}$ (°C)	$T_{max\ 1}$ (°C)	$T_{max\ 2}$ (°C)	$T_{max\ 3}$ (°C)	Residue at 800 °C (%)
F0	295	158	371	418	1 ± 1
F1	196	161	357	424	14.7 ± 0.5
F2	169	168	331	424	19.6 ± 0.5
TAEP	113	132	307	436	30.6 ± 0.2

The thermal decomposition of the pure monomer TAEP happened in three distinct steps, with maximum degradation rates at approximately 160 °C, 370 °C and 420 °C. A similar behaviour was reported in the literature for the thermogravimetric analysis of TAEP homopolymer in air [26]. Three main decomposition steps were described and assigned to the decomposition of the phosphate groups (from 160 °C to 270 °C), the decomposition of the ester groups (at approximately 330 °C) followed by the decomposition of unstable structures in the char (above 500 °C). In our study, the analyses showed that the pristine coating underwent a three-stage decomposition process, with a first step around 160 °C, and then two overlapping steps between 300 °C and 475 °C. Even though the F0 formulation was more thermally stable under 400 °C compared to F1 and F2, there was no residue at 800 °C. The major F0 degradation range corresponded to a weight loss of approximately 90%, a percentage greatly higher compared to the ones of phosphorus-containing formulations. F1 and F2 coatings also displayed a several stages of decomposition but with better distinguishable steps. F1 coating first had thermal degradation at a higher temperature range followed by another thermal degradation at approximately 420 °C, which corresponds to the final decomposition of the oligomer and the monomer. For both phosphorus-containing coatings, it was observed that the thermal pyrolysis of unstable structures allowed the formation of a stable char above 450 °C. The major mass loss of F2 cured films occurred between 300 °C and 360 °C. The percentage of residue at 800 °C reached 14.7% for F1 and 19.6% for F2.

Thus, adding phosphorus compounds in the formulation shifts the thermal degradation to a lower temperature but leads to the formation of stable residues at 800 °C. The development of a char could be a benefit in terms of fire performance; the formation of such a protective layer was verified with the cone calorimeter tests.

3.3. Coating Adhesion on Wood

Good adhesion between the coating and the wooden substrate is mandatory to ensure good coating properties and durability [39]. Pull-off tests is a common method to assess the coating adhesion to the substrate that provides a quantitative result. The results of the pull-off strength of the different coatings applied on wood are presented in Figure 8. During the tests, only adhesive failure between the substrate and the first coat happened. An increasing phosphorus content in the coating led to an increasing fibre deposit on the dollies after the test. Thus, the cohesive strength of the coatings was good and the phosphorus coatings seemed to offer a better mechanical anchoring with a higher-wood fibres wetting. The adhesion strength also varied with the phosphorus content. Indeed, while the mean value of the adhesion strength of the F0 coating was under 1.00 MPa, it attained 1.01 and 1.12 MPa for F1 and F2, respectively. The statistical analysis conducted on the results revealed a significant difference in adhesion strength between the coating without phosphorus monomer and the one with the highest content. Thus, above a certain P-monomer addition in the formulation, the pull-off strength was slightly enhanced. Phosphorus derivatives are well-known adhesion promoters [43]. For instance, Maege et al. studied the adsorption of phosphorus derivatives on metallic substrates as an anchor to the coating system, offering an improved adhesion [44]. Wang et al. suggested that a phytic acid-based compound enhanced the adhesion of UV-curing coating to metal by the formation of P-O-Fe chemical bonds [45]. Even though the presence of phosphorus additives enhancing the adhesion with wood for coatings or adhesives has been reported in different studies [15,46,47], the mechanism explaining this improvement is not well detailed.

The difference in pull-off test results could be explained by several factors including the limited coating penetration into wood [48] or the UV coating shrinkage through curing [49]. During the photopolymerisation of acrylate systems, the weak intermolecular Van der Waals interactions in the liquid coating are replaced by strong and short covalent bonds between the carbon atoms in the solid film [50]. The distance between carbon atoms is reduced, building up internal stress in the coating. The volume contraction induces a shrinkage that may lead to low coating adhesion and poor mechanical properties [49]. In order to obtain a better understanding, SEM analyses were performed on the sample cross-sections to observe the wood/coating interface (Figure 9). The F0 sample clearly displayed the coating detachment from the wooden substrate in different areas. The interface between the phosphorus coating and wood showed better bonding. The TAEP presence in the formulation seemed to induce lesser coating shrinkage through photopolymerisation, enhancing the coating adhesion.

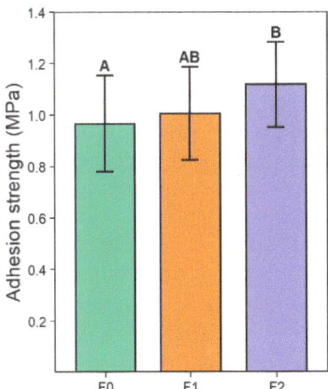

Figure 8. Pull-off strength of the coatings applied on wood. Letters above the bars correspond to the groups determined in the Tukey comparison test after the ANOVA analysis ($\alpha = 0.05$). Groups having a letter in common are not significantly different.

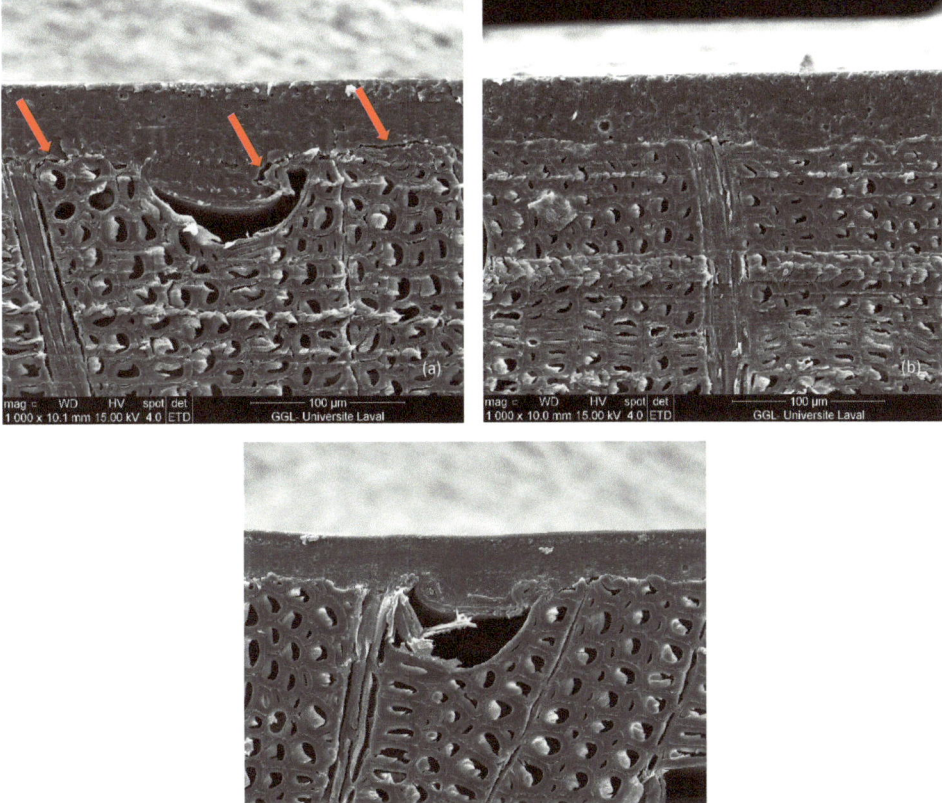

Figure 9. SEM images of the wood/coating interface, F0 (**a**), F1 (**b**) and F2 (**c**). Arrows show coating detachments of few microns wide from wood.

3.4. Fire Performance of the Coated Wood Samples

Heat release curves are displayed in Figure 10 and key values are summarised in Table 4. All coated panels present a heat release curve with two peaks, similar to what is usually observed in the literature for virgin wood [51]. The first peak is attained before charring. The heat release rate then decreases as a protective layer, i.e. a char, is formed. Then, the HRR curve illustrates the gradual burning through the thickness of the wood sample after the formation of the initial char. Once the carbonaceous protective layer crakes, a second HRR rise happens. The second peak also corresponds to the increased rate of volatile formation prior to the end of the flame burning [52].

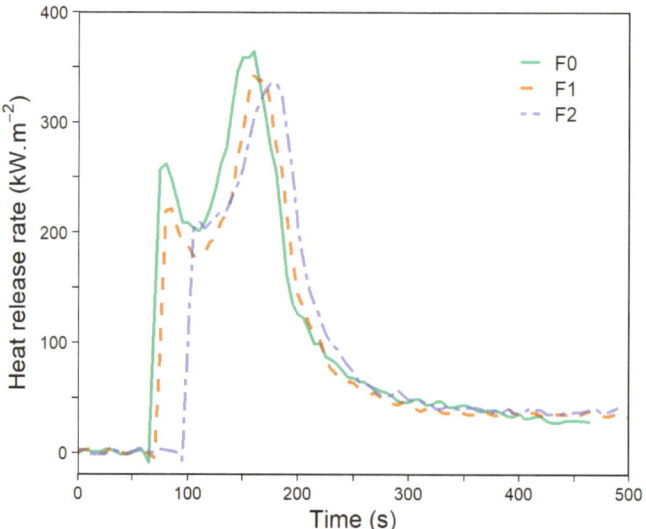

Figure 10. Cone calorimeter heat release rate curves of the coated wood samples.

Table 4. Main cone calorimeter results of the coated wood samples.

Parameter	F0	F1	F2
Phosphorus coating content * (wt%)	0.0	1.7	3.4
TTI (s)	71 ± 2	77 ± 9	90 ± 20
pHRR (kW·m^{-2})	390 ± 10	340 ± 10	340 ± 10
THR (MJ·m^{-2})	47 ± 1	44 ± 2	42 ± 1
Residue (wt%)	7.6 ± 0.3	9.8 ± 0.6	10.4 ± 0.6

* Phosphorus coating contents were calculated based on the phosphorus content value (7.89 wt%) of the TAEP monomer obtained by ICP-OES analysis.

F0 coating cracked within a few seconds after exposure to the heat flux. The coating delaminated from the surface, exposing the wooden substrate to the flame. A high first heat release rate peak was rapidly reached for the F0 sample. A second peak was observed attaining a value of 390 kW·m^{-2}. The higher the value of the HHR first peak for the F0 coating compared to the other samples was attributed to the greater organic coating contribution to the burning. The phosphorus coatings also showed some delamination but to a lesser extent compared to F0. This difference in coating adhesion to wood was already highlighted with pull-off tests. A good coating adhesion is critical in the beginning of the cone calorimeter test and decisive for the fire behaviour of the material. Some cracks also appeared on F1 and F2 coating surface after the ignition. Similarly to what was observed for the F0 coating, the HRR curve of F1-coated wood samples presented a first peak but with a lower value, suggesting the formation of a protective char layer.

The heat transfer and fuel supply to the flame were limited, slowing down the pyrolysis process. Cracks in the protective layer appeared, allowing the flame to reach the virgin wood. An increase in the pyrolysis then occurred, leading to a second heat release rate rise. For the highest phosphorus-FR-containing film (F2 coating), the curve depression was considerably flattened. It was assumed that the protective layer is formed more efficiently for the F2 sample as a plateau rather than a peak was observed on the HRR curve in the early stage of the test. In the case of phosphorus-containing coatings, the pHRR only reached 340 kW·m^{-2} and was delayed compared to F0. Adding a phosphorus monomer enhanced the residual mass percentage as well as reduced the THR, suggesting an improved wood charring behaviour. The time to ignition also varied with the FR loading with an upward tendency.

A reduction of 13% in the pHRR could appear as a moderate improvement. However, adding a phosphorus monomer positively affected other parameters such as the percentage of residues after the test. Wang et al. also reported a pHRR decrease of approximately 10% with their best treatment obtained in the cone calorimeter test performed at 50 kW·m^{-2} [15]. The wood coatings reported in this study contained the following FR at different ratios: a multifunctional bio-mercaptan of castor oil grafted thiol, 2,4,6-trimethyl-2,4,6-trivinylcyclotrisilazane and an organic phosphorus compound. The UV-curable coating was applied at 180 μm (versus 50 μm in the present study) on 4 mm thick Chinese fir samples. The pHRR curve followed a reverse tendency compared to our study, with a higher first peak compared to the second one. Higher pHRR decreases were observed for a non UV-curable waterborne intumescent wood coating [53] or for coating applied on wood with a thicker layer [54]. For the first study, Song et al. obtained a clear pHRR decrease of approximately 40% and a delayed TTI compared to pure wood, a 4 mm thick poplar panel. All the pHRR of the samples treated with the different intumescent coatings ranged from 140 kW·m^{-2} to 190 kW·m^{-2} after the cone calorimeter test. In the second study, epoxy coatings were applied on a poplar wood sample at a 5 mm thick layer. The authors reported a pHRR reduction of 20% with the addition of a phosphaphenanthrene-containing epoxy co-curing agent in the coating formulation compared to the epoxy coating reference.

The cone calorimeter residues were collected and analysed to better comprehend the charring mechanism. SEM analyses were performed on the residues collected after the test to confirm the action of phosphate moieties in the condensed phase (Figure 11). The mass percentages of different elements obtained from the EDS analysis performed on the same residues are gathered in Table 5.

Table 5. Elemental composition analysis performed on different areas (identified on Figure 11) of the cone calorimeter residues.

Analysis Area	C (wt%)	O (wt%)	P (wt%)	Ca (wt%)	Mg (wt%)	Mn (wt%)	K (wt%)
F0—Area 1	65.45	27.33	0.31	4.41	1.08	0.98	0.44
F0—Area 2	12.28	51.08	1.59	21.95	4.84	4.85	3.40
F1—Area 3	81.69	16.49	1.46	0.35	-	-	-
F1—Area 4	81.16	17.24	1.44	0.17	-	-	-
F2—Area 5	76.47	21.04	2.14	0.35	-	-	-
F2—Area 6	80.77	16.44	2.11	0.52	0.15	-	-

F0 residues displayed a very fragile fibrous structure with no visible coating deposit. No charring was observed, only ashes were formed. It is similar to wood sample morphology after being exposed to a flame as reported in the literature [55]. On the contrary, the flame-retarded samples clearly displayed some wooden charred structure demonstrating that a protective mechanism occurred. For both F1 and F2 samples, high temperature-resisting residues of the coating were noticeable. In the case of the F1 sample, a partial delamination of the coating was observed, revealing a thick char underneath. This solid structure is present on the right-hand side of Figure 11b. Fractures in the F1 residues were also visible. They went through the coating and the substrate. F2 residues also displayed a thick char formation. Almost no coating detachment was noticed but cracks were still observed. Interestingly, most of the cracks did not go through the substrate either. Indeed, Figure 11d revealed that most F2 coating fractures were protected by an underneath layer. The F2 burnt sample showed a surface morphology with hollowed areas and trapped bubbles underneath the top layer.

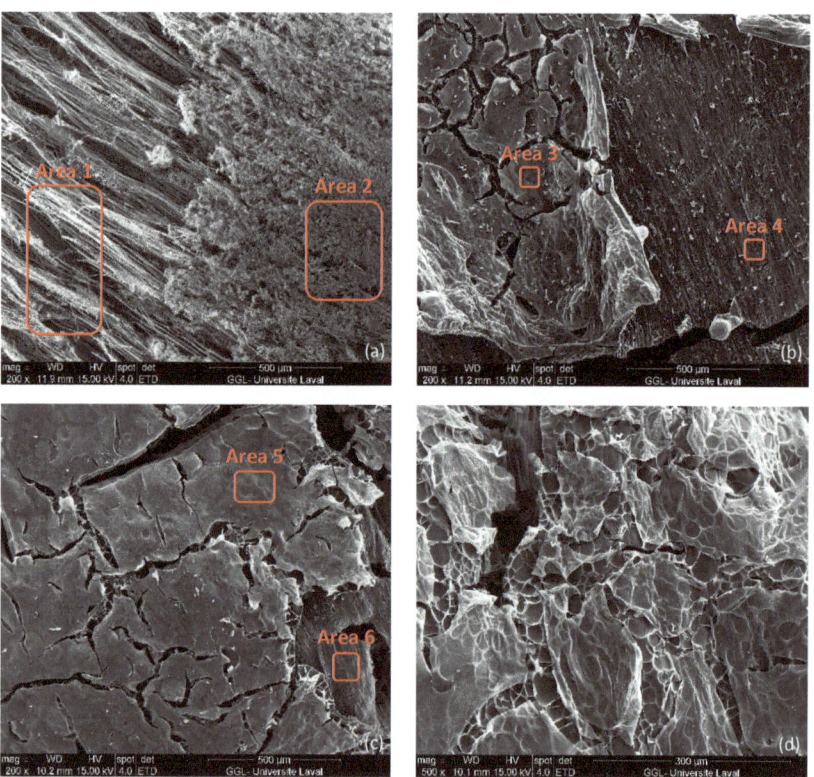

Figure 11. SEM images of cone calorimeter residues of F0 (**a**), F1 (**b**) and F2 (**c,d**).

Thus, the presence of a protective layer after the cone calorimetry tests on the residues of wood samples, either coated with F1 or F2 formulation, validates an action of the phosphate moieties in the condensed phase. The TAEP addition to the coating formulation helped the char formation, increasing the mass residue. An insulating barrier is formed, protecting the underneath substrate from fire, and preventing gas exchange between the condensed phase and the flame.

EDS analyses further endorsed the action of phosphorus compounds participating in the char formation. Whereas the analysis of the F0 residues revealed the presence of typical elements found in wood ash, the elemental composition significantly varied when a phosphorus-containing coating was applied on the wood. Misra et al. observed that the wood ash at 600 °C was mainly composed of calcium, potassium and magnesium, apart from carbon and oxygen elements [56]. Phosphorus, sulphur and manganese were also present to a minor extent. Wood decomposition forms calcium carbonate and oxide [57]. High O wt% and Ca wt% for F0 endorsed the presence of calcium derivatives. F1 and F2 surface analysis demonstrated higher C and P contents. The phosphorus coming from the FR coating application was still present in the residues after the cone calorimeter test. As expected, a higher P content was observed for F2 compared to F1. A significant amount of phosphorus induces an active role of the FR in a condensed phase mechanism. It is worth mentioning that even though the P content of area 2 of F0 sample and the one of F1 sample were in the same range, these were not comparable as the percentages of the other elements were completely different. EDS analysis gives the relative elemental composition of a local sample area based on the total detected elements in it. While F0 corresponded to an ashes area with a C content under 15%, the F1 sample charred, resulting in a high C content, above 80%.

We may conclude that a decrease in the pHRR combined with an increase in the residual mass indicates an effective action of the TAEP addition regarding flame retardancy. The cone calorimeter tests along with the observation of the residues also suggested a difference in coating adhesion depending on the phosphorus rate in the formulation. With a higher phosphorus content, the least coating delamination occurred, contributing to the improvement of the fire behaviour of the coated wood.

4. Conclusions

High-solid-content transparent UV-curable coatings were prepared at different TAEP percentages in an acrylate-based formulation. Coatings were applied on wood panels. The photopolymerisation kinetics was monitored by photo-DSC. It was shown that adding the phosphorus monomer did not impact the UV-curing behaviour as the conversion of the acrylate group was in the same range, i.e., over 65%. A change in coating hardness was observed and attributed to the fact that TAEP brought flexibility to the coating films. This study also showed that P-containing coatings displayed a higher water sorption for high relative humidity. The issue related to water affinity could be easily lowered by integrating the formulation as a sealer coat in a prefinished wood flooring system. The presence of the phosphorus monomer at the highest content significantly improved the coating adhesion on the wood. The wood coating fire efficacy was evaluated by cone calorimetry. A decreased pHRR as well as an enhanced residue percentage were obtained for all the phosphorus-containing coatings. SEM and EDS analyses performed on the collected residues confirmed the formation of a solid protective structure and the presence of phosphorus in the flame-retarded samples. The phosphorus FR had an effective action in the condensed phase, improving the fire retardancy of the coated wood. The formulation containing approximately 40% of TAEP clearly displayed the best FR performances compared to the non-FR coating. Furthermore, the major expected properties of a sealer coating were positively assessed. This study thus demonstrates the high potential of phosphorus-containing UV-curable acrylate system to improve the fire behaviour of wood flooring. Further investigations could include the introduction of another FR retardant in the formulation. Nitrogen or silicon compounds have demonstrated a synergistic behaviour along with the phosphorus FR. The F2 coating highlighted in this study could also be of interest for metal coating where TAEP could act as both a flame retardant and an adhesion promoter.

Supplementary Materials: The following supporting information can be downloaded at: https://www.mdpi.com/article/10.3390/coatings12121850/s1. Figure S1. ^{1}H NMR spectrum of tri(acryloyloxyethyl) phosphate (TAEP); Figure S2. FTIR spectrum of tri(acryloyloxyethyl) phosphate (TAEP); Figure S3. Appearance of the coatings applied on wood panels; Figure S4. Storage modulus and tan δ curves resulting from DMA analysis on the coating films.

Author Contributions: Conceptualisation, S.P., F.S., S.D. and V.L.; methodology, S.P., F.S., S.D. and V.L.; software, S.P.; validation, S.P.; formal analysis, S.P.; investigation, S.P.; writing—original draft preparation, S.P.; writing—review and editing, S.P., F.S., S.D. and V.L.; supervision, S.D. and V.L.; project administration, S.D. and V.L.; funding acquisition, V.L. All authors have read and agreed to the published version of the manuscript.

Funding: This work is part of the research program of Natural Sciences and Engineering Research Council of Canada (NSERC) Canlak Industrial Research Chair in Finishes for Interior wood products (CRIF) through programs CRD (Grant No. 500235) and Research Support Program, part 4: Support for

International Research and Innovation Initiatives (PSR-SIRII in French, Program No. 2590) subsidised by the Ministry for Economy, Science and Innovation (MESI) of Québec.

Institutional Review Board Statement: Not applicable.

Informed Consent Statement: Not applicable.

Data Availability Statement: The data presented in this study are available upon request from the corresponding author.

Acknowledgments: The authors are grateful to the NSERC-Canlak Industrial Research Chair in Finishes for Interior wood products (CRIF) industrial partners for their help, support, and product supply. The authors would like to acknowledge the help of Suzie Côté and the Research Centre for Advanced Materials (CERMA, Université Laval) for SEM imaging as well as Jenny Lapierre and the SEREX for the Karl Fischer titrations. The authors would also like to thank all the technical team at the Renewable Materials Research Centre (CRMR, Université Laval) and specifically Yves Bédard, Daniel Bourgault and Luc Germain, who provided technical support. The assistance provided by Julia B. Grenier for the phosphorus monomer synthesis was greatly appreciated.

Conflicts of Interest: The authors declare no conflict of interest. The funders had no role in the design of the study; in the collection, analyses or interpretation of data; in the writing of the manuscript, or in the decision to publish the results.

References

1. De Coninck, H.; Revi, A.; Babiker, M.; Bertoldi, P.; Buckeridge, M.; Cartwright, A.; Dong, W.; Ford, J.; Fuss, S.; Hourcade, J.-C.; et al. Strengthening and Implementing the Global Response. Available online: https://www.ipcc.ch/report/sr15/chapter-4-strengthening-and-implementing-the-global-response/ (accessed on 13 September 2022).
2. Cobut, A.; Blanchet, P.; Beauregard, R. Prospects for Appearance Wood Products Ecodesign in the Context of Nonresidential Applications. *For. Prod. J.* **2016**, *66*, 196–210. [CrossRef]
3. Lowden, L.; Hull, T. Flammability Behaviour of Wood and a Review of the Methods for Its Reduction. *Fire Sci. Rev.* **2013**, *2*, 4. [CrossRef]
4. Lu, S.-Y.; Hamerton, I. Recent Developments in the Chemistry of Halogen-Free Flame Retardant Polymers. *Prog. Polym. Sci.* **2002**, *27*, 1661–1712. [CrossRef]
5. Joseph, P.; Ebdon, J.R. Phosphorus-Based Flame Retardants. In *Fire Retardancy of Polymeric Materials*; Wilkie, C.A., Morgan, A.B., Eds.; CRC Press: Boca Raton, FL, USA, 2010; pp. 107–127. ISBN 978-1-4200-8399-6.
6. Schartel, B. Phosphorus-Based Flame Retardancy Mechanisms—Old Hat or a Starting Point for Future Development? *Materials* **2010**, *3*, 4710–4745. [CrossRef]
7. Levchik, S.V.; Weil, E.D. Developments in Phosphorus Flame Retardants. In *Advances in Fire Retardant Materials*; Woodhead Publishing Series in Textiles; Elsevier Science: Amsterdam, The Netherlands, 2008; pp. 41–66. ISBN 978-1-84569-507-1.
8. Lyons, J.W. Mechanisms of Fire Retardation with Phosphorus Compounds: Some Speculation. *J. Fire Flammabl.* **1970**, *1*, 302–311.
9. Braun, U.; Balabanovich, A.I.; Schartel, B.; Knoll, U.; Artner, J.; Ciesielski, M.; Döring, M.; Perez, R.; Sandler, J.K.W.; Altstädt, V.; et al. Influence of the Oxidation State of Phosphorus on the Decomposition and Fire Behaviour of Flame-Retarded Epoxy Resin Composites. *Polymer* **2006**, *47*, 8495–8508. [CrossRef]
10. Decker, C. High-Performance UV-Cured Acrylic Coatings. In *New Developments in Coatings Technology*; Zarras, P., Wood, T., Richey, B., Benicewicz, B.C., Eds.; American Chemical Society: Washington, DC, USA, 2007; Volume 962, pp. 176–189. ISBN 978-0-8412-3963-0.
11. Decker, C. Kinetic Study and New Applications of UV Radiation Curing. *Macromol. Rapid Commun.* **2002**, *23*, 1067–1093. [CrossRef]
12. Müller, B.; Poth, U. Radiation Curing. In *Coatings Formulation: An International Textbook*; European coatings tech files; Vincentz Network: Hanover, Germany, 2011; pp. 242–255. ISBN 978-3-86630-872-5.
13. Ma, G.; Wang, X.; Cai, W.; Ma, C.; Wang, X.; Zhu, Y.; Kan, Y.; Xing, W.; Hu, Y. Preparation and Study on Nitrogen- and Phosphorus-Containing Fire Resistant Coatings for Wood by UV-Cured Methods. *Front. Mater.* **2022**, *9*, 851754. [CrossRef]
14. Wang, T.; Liu, T.; Ma, T.; Li, L.; Wang, Q.; Guo, C. Study on Degradation of Phosphorus and Nitrogen Composite UV-Cured Flame Retardant Coating on Wood Surface. *Prog. Org. Coat.* **2018**, *124*, 240–248. [CrossRef]
15. Wang, T.; Li, L.; Cao, Y.; Wang, Q.; Guo, C. Preparation and Flame Retardancy of Castor Oil Based UV-Cured Flame Retardant Coating Containing P/Si/S on Wood Surface. *Ind. Crops Prod.* **2019**, *130*, 562–570. [CrossRef]
16. Huang, Y.; Ma, T.; Li, L.; Wang, Q.; Guo, C. Facile Synthesis and Construction of Renewable, Waterborne and Flame-Retardant UV-Curable Coatings in Wood Surface. *Prog. Org. Coat.* **2022**, *172*, 107104. [CrossRef]
17. Ma, T.; Li, L.; Liu, Z.; Zhang, J.; Guo, C.; Wang, Q. A Facile Strategy to Construct Vegetable Oil-Based, Fire-Retardant, Transparent and Mussel Adhesive Intumescent Coating for Wood Substrates. *Ind. Crops Prod.* **2020**, *154*, 112628. [CrossRef]

18. Mali, P.P.; Pawar, N.S.; Sonawane, N.S.; Patil, V.; Patil, R. UV Curable Flame Retardant Coating: A Novel Synthetic Approach of Trispiperazido Phosphate Based Reactive Diluent. *Pigment Resin Technol.* **2021**, *50*, 271–283. [CrossRef]
19. Chambhare, S.U.; Lokhande, G.P.; Jagtap, R.N. UV-Curable Behavior of Phosphorus- and Nitrogen-Based Reactive Diluent for Epoxy Acrylate Oligomer Used for Flame-Retardant Wood Coating. *J. Coat. Technol. Res.* **2016**, *13*, 703–714. [CrossRef]
20. Lokhande, G.; Chambhare, S.; Jagtap, R. Synthesis and Properties of Phosphate-Based Diacrylate Reactive Diluent Applied to UV-Curable Flame-Retardant Wood Coating. *J. Coat. Technol. Res.* **2017**, *14*, 255–266. [CrossRef]
21. Naik, D.; Wazarkar, K.; Sabnis, A. UV-Curable Flame-Retardant Coatings Based on Phosphorous and Silicon Containing Oligomers. *J. Coat. Technol. Res.* **2019**, *16*, 733–743. [CrossRef]
22. Mulge, S.; Mestry, S.; Naik, D.; Mhaske, S. Phosphorus-Containing Reactive Agent for UV-Curable Flame-Retardant Wood Coating. *J. Coat. Technol. Res.* **2019**, *16*, 1493–1502. [CrossRef]
23. Liang, H.; Huang, Z.; Shi, W. Different Effects of Tri(Acryloyloxyethyl) Phosphate on the Thermal Degradation of Photopolymerized Epoxy Acrylate and Polyurethane Acrylate Films. *J. Appl. Polym. Sci.* **2006**, *99*, 3130–3137. [CrossRef]
24. Yao, C.; Xing, W.; Ma, C.; Song, L.; Hu, Y.; Zhuang, Z. Synthesis of Phytic Acid-Based Monomer for UV-Cured Coating to Improve Fire Safety of PMMA. *Prog. Org. Coat.* **2020**, *140*, 105497. [CrossRef]
25. Wang, X.; Xing, W.; Song, L.; Yu, B.; Shi, Y.; Yang, W.; Hu, Y. Flame Retardancy and Thermal Properties of Novel UV-Curing Epoxy Acrylate Coatings Modified by Phosphorus-Containing Hyperbranched Macromonomer. *J. Polym. Res.* **2013**, *20*, 165. [CrossRef]
26. Liang, H.; Shi, W. Thermal Behaviour and Degradation Mechanism of Phosphate Di/Triacrylate Used for UV Curable Flame-Retardant Coatings. *Polym. Degrad. Stab.* **2004**, *84*, 525–532. [CrossRef]
27. Qi, L.; Wang, B.; Zhang, W.; Yu, B.; Zhou, M.; Hu, Y.; Xing, W. Durable Flame Retardant and Dip-Resistant Coating of Polyester Fabrics by Plasma Surface Treatment and UV-Curing. *Prog. Org. Coat.* **2022**, *172*, 107066. [CrossRef]
28. Andrzejewska, E.; Andrzejewski, M. Polymerization Kinetics of Photocurable Acrylic Resins. *J. Polym. Sci. Part Polym. Chem.* **1998**, *36*, 665–673. [CrossRef]
29. International Standard Organization. *Reaction-to-Fire Tests, Heat Release, Smoke Production and Mass Loss Rate*; ISO: Geneva, Switzerland, 2005.
30. Khudyakov, I.V.; Legg, J.C.; Purvis, M.B.; Overton, B.J. Kinetics of Photopolymerization of Acrylates with Functionality of 1−6. *Ind. Eng. Chem. Res.* **1999**, *38*, 3353–3359. [CrossRef]
31. Jiang, F.; Drummer, D. Curing Kinetic Analysis of Acrylate Photopolymer for Additive Manufacturing by Photo-DSC. *Polymers* **2020**, *12*, 1080. [CrossRef]
32. Norrish, R.G.W.; Smith, R.R. Catalysed polymerization of methyl methacrylate in the liquid phase. *Nature* **1942**, *150*, 336–337. [CrossRef]
33. Decker, C. Photoinitiated Crosslinking Polymerisation. *Prog. Polym. Sci.* **1996**, *21*, 593–650. [CrossRef]
34. Liang, H.; Asif, A.; Shi, W. Photopolymerization and Thermal Behavior of Phosphate Diacrylate and Triacrylate Used as Reactive-Type Flame-Retardant Monomers in Ultraviolet-Curable Resins. *J. Appl. Polym. Sci.* **2005**, *97*, 185–194. [CrossRef]
35. Huang, Z.; Shi, W. Synthesis and Properties of a Novel Hyperbranched Polyphosphate Acrylate Applied to UV Curable Flame Retardant Coatings. *Eur. Polym. J.* **2007**, *43*, 1302–1312. [CrossRef]
36. Scranton, A.B.; Bowman, C.N.; Peiffer, R.W. (Eds.) *Photopolymerization: Fundamentals and Applications*; ACS Symposium Series; American Chemical Society: Washington, DC, USA, 1997; Volume 673, ISBN 978-0-8412-3520-5.
37. Plazek, D.J.; Ngai, K.L. The Glass Temperature. In *Physical Properties of Polymers Handbook*; Mark, J.E., Ed.; Springer: New York, NY, USA, 2006; pp. 187–215. ISBN 978-0-387-31235-4.
38. Sato, K. The Hardness of Coating Films. *Prog. Org. Coat.* **1980**, *8*, 1–18. [CrossRef]
39. Bulian, F.; Graystone, J. *Wood Coating: Theory and Practice*; Elsevier Science: Amsterdam, The Netherlands, 2009; ISBN-10: 0444528407; ISBN-13: 978-0444528407.
40. Lambourne, R.; Strivens, T.A. *Paint and Surface Coatings*; Woodhead Publishing Limited: Sawston UK, 1999; ISBN 978-1-85573-348-0.
41. Maiti, S.; Banerjee, S.; Palit, S.K. Phosphorus-Containing Polymers. *Prog. Polym. Sci.* **1993**, *18*, 227–261. [CrossRef]
42. Thommes, M.; Kaneko, K.; Neimark, A.V.; Olivier, J.P.; Rodriguez-Reinoso, F.; Rouquerol, J.; Sing, K.S.W. Physisorption of Gases, with Special Reference to the Evaluation of Surface Area and Pore Size Distribution (IUPAC Technical Report). *Pure Appl. Chem.* **2015**, *87*, 1051–1069. [CrossRef]
43. Cassidy, P.E.; Yager, B.J. Coupling Agents as Adhesion Promoters. *J. Macromol. Sci. Part Rev. Polym. Process.* **1971**, *1*, 1–49. [CrossRef]
44. Maege, I.; Jaehne, E.; Henke, A.; Adler, H.-J.P.; Bram, C.; Jung, C.; Stratmann, M. Self-Assembling Adhesion Promoters for Corrosion Resistant Metal Polymer Interfaces. *Prog. Org. Coat.* **1998**, *34*, 1–12. [CrossRef]
45. Wang, X.; Zhang, J.; Liu, J.; Luo, J. Phytic Acid-Based Adhesion Promoter for UV-Curable Coating: High Performance, Low Cost, and Eco-Friendliness. *Prog. Org. Coat.* **2022**, *167*, 106834. [CrossRef]
46. Cheng, H.N.; Ford, C.; Dowd, M.K.; He, Z. Effects of Phosphorus-Containing Additives on Soy and Cottonseed Protein as Wood Adhesives. *Int. J. Adhes. Adhes.* **2017**, *77*, 51–57. [CrossRef]
47. Zhao, X.; Liang, Z.; Huang, Y.; Hai, Y.; Zhong, X.; Xiao, S.; Jiang, S. Influence of Phytic Acid on Flame Retardancy and Adhesion Performance Enhancement of Poly (Vinyl Alcohol) Hydrogel Coating to Wood Substrate. *Prog. Org. Coat.* **2021**, *161*, 106453. [CrossRef]

48. de Moura, L.F.; Hernández, R.E. Evaluation of Varnish Coating Performance for Two Surfacing Methods on Sugar Maple Wood. *Wood Fiber Sci.* **2005**, *37*, 355–366.
49. He, Y.; Yao, M.; Nie, J. Shrinkage in UV-Curable Coatings. In *Protective Coatings: Film Formation and Properties*; Wen, M., Dušek, K., Eds.; Springer International Publishing: Cham, Switzerland, 2017; pp. 195–223. ISBN 978-3-319-51625-7.
50. Park, J.-W.; Shim, G.-S.; Back, J.-H.; Kim, H.-J.; Shin, S.; Hwang, T.-S. Characteristic Shrinkage Evaluation of Photocurable Materials. *Polym. Test.* **2016**, *56*, 344–353. [CrossRef]
51. Schartel, B.; Hull, T.R. Development of Fire-Retarded Materials—Interpretation of Cone Calorimeter Data. *Fire Mater.* **2007**, *31*, 327–354. [CrossRef]
52. Grexa, O.; Lübke, H. Flammability Parameters of Wood Tested on a Cone Calorimeter. *Polym. Degrad. Stab.* **2001**, *74*, 427–432. [CrossRef]
53. Song, F.; Liu, T.; Fan, Q.; Li, D.; Ou, R.; Liu, Z.; Wang, Q. Sustainable, High-Performance, Flame-Retardant Waterborne Wood Coatings via Phytic Acid Based Green Curing Agent for Melamine-Urea-Formaldehyde Resin. *Prog. Org. Coat.* **2022**, *162*, 106597. [CrossRef]
54. Li, M.; Hao, X.; Hu, M.; Huang, Y.; Qiu, Y.; Li, L. Synthesis of Bio-Based Flame-Retardant Epoxy Co-Curing Agent and Application in Wood Surface Coating. *Prog. Org. Coat.* **2022**, *167*, 106848. [CrossRef]
55. Jiang, J.; Li, J.; Hu, J.; Fan, D. Effect of Nitrogen Phosphorus Flame Retardants on Thermal Degradation of Wood. *Constr. Build. Mater.* **2010**, *24*, 2633–2637. [CrossRef]
56. Misra, M.K.; Ragland, K.W.; Baker, A.J. Wood Ash Composition as a Function of Furnace Temperature. *Biomass Bioenergy* **1993**, *4*, 103–116. [CrossRef]
57. Etiégni, L.; Campbell, A.G. Physical and Chemical Characteristics of Wood Ash. *Bioresour. Technol.* **1991**, *37*, 173–178. [CrossRef]

Article

Quality of Oil- and Wax-Based Surface Finishes on Thermally Modified Oak Wood

Zuzana Vidholdová *, Gabriela Slabejová and Mária Šmidriaková

Faculty of Wood Sciences and Technology, Department of Wood Technology, Technical University in Zvolen, T.G. Masaryka 24, SK 96001 Zvolen, Slovakia; slabejova@tuzvo.sk (G.S.); smidriakova@tuzvo.sk (M.Š.)
* Correspondence: zuzana.vidholdova@tuzvo.sk; Tel.: +421-45-520-6389

Abstract: In this study, natural linseed oil, hard wax oil, and hard wax, commonly used as finishes for wooden furniture and parquet, were used for surface finishes on Turkey oak wood (*Quercus cerris* L.), thermally modified at temperatures of 175 °C and 195 °C for 4 h. Several resistance surface properties were investigated. The mechanical resistance properties of all surface finishes were very much allied to interactions between the finish and the type of substrate. The adhesion strength and impact resistance decreased if higher temperature was used for thermal modification of the substrate. The surface hardness and the resistance to abrasion were high and increased slightly with increasing temperature during thermal modification of wood. It was also found that surface adhesion, hardness and resistance to impact were very much related to interactions between the coating film and the substrate. The resistance properties of finishes, such as resistance to cold liquids and mold, were mainly influenced by the type of the surface finish. The resistance to cold liquids increased in the order: surface finish with hard wax < linseed oil < finish system of linseed oil + hard wax oil. The lowest resistance to cold liquids showed up in condensed milk and sanitizer. Resistance to *Aspergillus niger* and *Penicillium purpurogenum* was relatively weak, however apparently improved during the first 7 days of the fungal test; the surfaces were covered with a lower distribution density of fungal mycelium after 21 days of the fungal test. Individual surface performances of oil and wax-based surface finishes on native wood were different from thermally modified wood.

Keywords: coating; finishing; oak; oil; thermal modification; wax; wood

Citation: Vidholdová, Z.; Slabejová, G.; Šmidriaková, M. Quality of Oil- and Wax-Based Surface Finishes on Thermally Modified Oak Wood. *Coatings* **2021**, *11*, 143. https://doi.org/10.3390/coatings11020143

Academic Editor: Salim Hiziroglu
Received: 31 December 2020
Accepted: 25 January 2021
Published: 28 January 2021

Publisher's Note: MDPI stays neutral with regard to jurisdictional claims in published maps and institutional affiliations.

Copyright: © 2021 by the authors. Licensee MDPI, Basel, Switzerland. This article is an open access article distributed under the terms and conditions of the Creative Commons Attribution (CC BY) license (https://creativecommons.org/licenses/by/4.0/).

1. Introduction

Thermally modified wood (TMW) is considered as an available, dimensionally stable and durable material. TMW products are commonly used as nonstructural material for various indoor and above-ground outdoor applications, e.g., flooring, cladding or decking. It is known that different processes of thermal modification and operating conditions [1] causes different changes in chemical composition [2–4], physical, mechanical properties and durability [5–8], as well as surface energy and color of TMW [9,10]. One of the main advantages of thermal modification at a temperature above 180 °C is the dimensional stability improvement of wood. Wood is a hygroscopic material, which means it readily absorbs moisture. Thermal modification is reported to increase the dimensional stability of the wood significantly [11–13]. At that high temperature, hemicelluloses and lignin structures in the wood will go through irreversible changes that make this effect permanent in the wood [14,15].

Coating is necessary to protect the attractive appearance and color of thermally modified wood [16]. Higher dimensional stability of TMW translates to less swelling and shrinkage [17–19]; thus, in general, they should create fewer stresses for a coating's film [20]. Additionally, the hydrophilicity of TMW has great impact on the coating process and coating performance. Due to the migration of fats, resins and other non-polar or less polar substances contained in wood as well as due to the degradation of hydroxyl groups in wood polysaccharides, the wood surface becomes more hydrophobic [21,22]. Due to this fact, painting-based oil well wetted the surface of TMW. Conversely, when applying water-borne coatings, the time of curing and penetration is prolonged due to the lower water surface absorption and changed polarity of wood [23]. Petrič et al. [24] reported good wetting of oil-treated Scots pine wood by waterborne coatings. In the works of Altgen and Militz [25], the penetration of coating systems based on a solvent-borne oil, a waterborne alkyd-reinforced acrylate paint and a waterborne acrylate paint into Norway spruce and Scots pine TMW was analysed. They found that the penetration of coating systems into TMW did not differ from unmodified wood, although an excessive penetration of solvent-borne oil was found occasionally for TMW. The adhesion strength of waterborne coatings depended on the system that was used. Additionally, Nejad et al. [20] investigated the weathering performance of exterior penetrating stains when applied to oil-treated wood. They found better color stability and overall better general appearance ranking and lower moisture uptake of heat-treated woods than coated untreated wood samples. In the work of Herrera et al. [26], the changes in visual appearance, the surface topography, and the wood-coating interactions of UV-hardened solvent-free and water-borne polyurethane coating on ash TMW were observed due to artificial accelerated ageing in a UV chamber. In addition, the mechanical resistance of oils and waxes on beech TMW was also studied in our previous work [27].

Coatings based on natural and synthetic oils and waxes, or in their mixture with aqueous dispersions, belong to the group of "green coatings" or "ecological coatings". Oils fall into category of penetrating finishes which enhance the natural wood gain and appearance. Still, oils fill the wood lumens and cavities, not being chemically bonded to the wood cell walls. Vegetable oil such as linseed oil, tall oil and tung oil are very efficient against water uptake when used for treating wood surfaces. However, they have limiting quality characteristics such as low hardness, problematic light stability, and low resistance to liquids—detergents, food, and chemicals [28]. They are poorly resistant to application to the surface of stressed wood products such as kitchen worktop surfaces and wood flooring. The price ranges of these coating materials are different with respect to the chemical composition of the film-forming components. The most expensive oil-wax coatings contain pure natural products such as vegetable oils, beeswaxes, carnauba wax and others [29].

The presented work deals with the evaluation of selected mechanical and resistance properties of oil and wax wood surface finishes and their dependence on the type of the wooden substrate (native and thermally modified oak wood) that provide a comprehensive view of the quality and performance of these ones.

2. Materials and Methods

2.1. Material

In the experiments, Turkey oak (*Quercus cerris* L.) heartwood was used. The test samples were made from sound boards:

- wood without thermal modification (native) with drying temperature 100 °C and an average density of ρ_0 = 761 kg/m^3 and a moisture content of 8% ± 2%;
- wood thermally modified at 175 °C for 4 h (TMW 175 °C) with an average density of ρ_0 = 771 kg/m^3 and a moisture content of 5% ± 2%;
- wood thermally modified at 195 °C for 4 h (TMW 195 °C) with an average density of ρ_0 = 700 kg/m^3 and a moisture content of 4% ± 1%.

Wood and their thermal modification according to industrial production standards at given temperatures was provided by the company TECHNI—PAL (Polkanová, Staré Hory, Slovakia). The dimensions of boards were 1000 × 100 × 20 mm^3 (longitudinal × radial × tangential). For each surface finish and type of substrate, five boards were selected randomly. The surface of test specimens was grinded with sandpapers with grid numbers of 60 and 80.

2.2. Surface Finishing Process

The test specimens were coated with commercial coatings for interior such as:

- Linseed oil (Novochema, Levice, Slovakia) without siccative—used to make a base coat on wood or other absorbent surfaces, under coatings of oil or synthetic coating materials;
- Hard wax oil (Renojava s.r.o., Prešov, Slovakia)—a mixture of hard wax oil, siccative, and aliphatic solvent. It is used to finish all types of parquets and interior furniture with both normal and high loads;
- Hard wax (Adler-Werk Lackfabrik Johann Berghofer GmbH & Co KG, Schwaz, Austria)—hard wax, free of solvents, based on natural oils and wax. It contains linseed oil, bees wax, carnauba wax, and cobalt-zircone siccative.

The following surface finishes were made:

1. Linseed oil: one coat of Linseed oil with average film thickness of 30 ± 10 µm;
2. Linseed oil + oil: undercoat = one coat of Linseed oil + finishing coat = two coats of Hard wax oil—the average film thickness of 80 ± 10 µm;
3. Wax: one coat of Hard wax with average film thickness of 30 ± 10 µm.

The surface finishes were made according to the recommendations listed in the technical sheets. The coatings were applied and cured at a temperature of (23 ± 2) °C and a relative air humidity of (50 ± 5)%. The drying time between the first and second application was 24 h. Before the next application, the coating was ground by hand with sandpaper with grid number of P240. After finishing, the samples were conditioned at a temperature of (23 ± 2) °C and a relative humidity of (50 ± 5)% for 14 days. Then the mechanical resistance and properties were investigated. The coating film thickness was measured with Thickness Gauge (type PosiTector® 200 from DeFelsko Corporation, Ogdensburg, NY, USA) working on the ultrasonic principle.

2.3. Adhesion

The adhesion of the coating films on thermally modified wood and unmodified wood was evaluated by Pull-off test according to the standard STN EN ISO 4624 [30] and by Cross-cut test according to the standard STN EN ISO 2409 [31]. The adhesion was measured on 5 samples per surface finish and 2 measurements were performed on each sample.

The machine PosiTest® AT-M Pull-Off Adhesion Tester (DeFelsko Corporation, Ogdensburg, NY, USA) was used for Pull-off test. Small 20 mm diameter dollies were glued to the coating film using two-component epoxy resin (Pattex® Repair Epoxy, Henkel AG & Co. KGaA, Düsseldor, Germany). After 24 h of curing at 20 °C and a relative air humidity of 60%, perimeters of glued dollies were carefully incised to prevent the propagation of failures out the tested area. Pulling was applied at a rate of 1 mm/min up to separation of the dolly from the surface.

The disruption was also evaluated visually using a table magnifying glass. Classification of failure location (wood—coating film—glue joint—metal dolly) from the Pull-off strength test is shown in Figure 1.

The Cross-cut test was measured on 3 samples per surface finish and 3 measurements were performed on each sample. The Cross-cut test was done as follows: a crosshatch pattern was cut through the coating film to the substrate. The spacing between cuts was selected on the base of the thickness of the surface finish and on the type of substrate as follows: the thickness of the coating film up to 60 µm, the 2 mm spacing for soft wood substrates; the thickness of the coating film from 61 µm to 120 µm, the 2 mm spacing for both hard and soft substrates.

The adhesion of the coating film was classified into 6 grades according to Table 1.

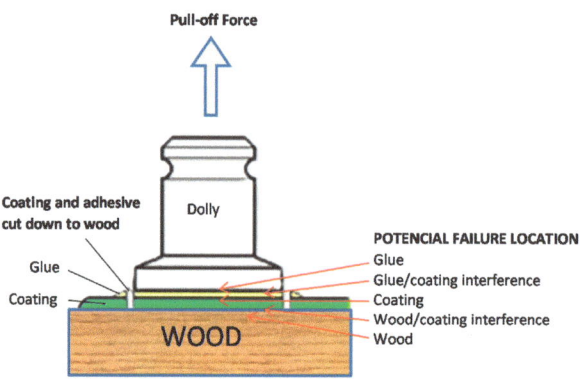

Figure 1. Classification of failure location for the Pull-off strength test.

Table 1. Classification of adhesion according to the Cross-cut test.

Classification	Grade 0	Grade 1	Grade 2	Grade 3	Grade 4	Grade 5
Surface of Cross-cut area from witch flaking has occurred	None	<5%	5–15%	15–35%	35–65%	>65%

2.4. Impact Resistance

The impact resistance of the surface finishes was determined according to the standard STN EN ISO 6272-2 [32]. The impact resistance of the surface finish was measured on 3 samples per surface finish. On each sample, 5 measurements were performed for each drop height—10 mm, 25 mm, 50 mm, 100 mm, 200 mm, and 400 mm.

The intrusion (a pinhole diameter) was measured for drop height of 400 mm with the measuring magnifying glass using 10-times magnification and the surface finish was evaluated subjectively according to Table 2.

Table 2. Impact resistance: degree and evaluation.

Degree	Visual Evaluation
1	No visible changes
2	No cracks on the surface and the intrusion was only slightly visible
3	Visible light cracks on the surface, typically one to two circular cracks around the intrusion
4	Visible large cracks at the intrusion
5	Visible cracks were also off—site of intrusion, peeling of the coating

2.5. Hardness

The film hardness was determined by the Pencil test according to the standard STN EN ISO 15184 [33]. The test started with the softest pencil (number 1 = 3B). The film hardness was measured on 3 samples per surface finish, and 3 measurements were performed on each sample.

The results of the test were evaluated according to the pencil that scratched the surface (Table 3).

Table 3. Degrees of film hardness by the Pencil test.

Pencil number	1	2	3	4	5	6	7	8	9	10	11	12	13
Pencil hardness	3B	2B	B	HB	F	H	3H	4H	5H	6H	7H	8H	9H

2.6. Abrasion Resistance

Evaluation of the surface finish resistance to abrasion was determined according to the standard STN EN ISO 7784-3 [34]. The abrasion resistance was measured on 3 samples per surface finish after 100 cycles. The sample dimensions were of $100 \times 100 \times 20$ mm.

The coefficient of the resistance to abrasion K_T was calculated according to the Equation (1):

$$K_T = (m_1 - m_2)/F_{COR} \tag{1}$$

where m_1—Specimen mass before sanding (g), m_2—Specimen mass after sanding (g), and F_{COR}—Correction coefficient of the used pair of abrasive papers (F_{COR} = 1.052).

2.7. Resistance to Cold Liquids

Surface resistance to cold liquids was determined according to the standard STN EN 12720+A1 [35]. The test was performed on two randomly selected boards per surface finish and wood substrate. Table 4 shows the selected cold liquids which are typical for everyday situations in household.

Table 4. Used cold liquids.

Cold Liquid	Characteristic
Acetic acid	10% (m/m) aqueous solution
Citric acid	10% (m/m) aqueous solution
Ethanol (p.a.)	96% (v/v)
vCondensed milk	10% fat content
Coffee	Dissolve 4 g of instant coffee, medium roasted, freeze—dried, in 100 mL boiling water
Wine	Blackcurrant wine
Black/Green tea	Infuse 1.75 g of tea leaves in 175 mL of boiling water leach for 5 min without stirring, and then carefully decant
Sanitizer	Chloramine T, 2.5% aqueous solution

After 16 h exposure to a cold liquid, the surfaces was cleaned by gently wiping with absorbent cloth, soaked first in a cleaning solution and then in water. Finally, the surfaces were carefully dried with a dry cloth. Damage to the surface, i.e., discoloration, changes in gloss and color and other defects were visually evaluated in an observation box with direct light and graded according to Table 5.

Table 5. Surface resistance to cold liquids: degree and evaluation.

Degree	Description
5	No visible changes (no damage)
4	Slight change in gloss—visible only in reflection of light source
3	Slight traces of damage (gloss)—visible from different directions
2	Strong traces of damage usually without changing the structure of varnish
1	Strong damage with change in varnish structure

2.8. Resistance to Molds

That surface resistance molds are more likely to grow in the interior environment was determined according to the standard STN EN 15457 [36]. Pure cultures of microscopic fungi *Aspergillus niger* Tiegh. (strain BAM 122) and *Penicillium purpurogenum* Dierckx (strain BAM 24) were grown and maintained on 2% Czapek-Dox agar at a temperature of

27 ± 1 °C and a relative humidity of 90%. The cultures were obtained from the Mycological Laboratory at the Faculty of Wood Sciences and Technology of the Technical University in Zvolen, Slovakia. A mixed spore suspension of the test fungi was prepared by washing the surface of individual 2-week-old Petri plate cultures with 10–15 mL of sterile demineralized water. Washings were combined in a spray bottle and diluted to approximately 100 mL with demineralized water to yield approximately 3×10^7 spores·mL^{-1}. The test was performed on 3 circular samples (with diameter of 55 mm) per surface finish. The samples placed individually on the surface of Czapek—Dox agar medium in Petri dishes (100 mm diameter and an outside height of 50 mm, one sample per Petri dish) were coated by brush with 1 mL of mixed mold spore suspension and incubated at 24 ± 2 °C and 80 ± 5% RH for 21 days.

Following incubation, the specimens were visually rated using a scale from 0 to 4 according to Table 6 and using a light microscope Olympus BX43F (Olympus Corporation, Tokyo, Japan) using 4× magnification.

Table 6. Surface resistance to mold growth: degree and evaluation.

Degree	Description
0	0 is no mycelium on the top surface of the specimen
1	growth up to 10% on the top surface of the specimen
2	growth more than 10% up to 30% on the top surface of the specimen
3	growth more than 30% up to 50% on the top surface of the specimen
4	more than 50% on the top surface of the specimen

2.9. Statistival Analysys

The measured data of the adhesion (σ), the abrasion, and the resistance to molds were evaluated for individual groups of wood samples on the basis of mean values and standard deviations.

The relationship between adhesion (σ) and the modification temperature (t) was evaluated using (1) a linear correlation "$\sigma = a + b \cdot t$" with a coefficient of determination (R^2), and levels of significance (p-level), and (2) a post-hoc Duncan test with a level of statistical significance (p-level) and a statistical difference between these values at 95% statistical significance by the ANOVA statistical test. A statistical difference of mean for abrasion resistance (K_T) was evaluated using a post-hoc Duncan test.

3. Results and Discussion

3.1. Adhesion

The adhesion was found to be dependent on the type of the wood substrate; the adhesion was higher on native wood than on the thermally modified wood. The highest adhesion was reached by the linseed oil and the lowest one by wax. The differences in adhesion of the surface finishes between native wood and TMW were statistically significant (Table 7—see average values, Duncan test, and p-level of significance).

Table 7. Adhesion of the surface finish.

Substrate	Adhesion (MPa)									
	Linseed Oil			Linseed Oil + Oil			Wax			
	Average	SD	Duncan Test (p-Level)	Average	SD	Duncan Test (p-Level)	Average	SD	Duncan Test (p-Level)	
Native	5.29	0.45	-	4.57	0.29	-	3.45	0.37	-	
TMW 175 °C	4.18	0.39	a (0.000)	3.65	0.46	a (0.000)	2.91	0.24	b (0.001)	
TMW 195 °C	3.11	0.27	a (0.000)	3.26	0.32	a (0.000)	2.42	0.21	a (0.000)	

Notes: Average—mean values from 10 replicates; SD—standard deviation; a, b—indexes of the Duncan test characterizing the significance level of adhesion strength in relation to the reference (native wood) (a—very significant decrease > 99.9%, b—significant decrease > 99%).

The linear decreases in the adhesion in dependence of the increased temperature during thermal modification of Turkey oak wood and of the used type finishes are shown in Figure 2. A significantly negative effect of the increased modification temperature was confirmed by the coefficient of determination R^2 of the linear correlation "$\sigma = a + b \cdot t$", of 0.780, 0.710, and 0.653. The analysis indicated a statistically significant relationship at the 95% confidence level (the *p*-value of the linear regression analysis was less than 0.05).

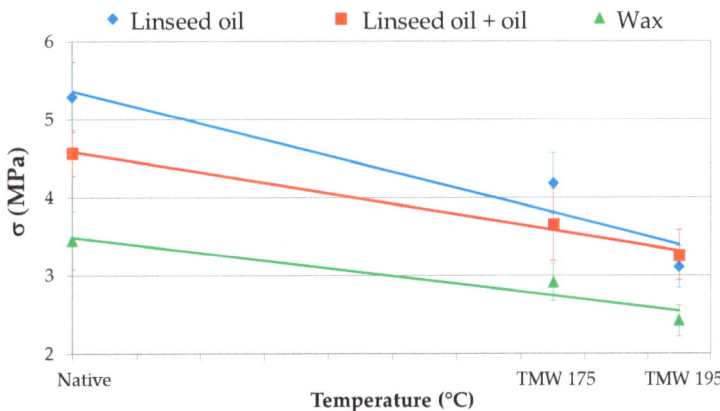

Regresion equation for Linseed oil: $\sigma = 7.437 - 0.021 \cdot t$; $R^2 = 0.780$; $p = 0.000$
Linseed oil + oil: $\sigma = 5.911 - 0.013 \cdot t$; $R^2 = 0.710$; $p = 0.000$
Wax: $\sigma = 4.473 - 0.010 \cdot t$; $R^2 = 0.653$; $p = 0.000$

Figure 2. Linear correlation between the adhesion (σ) of surface finishes and the temperature used for thermal modification of Turkey oak wood.

Not only the measured values of adhesion strength are important, but also the analysis of the area damaged after the Pull-off strength test. The damaged area of the wood surface layer on native wood with the linseed oil surface finish was the cohesive type of fracture of wood, the area of which was less than 25% of the tested area (Figure 3), the rest of the area showed as adhesive type of fracture.

We can state that the measured value means the adhesion of the coating film to the surface. Greater damage of the wood surface layer (up to 40%) was recorded on TMW 175 °C. After the Pull-off testing on TMW 195 °C, the damage of wood surface layer was 100%. We can assume that the adhesion of the coating film was greater than cohesion of wood surface layer.

After Pull-off testing on the surface of native wood with the surface finish of linseed oil + oil (Figure 3), the damage occurred in the interface of coating film-wood. It can be stated that the measured value is a value of adhesion. In the case of TMW 175 °C, 25% of the tested area was the cohesive fracture of wood. In the case of TMW 195 °C, the cohesive fracture was up to 100% (Figure 3). We assume that the measured value on TMW 195 °C means the value of cohesion of wood surface layer and the adhesion was higher. In previous work [37,38] we described an adhesion of an oil surface finish on aged wood and wood attacked by fungi. On the aged wood with an oil surface finish, the cohesive break occurred mainly in the coating film. On the wood attacked by fungi, the breakage occurred in the wood surface layer. The wood surface layers were weakened and the breakage occurred in the oil-impregnated layer of wood. In this experiment, the wood surface layers were weakened by the heat treatment.

After Pull-off testing on the surface of native wood with the wax surface finish, the damage occurred in the interface of coating film-wood. In the case of TMW 175 °C, the damage occurred partially in wood surface layers and was classified as the cohesive fracture (25% of the area); the damage on the rest of the tested area was classified as adhesive type of fracture. On TMW 195 °C, the damage was classified as the adhesive type of fracture.

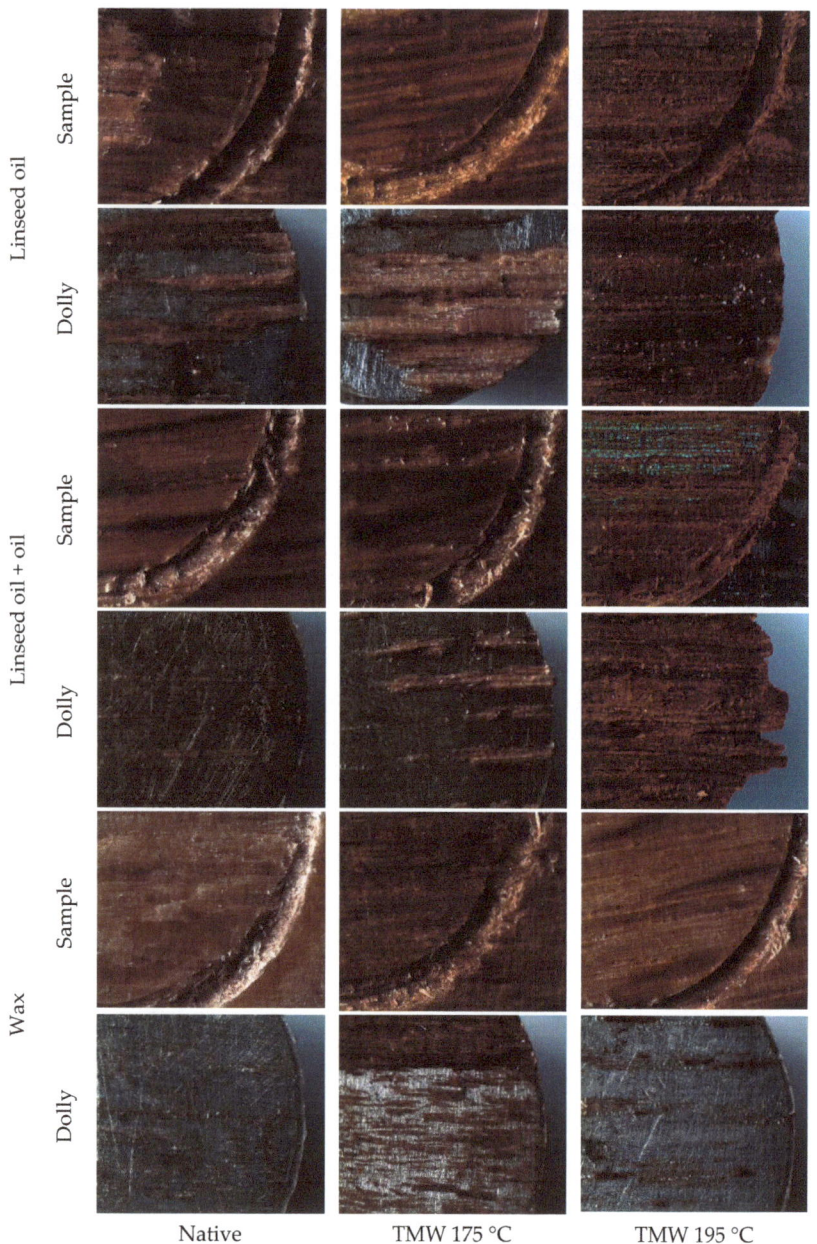

Figure 3. The failure location for wood surfaces and dolly after Pull-off testing film.

If comparing both oil surface finishes with the wax surface finish on TMW, it is seen that the adhesion of oil surface finishes was higher than the cohesion of wood surface layers, but the adhesion of the wax surface finish was lower than the cohesion of wood surface layers.

Sanding could improve the adhesion [39] of the coatings by increasing its wettability.

Table 8 also shows the Cross-cut adhesion of all tested coating films. The surface finish of linseed oil on native wood showed the adhesion classified as degree 2 and on TMW 175 °C and on TMW 195 °C as degree 3. The surface finish of linseed oil + oil on native wood and on TMW 175 °C showed the adhesion classified as degree 2; on TMW 195 °C showed the adhesion classified as 3. The surface finish of wax on native wood was classified as 2; on TMW 175 °C as degree 3 and on TMW 195 °C as degree 4 (Figure 4).

Table 8. Cross-cut adhesion, abrasion and hardness of the surface finish.

Surface Finish	Substrate	Cross-Cut Adhesion	Abrasion K_T (−)			Hardness
		Degree	Average	SD	Duncan Test (p-Level)	Degree
Linseed oil	Native	2	0.053	0.024	-	6
	TMW 175 °C	3	0.032	0.000	d 0.184	7
	TMW 195 °C	3	0.042	0.005	d 0.326	9
Linseed oil + oil	Native	2	0.052	0.033	-	5
	TMW 175 °C	2	0.021	0.000	d 0.208	6
	TMW 195 °C	3	0.027	0.025	d 0.259	8
Wax	Native	2	0.084	0.076	-	7
	TMW 175 °C	3	0.063	0.060	d 0.170	7
	TMW 195 °C	4	0.042	0.025	d 0.197	8

Notes: SD—standard deviation; d—index of the Duncan test characterizing the insignificance difference.

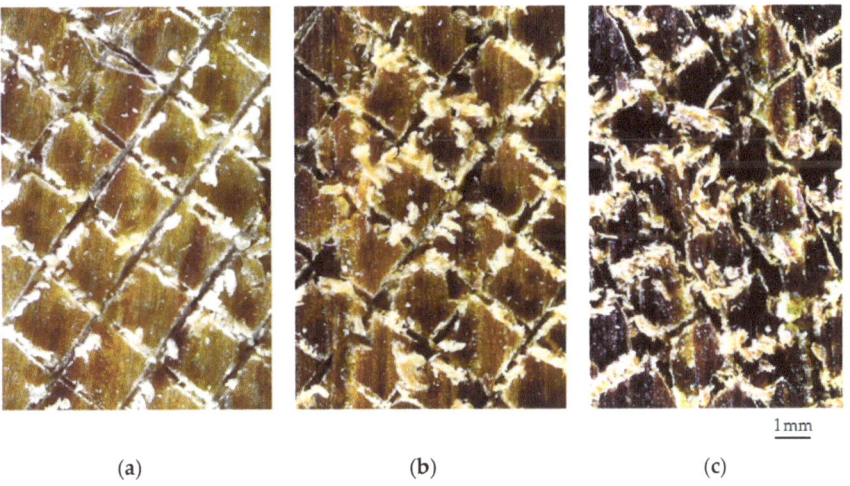

(a) (b) (c)

Figure 4. Adherence of wax surface finish to substrates: (**a**) on native wood (degree 2), (**b**) on thermally modified wood (TMW) 175 °C (degree 3) and (**c**) on TMW 195 °C.

3.2. Impact Resistance

Table 9 shows the grades of the changes on surface finishes after the impact resistance testing. Falling-weight test (small-area indenter) was done at the different drop height. In the last column in Table 8, the diameters of pinholes (mm) are recorded.

Table 9. Impact resistance of the surface finish: degree and evaluation.

Surface Finish	Substrate	Impact Resistance Degree (at the Different Drop Height (mm))					
		10	25	50	100	200	400 *
Linseed oil	Native	1	1	2	2	2	2 (4)
	TMW 175 °C	1	1	2	2	3	3 (3)
	TMW 195 °C	1	2	2	2	3	3 (6)
Linseed oil + oil	Native	1	1	2	2	2	2 (4)
	TMW 175 °C	1	2	2	2	3	3 (5)
	TMW 195 °C	1	2	2	2	3	3 (5)
Wax	Native	1	2	2	2	3	3 (5)
	TMW 175 °C	1	2	2	2	3	3 (5)
	TMW 195 °C	1	2	2	3	3	3 (5)

Note: * In parentheses is the diameter of the pinhole intrusion in mm.

The biggest intrusion (diameter of 6 mm) was measured on the surface finish of linseed oil on TMW 195 °C. The smallest intrusion (diameter of 3 mm) was measured on the same surface finish (linseed oil) on TMW 175 °C.

The changes on the surface finishes were graded as 3 maximally; only on two the surface finishes the changes on the surface finishes were graded as 2: linseed oil on native wood, and linseed oil + oil on native wood (Figure 5).

The similar diameters of intrusions bead at the height of 400 mm were recorded by [40], on pigmented surface finish (polyester-polyurethane) on MDF veneered with beech veneer. The diameter of intrusions measured on European oak veneer with silicone coating at the same test conditions was 3–3.5 mm [41]. The impact resistance of a surface finish is influenced by hardness of the substrate as well as the film's brittleness and elasticity. Additionally Pavlič et al. [42] recorded the influence of substrate on impact resistance of the coated surfaces. Tesařová et al. [43] reported a hypothesis about the relationship between the physical—mechanical properties of lacquers surface finishes and the ultimate tensile stress of free coating films.

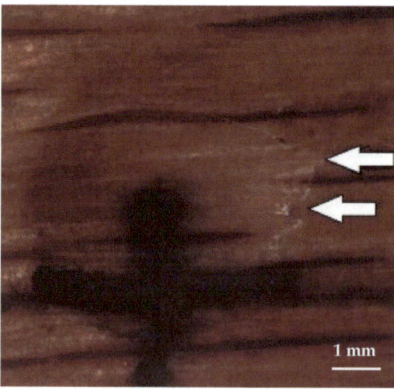

Figure 5. Cracking visible on the surface with surface finish of linseed oil + oil on thermally modified wood (TMW 195 °C) after impact resistance testing at the drop height of 400 mm.

The results indicate that the surface of European oak TMW at a higher temperature is more susceptible to surface cracking when being struck. The cracking formed in wood is spreading into the coating film. Similar results on thin silicone coatings were recorded by Slabejová et al. [41]; the cracks were developed, and the damage was graded up to 4 and 5. Additionally, Slabejová and Šmidriaková [40] tested polyester-polyurethane surface finish on beech wood; the damage was graded as 4 and 5.

3.3. Hardness

Table 8 summarizes the surface hardness of the tested oil and wax surface finishes. The linseed oil surface finish on TMW 195 °C was the hardest (degree 9); while the surface finish of linseed oil + oil on native wood showed only the degree of 5. The surface hardness of all three surface finishes was increasing with increasing temperature of thermal treatment. On native wood surface, the highest hardness was measured for wax surface finish. Nejad et al. [44] reported similar results of higher pencil hardness for polyurethane and acrylic-based coatings on oil-treated wood if compared with the hardness on native wood. In previous work [27], thermally modified beech wood showed the increased surface hardness of these oil and wax surface finishes. The surface hardness is a property of the coating film, but it was proven that the determined hardness was also influenced by the substrate [42].

3.4. Resistance to Abrasion

In the application of coated wood in interior, good resistance to abrasion is desired (which is presented by the lower value of K_T). The abrasion resistance also depended on the type of wood substrate; however, it was statistically insignificantly lower on native wood than on thermally modified wood (Table 8—see Duncan test, and p-level of significance, Figure 6). We assume that the calculated average value of the coefficient K_T differs because of some accidental influences. The lowest resistance to abrasion was measured for the wax surface finish; the greatest abrasion was observed on native wood (the highest value of $K_T = 0.084$). The highest resistance to abrasion (the lowest value of $K_T = 0.021$) was measured on the surface finish of linseed oil + oil on TMW 175 °C. Generally, the resistance to abrasion of all three surface finishes was satisfactory when compared with the technical requirements according to the standard STN 91 0102 [45]. The resistance to abrasion of the linseed oil, linseed oil + oil, and wax surface finishes was higher than the resistance of polyester—polyurethane surface finish [40] and a waterborne UV—hardened surface finish after 100 cycles [46,47].

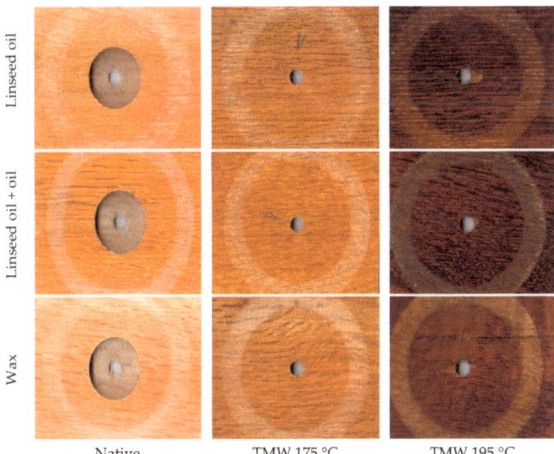

Figure 6. Grinded surface finishes after resistance to abrasion testing (after 100 cycles).

3.5. Resistance to Cold Liquids

A more detailed look at the changes in resistance of cold liquids for linseed oil, linseed oil + oil, wax surface finish and also for uncoated surface clearly shows Figures 7–10.

The surface finish of linseed oil reached the best resistance to cold liquids on darkest TMW 195 °C. Linseed oil surface finish showed the lowest resistance to condensed milk and sanitizer (Figure 7). The low resistance to condensed milk was because of the filter paper saturated with milk remained adhered to the surface. Sanitizer caused a distinct change in color on the surface. The changes in color visible to the naked eye were also on the surfaces of TMW 175 °C after exposure to green tea and black tea. The surface finish created by the system of two oils (linseed oil + oil) reached the best resistance of all tested surfaces (Figure 8). This system of surface finish also showed the lowest resistance to condensed milk due to the filter paper remained adhered to the tested surface. The wax surface finish showed the best resistance to cold liquids on TMW 175 °C than on TMW 195 °C. Additionally, in this case, the condensed milk and sanitizer caused the greatest surface damage (Figure 9).

Based on the experimental results, it can be stated that oil and wax surface finishes showed significantly lower resistance to cold liquids when compared to the resistance of polyurethane surface finish [40], UV-hardened finishes [48], and waterborne surface finishes [44,46].

The Turkey oak wood without any surface finish (Figure 10) shows that the resistance to cold liquids of native wood surface and also the TMW surface was low. The liquids commonly used in household (condensed milk, coffee, blackcurrant wine, black tea, green tea, and sanitizer) left color changes visible to the naked eye on the tested surfaces. Slabejová et al. [41] reported that even a silicone coating on the veneer surface did not provide a good resistance to cold liquids and the color stains were formed on the coating just like on the wood surface without any finish.

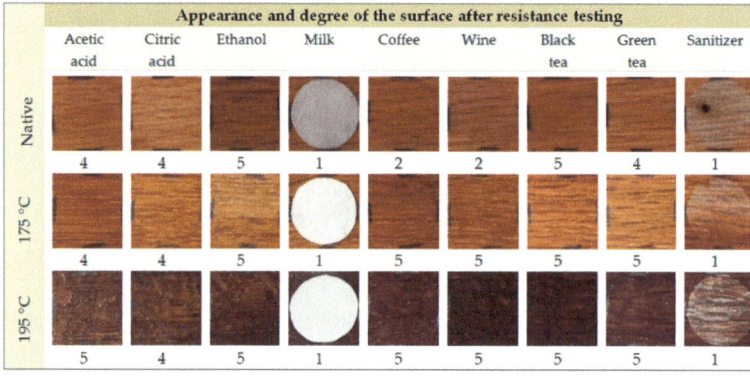

Figure 7. Resistance to cold liquids (degree): linseed oil surface finish.

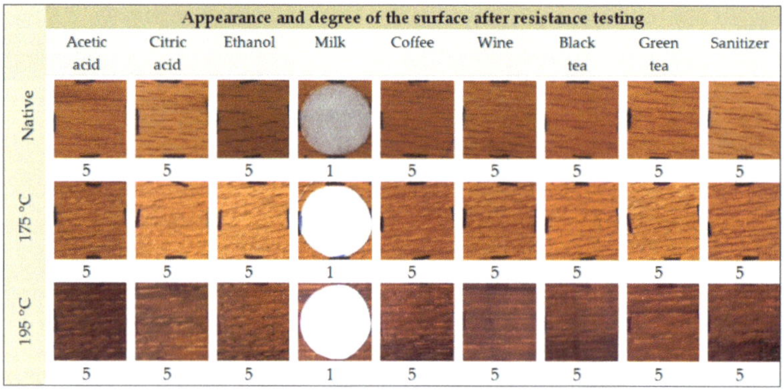

Figure 8. Resistance to cold liquids (degree): surface finish of linseed oil + oil.

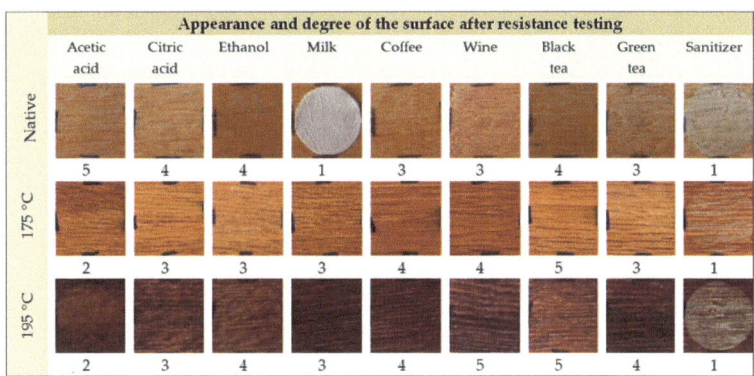

Figure 9. Resistance to cold liquids (degree): wax surface finish.

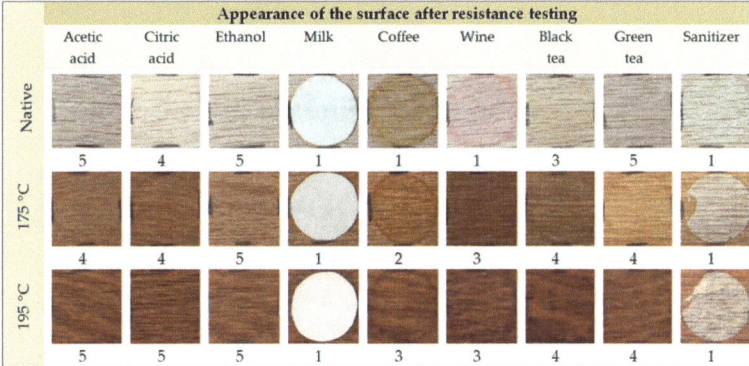

Figure 10. Resistance to cold liquids (degree): uncoated surface of native wood and TMW.

3.6. Resistance to Molds

The resistance to mold growth of the microscopic fungi *Aspergillus niger* and *Penicillium purpurogenum* on the top surfaces' tested surface finishes is summarized in Table 10. A more detailed look at the changes in mold growth for surface finish and also for uncoated surface clearly shows in the microscopic images of surface in Figure 11. The mold growth on the surfaces was evident from the first 7 days of the mold test, whereas on the final 21 days after inoculation, 30% and more of their area was covered with the mold hyphae with the sporangium (degree ranged from 2 to 4). However, all surfaces were covered with less distribution density of mold fungi as compared to the unmodified surfaces (see in Figure 11 white fibers and dark colorless conidiophores and spores on top surfaces). These tendencies could be explained by the water-repellent nature of the finishes which generates the diminished growth of molds. Similarly to previous work [49], the mold growth on surfaces coated with wax—oil or polyurethane—acrylate-based coating was higher than mold growth on the surface with polyurethane coating. The mold growth was decreasing with increasing number of coats of polyurethane or waterborne polyurethane—acrylate coating. Not all used surface finishes were designed for protecting against molds. However, when combined with antimicrobial agents, they may achieve the improved microbial durability [50,51].

Table 10. Resistance of the surface finishes to mold during 0–21 days of mold test.

Substrate	Statistical Characteristics	Mold Resistance Degree (−)											
		Linseed Oil			Linseed Oil + Oil			Wax			Without Finish		
		7th	14th	21th	7th	14th	21th	7th	14th	21th	7th	14th	21th
Native	Median	1	3	3	0	3	3	0	2	3	3	4	4
	SD	0.58	0.58	0.00	0.00	0.58	0.00	0.58	0.58	0.58	0.58	0.58	0.00
TMW 175 °C	Median	0	3	3	0	3	3	1	2	3	2	4	4
	SD	0.58	0.58	0.58	0.58	0.58	0.58	0.58	0.00	0.58	0.00	0.00	0.00
TMW 195 °C	Median	1	3	4	1	2	4	1	1	3	1	4	4
	SD	0.58	0.58	0.58	0.58	0.00	0.58	0.58	0.00	0.00	0.58	0.58	0.00

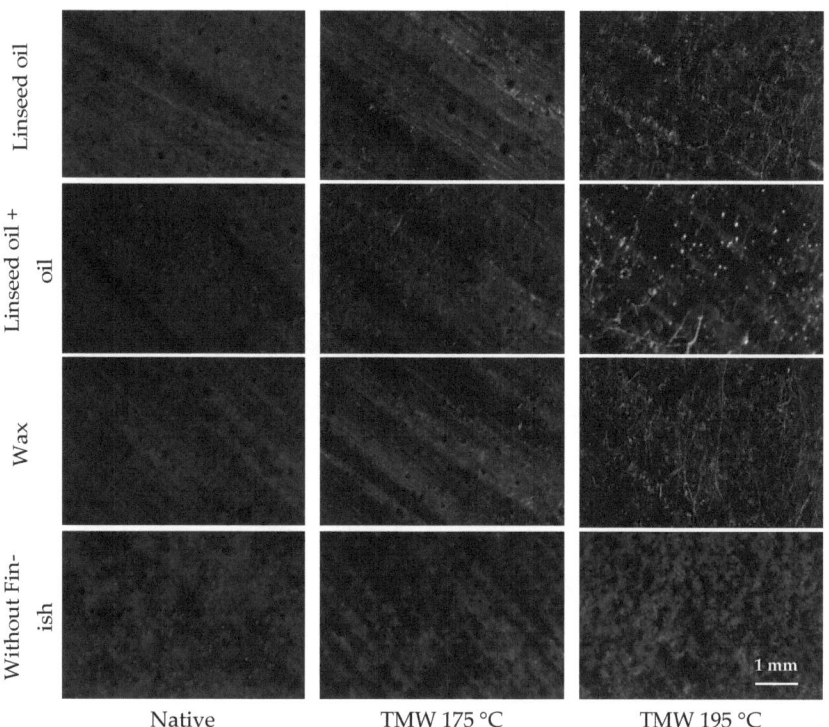

Figure 11. Resistance to molds after incubation for 21 days: surface finish with linseed oil, linseed oil + oil, wax and surface without finish.

4. Conclusions

Three surface finishes—linseed oil, finish system of linseed oil and hard wax oil, and wax were applied on Turkey oak wood to study their performances (adhesion, impact resistance, hardness, and resistances to abrasion, cold liquids and mold) with the dependence on type of the wooden substrate (native wood and wood thermally modified at the temperature of 175 °C and 195 °C for 4 h). Both adhesion and impact resistance were dependent on the type of wood substrate; these were higher on native wood than on thermally modified wood. The abrasion resistance, hardness, and resistance to cold liquids were also depended on the type of wood substrate; however, these were lower on native wood than on thermally modified wood. The coating performance such as abrasion resistance and resistance to cold liquids were also depended on the type of surface finish. Resistance to *Aspergillus niger* and *Penicillium purpurogenum* was relatively weak, however apparently improved during the first 7 days of the fungal test; and the surfaces were covered with a lower distribution density of fungal mycelium after 21 days of the fungal test.

Mechanical properties together with the resistance properties of surface finish must be considered when oil- and wax-based surface finishes were selected for application on wood. Their surface performances on native wood or thermally modified wood are different.

Author Contributions: Conceptualization, Z.V. and G.S.; methodology, Z.V. and G.S.; software, Z.V. and G.S.; validation, Z.V. and G.S.; formal analysis, Z.V., M.Š., and G.S.; investigation, Z.V. and G.S.; resources, Z.V. and G.S.; data curation, Z.V. and G.S.; writing—original draft preparation, Z.V., M.Š., and G.S.; writing—review and editing, Z.V. and G.S.; visualization, Z.V.; supervision, G.S.; project administration, Z.V.; funding acquisition, Z.V., G.S., and M.Š. All authors have read and agreed to the published version of the manuscript.

Funding: This work was supported by the Slovak Research and Development Agency under the contract No. APVV-16-0177 and APVV-17-0583, and the Scientific Grant Agency of the Ministry of Education SR Grant No. 1/0729/18.

Institutional Review Board Statement: Not applicable.

Informed Consent Statement: Not applicable.

Data Availability Statement: Data sharing is not applicable to this article.

Acknowledgments: We would like to thank, for providing some of the materials for this research, the company TECHNI—PAL (Polkanová, Staré Hory, Slovakia). This publication is also the result of the following project implementation: Progressive research of performance properties of wood-based materials and products (LignoPro), ITMS 313011T720 supported by the Operational Programme Integrated Infrastructure (OPII) funded by the ERDF.

Conflicts of Interest: The authors declare no conflict of interest.

References

1. Hill, C.A.S. *Wood Modification: Chemical, Thermal and Other Processes*, 1st ed.; John Wiley & Sons Ltd.: Chichester, UK, 2007; p. 239.
2. Čabalová, I.; Fačík, F.; Lagaňa, R.; Výbohová, E.; Bubeníková, T.; Čaňová, I.; Ďurkovič, J. Effect of thermal treatment on the chemical, physical, and mechanical properties of pedunculate Oak (*Quercus robur* L.) wood. *BioResources* **2018**, *13*, 157–170. [CrossRef]
3. Hrčka, R.; Kučerová, V.; Hýrošová, T. Correlations between Oak Wood Properties. *BioResources* **2018**, *13*, 8885–8898. [CrossRef]
4. Gaff, M.; Kačík, F.; Sandberg, D.; Babiak, M.; Turčani, M.; Niemz, P.; Hanzlík, P. The effect of chemical changes during thermal modification of European oak and Norway spruce on elasticity properties. *Compos. Struct.* **2019**, *220*, 529–538. [CrossRef]
5. Esteves, B.M.; Pereira, H.M. Wood modification by heat treatment: A review. *BioResources* **2009**, *4*, 370–404. [CrossRef]
6. Reinprecht, L.; Vidholdová, Z. *Thermowood*, 1st ed.; Šmíra–Print, s.r.o.: Ostrava, Czech Republic, 2011; p. 89.
7. Sandberg, D.; Kutnar, A.; Mantanis, G. Wood modification technologies-A review. *i-Forest* **2017**, *10*, 895–908. [CrossRef]
8. Lee, S.H.; Ashaari, Z.; Lum, W.C.; Halip, J.A.; Ang, A.F.; Tan, L.P.; Chin, K.L.; Tahir, P.M. Thermal treatment of wood using vegetable oils: A review. *Constr. Build Mater.* **2018**, *181*, 408–419. [CrossRef]
9. Poncsak, S.; Koceafe, D.; Younsi, R. Improvement of the heat treatment of Jack pine (*Pinus banksiana*) using ThermoWood technology. *Eur. J. Wood Wood Prod.* **2011**, *69*, 281–286. [CrossRef]
10. Dzurenda, L.; Dudiak, M. The effect of the temperature of saturated water steam on the colour change of wood *Acer pseudoplatanus* L. *Acta Fac. Xylologiae Zvolen* **2020**, *62*, 19–28. [CrossRef]
11. Priadi, T.; Hiziroglu, S. Characterization of heat treated wood species. *Mater. Des.* **2013**, *49*, 575–582. [CrossRef]
12. Aytin, A.; Korkut, S.; Ünsal, Ö.; Çakicier, N. The effect of heat treatment with the ThermoWood method on the equilibrium moisture content and dimensional stability of wild cherry wood. *BioResources* **2015**, *10*, 2083–2093. [CrossRef]
13. Kúdela, J.; Lagaňa, R.; Andor, T.; Csiha, C. Variations in beech wood surface performance associated with prolonged heat treatment at 200 °C. *Acta Fac. Xylologiae Zvolen* **2020**, *62*, 5–17. [CrossRef]
14. Nuopponen, M.; Vuorinen, T.; Jämsä, S.; Viitaniemi, P. Thermal modifications in softwood studied by FT-IR and UV resonance Raman spectroscopies. *J. Wood Chem. Technol.* **2004**, *24*, 13–26. [CrossRef]
15. Jirouš-Rajković, V.; Miklečić, J. Heat-treated wood as a substrate for coatings, weathering of heat-treated wood, and coating performance on heat-treated wood. *Adv. Mater. Sci. Eng.* **2019**. [CrossRef]
16. Vidholdová, Z.; Sandak, A.; Sandak, J. Assessment of the chemical change in heat treated pine wood by near infrared spectroscopy. *Acta Fac. Xylologiae Zvolen* **2019**, *61*, 31–42. [CrossRef]

17. Tjeerdsma, B.F.; Boonstra, M.; Pizzi, A.; Tekely, P.; Militz, H. Characterisation of thermally modified wood: Molecular reasons for wood performance improvement. *Holz Roh Werkst.* **1998**, *56*, 149. [CrossRef]
18. Srinivas, K.; Pandey, K.K. Effect of heat treatment on color changes, dimensional stability, and mechanical properties of wood. *J. Wood Chem. Technol.* **2012**, *32*, 304–316. [CrossRef]
19. Čermák, P.; Vahtikari, K.; Rautkari, L.; Laine, K.; Horáček, P.; Baar, J. The effect of wetting cycles on moisture behaviour of thermally modified Scots pine (*Pinus sylvestris* L.) wood. *J. Mater. Sci.* **2016**, *51*, 1504–1511. [CrossRef]
20. Nejad, M.; Dadbin, M.; Cooper, P. Coating performance on exterior oil-heat treated wood. *Coatings* **2019**, *9*, 225. [CrossRef]
21. Hakkou, M.; Pétrissans, M.; Zoulalian, A.; Gérardin, P. Investigation of wood wettability changes during heat treatment on the basis of chemical analysis. *Polym. Degrad. Stab.* **2005**, *89*, 1–5. [CrossRef]
22. Pétrissans, M.; Gérardin, P.; Serraj, M. Wettability of heat-treated wood. *Holzforschung* **2003**, *57*, 301–307. [CrossRef]
23. Kesik, H.I.; Akyildiz, M.A. Effect of the heat treatment on the adhesion strength of water based wood varnishes. *Wood Res. Slovak.* **2015**, *60*, 987–994.
24. Petrič, M.; Knehtl, B.; Krause, A.; Militz, H.; Pavlič, M.; Pétrissans, M.; Rapp, A.; Tomažič, M.; Welzbacher, C.; Gérardin, P. Wettability of waterborne coatings on chemically and thermally modified pine wood. *J. Coat. Technol. Res.* **2007**, *4*, 203–206. [CrossRef]
25. Altgen, M.; Militz, H. Thermally modified Scots pine and Norway spruce wood as substrate for coating systems. *J. Coat. Technol. Res.* **2017**, *14*, 531–541. [CrossRef]
26. Herrera, R.; Sandak, J.; Robles, E.; Krystofiak, T.; Labidi, J. Weathering resistance of thermally modified wood finished with coatings of diverse formulations. *Prog. Org. Coat.* **2018**, *119*, 145–154. [CrossRef]
27. Slabejová, G.; Vidholdová, Z.; Šmidriaková, M. Surface finishes for thermally modified beech wood. *Acta Fac. Xylologiae Zvolen* **2019**, *61*, 41–50. [CrossRef]
28. Bulian, F.; Graystone, J.A. *Wood Coatings-Theory and Practice*; Elsevier: Amsterdam, The Netherlands, 2009.
29. Ružinská, E. The analysis determining of surface treatment of wood for optimization of manufacturing processes finalization of wood products. *Mladá Veda* **2018**, *6*, 162–173.
30. *STN EN ISO 4624: Paints and Varnishes-Pull-off Test for Adhesion*; Slovak Office of Standards, Metrology and Testing: Bratislava, Slovakia, 2016.
31. *STN EN ISO 2409: Paints and Varnishes-Cross-Cut Test*; Slovak Office of Standards, Metrology and Testing: Bratislava, Slovakia, 2013.
32. *STN EN ISO 6272-2: Paints and Varnishes-Rapid-Deformation (Impact Resistance) Tests-Part 2: Falling-Weight Test, Small-Area Indenter*; Slovak Office of Standards, Metrology and Testing: Bratislava, Slovakia, 2011.
33. *STN EN ISO 15184: Paints and Varnishes-Determination of Film Hardness by Pencil Test*; Slovak Office of Standards, Metrology and Testing: Bratislava, Slovakia, 2012.
34. *STN EN ISO 7784-3: Paints and Varnishes-Determination of Resistance to Abrasion-Part 3: Method with Abrasive-Paper Covered Wheel and Linearly Reciprocating Test Specimen*; Slovak Office of Standards, Metrology and Testing: Bratislava, Slovakia, 2016.
35. *STN EN 12720+A1: Furniture-Assessment of Surface Resistance to Cold Liquids*; Slovak Office of Standards, Metrology and Testing: Bratislava, Slovakia, 2014.
36. *STN EN 15457: Paints and Varnishes-Laboratory Method for Testing the Efficacy of Film Preservatives in a Coating against Fungi*; Slovak Office of Standards, Metrology and Testing: Bratislava, Slovakia, 2015.
37. Slabejová, G.; Vidholdová, Z. Adhesion of coating films on the weathered wood. *TZB Info* **2019**, *10*. Available online: https://stavba.tzb-info.cz/drevostavby/19533-adhezia-naterovych-filmov-na-poveternostne-starnutom-dreve (accessed on 9 September 2019).
38. Vidholdová, Z.; Slabejová, G. Adhesion of surface treatments on rot wood. *TZB Info* **2019**, *11*. Available online: https://stavba.tzb-info.cz/drevene-konstrukce/19857-adhezia-naterovych-filmov-na-hnilom-dreve (accessed on 13 November 2019).
39. Kúdela, J.; Javorek, Ľ.; Mrenica, L. Influence of milling and sanding on beech wood surface properties. Part II. Wetting and thermo-dynamical characteristics of wood surface. *Ann. WULS-SGGW For. Wood Technol.* **2016**, *95*, 154–158.
40. Slabejová, G.; Šmidriaková, M. Quality of pigmented gloss and matte surface finish. *Acta Fac. Xylologiae Zvolen* **2018**, *60*, 105–113. [CrossRef]
41. Slabejová, G.; Šmidriaková, M.; Pánis, D. Quality of silicone coating on the veneer surfaces. *BioResources* **2018**, *13*, 776–788. [CrossRef]
42. Pavlič, M.; Žigon, J.; Petrič, M. Wood Surface Finishing of Selected Invasive Tree Species. *Wood Ind.* **2020**, *71*, 271–280. [CrossRef]
43. Tesařová, D.; Čech, P.; Hlavatý, J. Influence of coating formulation on physical-mechanical properties. In *Wood Science and Engineering in the Third Millenium: Proceedings of the International Conference (ICWSE 2017)*; Universitatea Transilvania din Brasov: Brasov, Romania, 2017; pp. 486–493. ISSN 1843-2689. Available online: http://www.unitbv.ro/il/Conferinte/ICWSE2017.aspx (accessed on 15 December 2020).
44. Nejad, R.; Shafaghi, H.; Ali, H.; Cooper, P. Coating performance on oil-heat treated wood for flooring. *BioResources* **2013**, *8*, 1881–1892. [CrossRef]
45. *STN 91 0102: Furniture. Surface Finishing of Wooden Furniture. Technical Requirements*; Slovak Office of Standards, Metrology and Testing: Bratislava, Slovakia, 1986.

46. Tesařová, D.; Chladil, J.; Čech, P.; Tobiášová, K. *Ecological Surface Treatment*, 1st ed.; Mendel University in Brno: Brno, Czech Republic, 2010; p. 126.
47. Slabejová, G. *Roughness and Adhesion in the Surface Treatment of Beech Wood with Water-Borne Paint Materials*, 1st ed.; Technical University in Zvolen: Zvolen, Slovakia, 2016; p. 96.
48. Kaygin, B.; Akgun, E. A nano-technological product: An innovative varnish type for wooden surfaces. *Sci. Res. Essays* **2009**, *4*, 1–7.
49. Vidholdová, Z.; Slabejová, G. Environmental valuation of selected transparent wood coatings from the view of fungal resistance. *Ann. Wars. Univ. Life Sci. SGGW For. Wood Technol.* **2018**, *103*, 164–168.
50. Reinprecht, L.; Vidholdová, Z. Growth inhibition of moulds on wood surfaces in presence of nano-zinc oxide and its combinations with polyacrylate and essential oils. *Wood Res.* **2017**, *62*, 37–44.
51. Lu, K.T.; Chang, J.P. Synthesis and antimicrobial activity of metal-containing linseed oil-based waterborne urethane oil wood coatings. *Polymers* **2020**, *12*, 663. [CrossRef]

Article

Superhydrophilic Coating of Pine Wood by Plasma Functionalization of Self-Assembled Polystyrene Spheres

Sebastian Dahle [1,2,*], John Meuthen [1], René Gustus [3], Alexandra Prowald [4], Wolfgang Viöl [5] and Wolfgang Maus-Friedrichs [3]

1. Institute of Energy Research and Physical Technologies, Clausthal University of Technology, Leibnizstr. 4, 38678 Clausthal-Zellerfeld, Germany; john.meuthen@gmail.com
2. Department of Wood Science and Technology, Biotechnical Faculty, University of Ljubljana, Jamnikarjeva ulica 101, 1000 Ljubljana, Slovenia
3. Clausthal Centre for Material Technology, Clausthal University of Technology, Agricolastr. 2, 38678 Clausthal-Zellerfeld, Germany; rene.gustus@tu-clausthal.de (R.G.); wolfgang.maus-friedrichs@tu-clausthal.de (W.M.-F.)
4. Institute of Electrochemistry, Clausthal University of Technology, Arnold-Sommerfeld-Str. 6, 38678 Clausthal-Zellerfeld, Germany; alexandra.prowald@tu-clausthal.de
5. Faculty of Engineering and Health, University of Applied Sciences and Arts, Von-Ossietzky-Str. 99, 37085 Göttingen, Germany; wolfgang.vioel@hawk.de
* Correspondence: sebastian.dahle@bf.uni-lj.si or sebastian.dahle@tu-clausthal.de; Tel.: +386-1-320-3618

Citation: Dahle, S.; Meuthen, J.; Gustus, R.; Prowald, A.; Viöl, W.; Maus-Friedrichs, W. Superhydrophilic Coating of Pine Wood by Plasma Functionalization of Self-Assembled Polystyrene Spheres. *Coatings* 2021, 11, 114. https://doi.org/10.3390/coatings11020114

Received: 11 December 2020
Accepted: 12 January 2021
Published: 20 January 2021

Publisher's Note: MDPI stays neutral with regard to jurisdictional claims in published maps and institutional affiliations.

Copyright: © 2021 by the authors. Licensee MDPI, Basel, Switzerland. This article is an open access article distributed under the terms and conditions of the Creative Commons Attribution (CC BY) license (https://creativecommons.org/licenses/by/4.0/).

Abstract: Self-assembling films typically used for colloidal lithography have been applied to pine wood substrates to change the surface wettability. Therefore, monodisperse polystyrene (PS) spheres have been deposited onto a rough pine wood substrate via dip coating. The resulting PS sphere film resembled a polycrystalline face centered cubic (FCC)-like structure with typical domain sizes of 5–15 single spheres. This self-assembled coating was further functionalized via an O_2 plasma. This plasma treatment strongly influenced the particle sizes in the outermost layer, and hydroxyl as well as carbonyl groups were introduced to the PS spheres' surfaces, thus generating a superhydrophilic behavior.

Keywords: X-ray photoelectron spectroscopy; atomic force microscopy; confocal laser scanning microscopy; scanning electron microscopy

1. Introduction

The deposition of self-assembling films has been developed from colloidal particle films [1,2]. The thickness of these films can be controlled between monolayers and several microns by varying the deposition parameters [3,4]. It has been found that monodisperse colloidal spheres are especially useful to deposit self-assembled films, sometimes enhanced by surfactants [5]. Such self-assembled films offer a great potential for many manufacturing methods due to their periodicity or crystallinity, and they are often referred to as nanosphere lithography or colloidal lithography [6]. Even the use of different particles and spheres can be carried out, thus gaining multicolloidal crystals [7]. The applications of these crystals include biosensors [8], chemical sensors [9], high-strength ceramics, optical sensors and templates for macroporous structure synthesis [10,11]. Polystyrene colloidal crystals can be used for the preparation of macroporous films [12] and conducting polymers [13]. The self-assembly induces periodicities in dimensions high enough to form photonic crystals [10,14]. One critical issue for many of the given applications is the roughness of the substrate's surface, onto which the layers are applied, because it accounts for the degree of ordering. The rougher the surface, the less perfect the arrangement of the spheres becomes [15]. For patterning nonplanar surfaces, colloidal lithography is also applicable [16].

The highly ordered structures obtained by self-assembly of spheres strongly influence the surface wettability, ranging from superhydrophilic to superhydrophobic behavior [17].

Lotus leaves with their self-cleaning ability show a multiscale roughness [18,19] together with superhydrophobic properties [20]. Several different approaches have been made to yield such a roughened texture [21,22]. First successful attempts to obtain films of self-assembled spheres on rough surfaces have been made by Allard et al., depositing polymethyl methacrylate spheres onto paper substrates via spray coating [23]. However, the method used by Allard et al. depends on the draining of the solvent through the substrate. For impermeable substrates with similar roughness, no successful application has been published until now. Nevertheless, the colloidal lithography can be considered as an easy method to protect damageable organic surfaces. One example for rough organic surfaces with the need of protection against soaking with water are wood surfaces, since the decay of wood is mainly driven by fungi and bacteria, which can only survive in damp wood. Such self-assembling colloidal arrays have been widely investigated for usage as photonic crystals [24,25]. The fabrication of these films can be scaled up to large areas [26,27] for applications such as color patterns and color-changing paints, even in the automotive industry [28], thin-film optics [29], electric applications, such as solar cells [28,30], or solarthermal applications [31]. Among further applications of colloidal crystallization techniques is the fabrication of superhydrophobic, superamphiphobic, or superhydrophilic surfaces [32,33].

Similar superhydrophobic characteristics have been creating on wood and lignocellulosic substrates using soot from flames [22,33–35]. However, this suffered from the visual appearance and required further steps for fixation of the particles to achieve mechanical durability [36,37]. Moreover, for wood and other roughly structured substrates, this typically requires varnishes or primer films to smoothen the surface [38,39], with some authors even concluding a direct deposition of such coatings from suspensions onto wood were not possible due to the substrates natural structure [40]. On wood substrates precoated e.g., with a polydimethylsiloxane (PDMS) film though, spray-coated nanocomposites were superhydrophobic, mechanically durable and self-healing [41].

Other approaches include fluorine functionalized superamphiphobic latex-type coatings that are cured at elevated temperatures [42]. Similar coatings were reported with improved mechanic durability and high transparency at the cost of a reduction of the superhydrophobicity [43]. Even more elaborate is the production of superhydrophobic or superamphiphobic films on lignocellulosic substrates by magnetron sputtering in high vacuum [44,45].

Another possible, but much less investigated application for self-assembling colloidal arrays are superhydrophilic films [46]. The concept of superhydrophilic coatings is similarly connected to hierarchical structures, but includes a different chemical functionality [47,48]. Superhydrophilic surfaces are an alternative approach to self-cleaning surfaces, which in some cases has been shown to outperform superhydrophobic surfaces [49]. Both, opaline pigments [23] and self-cleaning coatings are promising for applications in the wood sectors, although requiring an additional fixation of the colloidal arrays. Moreover, wettability generally plays an important role for many application on wooden work pieces [50], including for coatings [51,52], adhesive bonding [53] and fabrication of composites [54], particularly in the prevention of interfacial voids [55].

In this work, we demonstrate the applicability of self-assembling colloidal arrays on wood as a surface with initially high chemical and structural heterogeneities. Further, the etching of the colloidal arrays by means of an air plasma is shown, which allows to set up colloidal templates and photonic crystals with defined geometric dimensions. Moreover, the functionalization that occurs during the plasma etching yields a chemical functionalization that leads to a superhydrophilic film with potential applications as base coat, adhesion promoter or basis for self-cleaning surfaces.

2. Materials and Methods

A Digital Microscope (Keyence VHX-1000D with VH-Z100UW objective) is used for Light Microscopy (LM). The microscope allows a magnification from 100× to 1000×. The depth of sharpness is increased by merging several pictures.

Confocal Laser Scanning Microscopy (CLSM) was employed to study surface topographies previous and subsequent to the plasma treatments, thus allowing for a verification of the SEM results, which possibly may be affected by surface charging. Due to the roughness of the surfaces, no AFM images with similar fields of view as the SEM images could be obtained. The Keyence VK-X210 microscope with VK-X200K controller has a total magnification up to 24,000×, this consists of up to 150× objective lens magnification (M_o), up to 8× optical zoom and further digital magnification. A laser with a wavelength of 408 nm is used for the illumination. For all images shown here, objectives with M_o = 10× (N.A. = 0.3) and M_o = 150× (N.A. = 0.95) were used. The images are overlaid with wide field microscopic and laser intensity measurements.

A Scanning Auger Electron Microscope (Omicron NanoSAM, Uppsala, Sweden) with a base pressure below 10^{-10} hPa is used for Scanning Electron Microscopy (SEM) (c.f. successor equipment Scienta Omicron NanoScan Lab [56]). No conductive coating has been applied to the samples prior to the SEM measurements, thus enabling us to retrieve images of the same sample before and after plasma functionalization. However, no Auger electron spectroscopy measurements can be conducted due to charging of the surface. The beam current during operation was 200 pA at a primary electron energy of 2 keV with a typical spatial resolution of 3 nm for SEM.

Atomic Force Microscopy (AFM) was carried out using a Veeco Dimension 3100 SPM (Veeco Instruments Inc. 112 Robin Hill Road Santa Barbara CA 93117, Goleta, CA, USA). All measurements are performed in Tapping Mode with Al-coated silicon cantilevers (NSC15, Micromasch). The typical resonance frequencies of this series are about 325 kHz, typical spring constants are in the range of 40 N/m. The radius of the tip curvature is less than 10 nm. All images consist of 512 lines each containing 512 pixels. They are recorded with a line-scan frequency of 0.5 Hz. SPIP (Image Metrology A/S) is used for the depiction of the AFM images and the calculation of the root mean square (RMS) surface roughness according to ISO 4287/1 [57].

An Ultra High Vacuum (UHV) apparatus with a base pressure of 5×10^{-11} hPa was used to carry out the spectroscopic experiments [58]. Scanning Electron Microscopy was performed in another UHV apparatus [59]. Atomic Force Microscopy and Confocal Laser Scanning Microscopy were carried out at atmospheric air. All measurements were performed at room temperature.

X-ray Photoelectron Spectroscopy (XPS) was performed using a hemispherical analyser (VSW HA100, VSW Atomtech Ltd., Witney, Oxfordshire, UK) in combination with a commercial non-monochromatic X-ray source (Specs RQ20/38C). During XPS, X-ray photons irradiate the surface under an angle of 80° to the surface normal, illuminating a spot with a diameter of several mm. For all measurements presented here the Al Kα line (photon energy 1486.6 eV) is used. Electrons were recorded by the hemispherical analyzer with an energy resolution of 1.1 eV emitted under an angle of 10° to the surface normal. All XPS spectra are displayed as a function of binding energy with respect to the Fermi level.

For quantitative XPS analysis, photoelectron peak areas were calculated via mathematical fitting with Gauss-type profiles using CasaXPS to achieve the best agreement possible between experimental data and fit. To optimize our fitting procedure, Voigt-profiles were applied to various oxidic and metallic systems previously, but for most systems the Lorentzian contribution converged to zero. Therefore, all XPS peaks were fitted with Gaussian shapes. For the analysis of all of the spectra, a Shirley-type background has been used. Photoelectric cross sections as calculated by Scofield [60] with asymmetry factors after Powell and Jablonski [61], taking into account asymmetry parameters after Reilman et al. [62] and Jablonski [63] as well as inelastic mean free paths from the NIST database [64] (using the database of Tanuma, Powell and Penn for elementary contributions and the

TPP-2M equation for molecules), as well as the energy dependent transmission function of our hemispherical analyser are taken into account when calculating the stoichiometry.

Pine wood veneer with a thickness of about 0.7 mm were cut into pieces of approximately 10×100 mm^2 prior to the PS sphere dip coating process. The coated pine wood sample was then cut into chips of 10×10 mm^2 for further investigations.

The self-assembled polystyrene structures are prepared on the pine wood veneer chips by dipping in a suspension of polystyrene (PS) spheres (OptiBind®, d = 600 nm, previous catalog no. 11001397100390 at Seradyn, Inc. (Indianapolis, IN, USA), now catalog no. 91001397100390 at Thermo Fisher). The initial suspension was washed several times with water to remove commercial stabilizers and additives, after each washing step the spheres were centrifuged, and at last dispersed at an amount of 10 vol. % in ethanol. The film deposition was carried out at 40 °C by dipping of the substrate into the dispersion, resting in the dispersion for 1 s, and subsequent manual pulling out at an angle of approx. 40°–45° and a velocity of approx. 5 mm/s [65]. Similar flow-induced processes reported in literature result in well defined multilayer colloidal crystals [4,25,66]. The self-assembly of layers occurs, as capillary forces drive the dispersed polymer spheres into the wetting meniscus during the lifting process. The concentration of spheres lifting speed are main parameters that define the film thickness or number of layers of particles [32], further influences are presented by the wetting behavior of the dispersion on the substrate's surface and the solvent's evaporation rate. The self-assembled films were not subjected to any additional rinsing in order to exclude any impact.

Plasma treatments were carried out employing a dielectric barrier discharge in 200 hPa oxygen (99.99%, Westfalen AG). The plasma source has been described in detail elsewhere [67]. A brass-filled sealed quartz glass tube with a wall thickness of 2.4 mm is used as isolated high voltage electrode, which is mounted to a preparation chamber with a base pressure of 5×10^{-8} hPa which is connected directly to the UHV recipient via a common transfer system. An alternating high voltage pulse generator with a pulse duration of 0.6 µs and a pulse repetition rate of 10 kHz is connected to the dielectric isolated electrode, while the sample forms the grounded counter electrode. The discharge gap was set to about 1 mm and the discharge area is about 2 cm^2. During the plasma treatment, a voltage of 11 kV (peak) is measured. The high voltage supply delivers a power of 2 W, the plasma power density can be calculated to 1 W/cm^2 and with a plasma treatment time of 300 s, an energy density of 300 J/cm^2 is applied to the sample. The increase of the sample temperature during the plasma treatment does not exceed 10 K [34,68]. Typical electron energies with this setup amounted to approx. 7 eV [7,69].

Contact angles have been measured via the sessile drop method, using a Nikon D60 on a tripod together with a AF-S DX NIKKOR 18–105 mm/3.5–5.6G ED VR zoom lens. Deionized water drops were deposited by a commercial syringe. Contact angles have been estimated using the LB-ADSA method of the Drop Analysis plugin for the ImageJ image processing program [70]

3. Results and Discussion

The pine wood samples coated with self-assembling PS spheres have been characterized by a variety of microscopic tools, thus overcoming the single disadvantages of every of these methods. Since the optical appearance of the coating is of interest for many of the possible applications, the presented results begin with a LM image. CLSM has been used to study the structuring of the films over large scales. Scanning Electron Microscopy has been employed without any conductive coating. This allows to take images of the very same sample before and after plasma treatments, even though the occurrence of charging might affect the obtained images. However, it is a powerful tool to overcome the wavelength limitation of the Confocal Laser Scanning Microscope's field of view and thus closing the gap to the field of view for the Atomic Force Microscope. These microscopic studies of the coating and the influence of an air plasma treatment are then compared to the results on their chemical structure.

3.1. Microscopic Characterization

Figure 1 shows a light microscopic image of the PS sphere coated pine wood sample. The coating follows the natural structure of the wood substrate quite well on a rather macroscopic scale. The film thickness however seems to be unsteady. At some parts of the sample, the color of the substrate is still visible, whereas most parts are colored completely white by the PS spheres. The color, however, originates in the latex-type structure of the film, which causes a diffuse reflection of the incident light while actually the spheres are transparent [71]. Thus, the film thickness can hardly be estimated from the color of the film. The films obtained with the employed deposition method tend to have a film thickness of about 10–12 layers on smooth surfaces [72,73]. However, the roughness of the wood substrate might give rise to slightly higher film thicknesses.

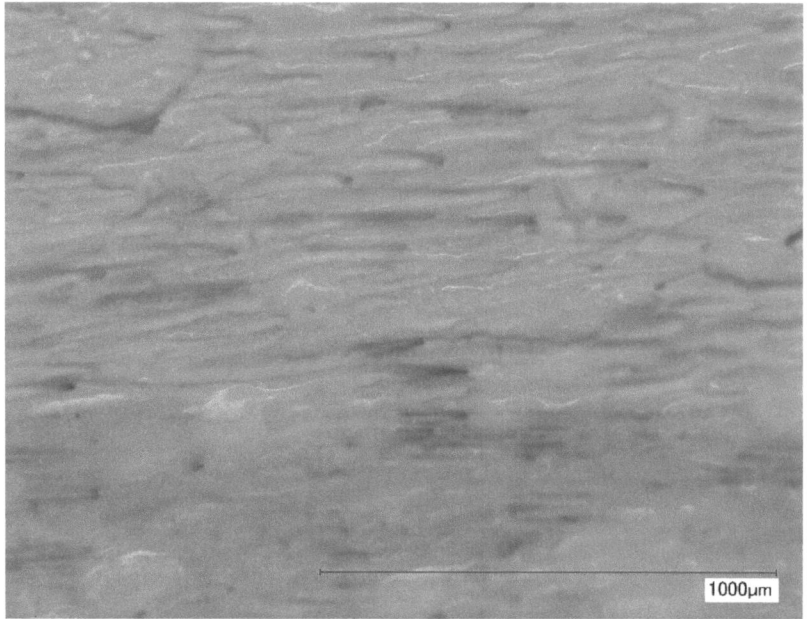

Figure 1. Light microscopic image of the PS sphere coated pine wood sample.

Figure 2 displays CLSM images of the PS sphere coated pine wood sample before (a, left image) and after O_2 plasma treatment (b, right image) with a magnification of 1500× (95 × 71 µm^2) in both images. The red bar indicates a horizontal length of 10 µm. Main parts of the obtained film of PS spheres are polycrystalline with typical domain sizes of about 5–15 single spheres, where the PS spheres are mainly resembling a face-centered cubic (FCC) crystal lattice. In the top image there is a rift visible which has not been filled up with PS spheres. Around this rift, the PS spheres assemble in an amorphous way, at some parts even revealing incomplete covering. The right image (b) pictures a rift, which is fully filled with PS spheres in an amorphous structure. However, no difference between the periodicity of the PS sphere film's structure of the plasma treated sample (b) and the untreated sample (a) is visible. The root mean square roughness R_q over the ordered domains amounts to approx. 4.4 µm before plasma and 2.7 µm after plasma. This likely originates in the reduced height of the PS spheres after the plasma etching, which yields a reduction in observable height differences within the CLSM data.

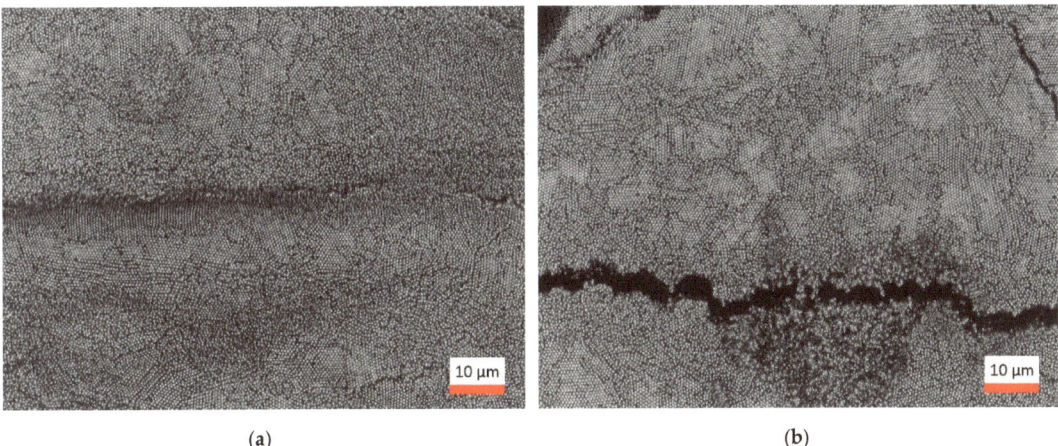

Figure 2. CLSM images of the PS sphere coated pine wood sample before (**a**) and after O$_2$ plasma treatment (**b**) with a magnification of 1500×.

Figure 3 exhibits SEM images of the PS sphere coated pine wood sample before (a, left image) and after O$_2$ plasma treatment (b, middle image) with image sizes of 16 × 21 µm^2 and 11 × 15 µm^2 respectively, as well as a CLSM image after the O$_2$ plasma treatment (c, right image) with a magnification of 12,000× (11.9 × 8.9 µm^2). The red bar indicates a horizontal length of 5 µm. The SEM image reveal that the spheres are densely packed prior to the plasma treatment. After the plasma treatment, the SEM image shows gaps between the PS spheres, which are still arranged in a FCC-like manner, even though the spheres are now shrunk to about half of their size. The CLSM image after O$_2$ plasma treatment (c) shows just the same influence of the plasma treatment as the SEM image, thus excluding misleading results due to possible charging effects. It is noteworthy, that the distances between PS spheres remains the same throughout the plasma etching, whereas the diameter of the outermost layers is significantly reduced. This indicates a minor influence of repulsive forces, such as electrostatic repulsion or hydration forces. This is consistent to earlier plasma applications on similar self-assembled PS layers [39,72]. Additional SEM images after O$_2$ plasma treatment are available in the (supplemental materials Figure S1).

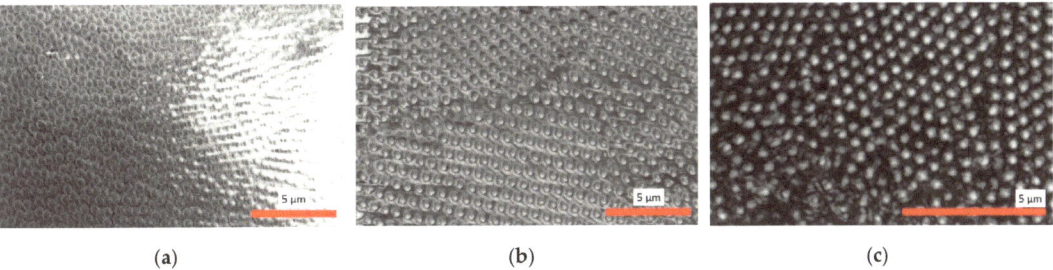

Figure 3. SEM images of the PS sphere coated pine wood sample before (**a**) and after O$_2$ plasma treatment (**b**), as well as a CLSM image after the O$_2$ plasma treatment (**c**) with a magnification of 12,000×.

Figure 4 presents AFM images of the PS sphere coated pine wood sample with a field of view of 10 × 10 µm^2 (a, left image) and 1 × 1 µm^2 (b, right image), respectively. Near surface parts with higher roughness, stacking errors and even amorphous phases are visible. The size of the spheres amounts to about 0.61 µm which fits very well to the manufacturer information and thus excludes broadening effects due to the AFM tip. The

AFM results after the plasma treatment (not shown) resemble exactly the images prior to the plasma treatment (see Figure 4), in contrast to the SEM images clearly showing etched layers of spheres. This indicates a lack of mechanical strength of the sphere-based coating, as the affected layers are apparently removed by performing contact AFM measurements.

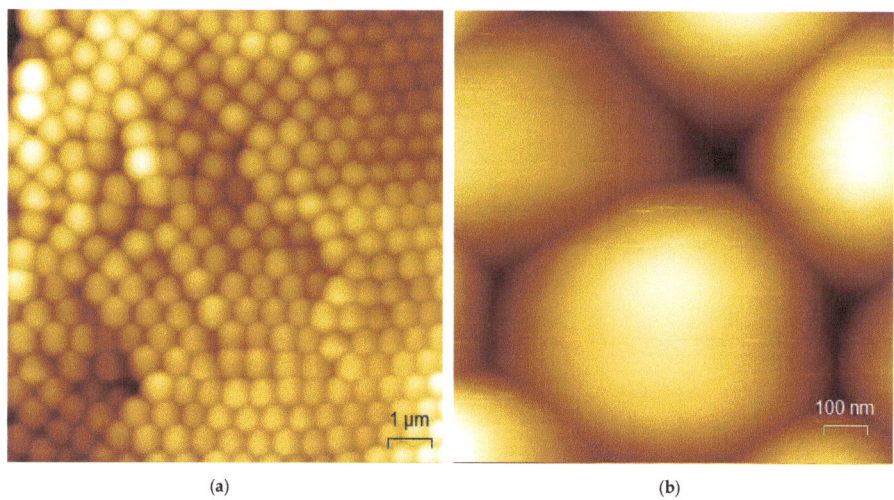

Figure 4. AFM images of the PS sphere coated pine wood sample on an area of 10×10 µm² (**a**) and 1×1 µm² (**b**).

3.2. Spectroscopic Characterization

Figure 5 shows XPS C 1s spectra of the PS sphere coated pine wood sample before (a, left image) and after O_2 plasma treatment (b, right image). The black points represent the original data, the Gaussian peaks of the individual components are displayed using green lines, while the sum for each spectrum is shown as red line. During the fitting procedure, the relative binding energies of the features have been set after Beamson and Briggs [74] as summarized in Table 1. Before the plasma treatment, the C 1s spectrum consists solely of aliphatic and aromatic groups. Due to the resolution limits, the ratios for peak areas and full widths at half mean (FWHM) have also been fixed for this fit. After the plasma treatment, some additional peaks are apparent, indicating an amount of oxidized carbon atoms of about 22%.

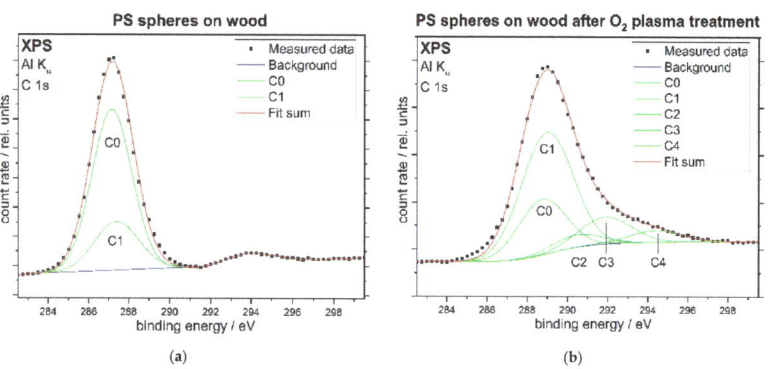

Figure 5. XPS C 1s spectra of the PS sphere coated pine wood sample before (**a**) and after O_2 plasma treatment (**b**).

Table 1. XPS binding energies according to [41].

Marking	Correlation	Binding Energy	Chemical Shift
C0	aromatic	284.8 eV	−0.2 eV
C1	aliphatic	285.0 eV	+0.0 eV
C2	C–O	286.5 eV	+1.5 eV
C3	C=O, O–C–O	287.8 eV	+2.8 eV
C4	O–C=O	288.8 eV	+3.8 eV

Figure 6 displays XPS O 1s spectra of the PS sphere coated pine wood sample before (a, left image) and after O_2 plasma treatment (b, right image). The spectrum before plasma treatment is enlarged due to its small intensity, which solely originates from the sample holder. After the plasma treatment, the oxide state from the sample holder is still present, but a larger peak has appeared at higher binding energies due to the hydroxyl, alkoxy, carbonyl and carboxyl groups that have been generated on the polystyrene spheres by the plasma treatment.

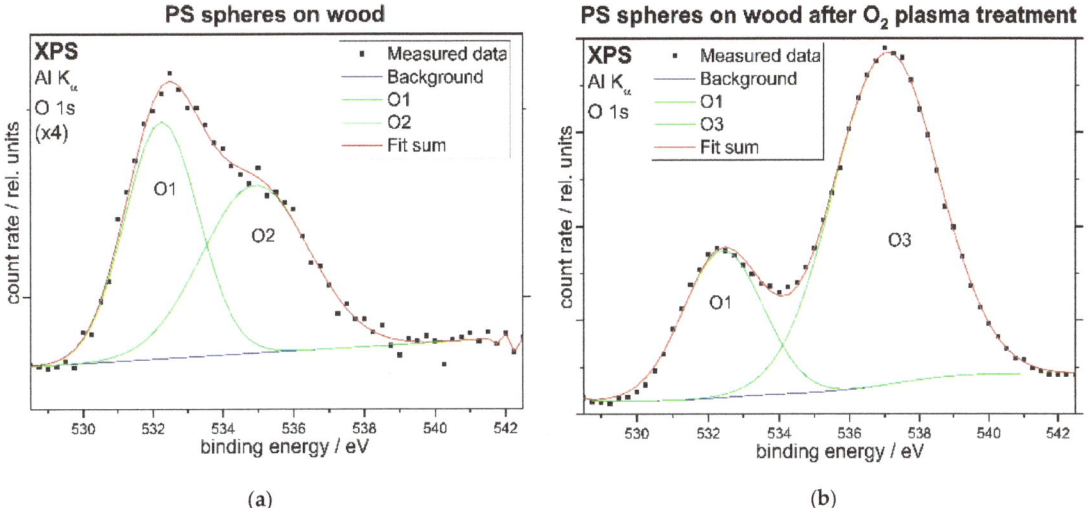

Figure 6. XPS O 1s spectra of the PS sphere coated pine wood sample before (**a**) and after O_2 plasma treatment (**b**).

All XPS C 1s and O 1s results from Figures 5 and 6 are summarized in Table 2. Contact angles estimated via the Sessile drop method were 84° for the uncoated wood substrate and 110° for the PS sphere coated wood prior to the plasma treatment, while the contact angle of the PS sphere coated wood sample after the plasma treatment was well below 10° and thus incapable of measurement. Water contact angles are exemplarily displayed in Figure 7 before (a) and after plasma treatment (b), as opposed to water contact angles on the uncoated pine wood substrate (c). One factor that has been introduced with high relevance to amphiphilic behaviours is the hydration state of layers of polymer particles and molecules encompassing different functionalities [75]. Although the reduced roughness and possible hydration effects may play a role, the hydrophilic effect is most likely to be caused foremost through the addition of the polar oxygen-containing groups to the previously non-polar, hydrophobic polystyrene. However, the porous structure and hydration effects may well add to this, thus giving rise to the particularly strong wetting that was observed on the plasma-treated surfaces.

Table 2. Summarized XPS results.

Sample	Region	Fraction	Binding Energy	FWHM	Rel. Intensity	Correlation	Mark
Before plasma	C 1s	0.94	287.1 eV	2.35 eV	0.75	aromatic	C0
			287.4 eV	2.58 eV	0.25	aliphatic	C1
	O 1s	0.06	532.2 eV	2.47 eV	0.50	Mo–O	O1
			534.9 eV	3.55 eV	0.50	Mo–CO$_3$	O2
After plasma	C 1s	0.69	288.7 eV	2.73 eV	0.23	aromatic	C0
			288.9 eV	3.00 eV	0.54	aliphatic	C1
			290.4 eV	2.90 eV	0.05	C–O	C2
			291.8 eV	3.00 eV	0.12	C=O, O–C–O	C3
			294.3 eV	3.00 eV	0.05	O–C=O	C4
	O 1s	0.31	532.4 eV	2.65 eV	0.25	Mo–O	O1
			537.1 eV	3.57 eV	0.75	C–O	O3

(a)

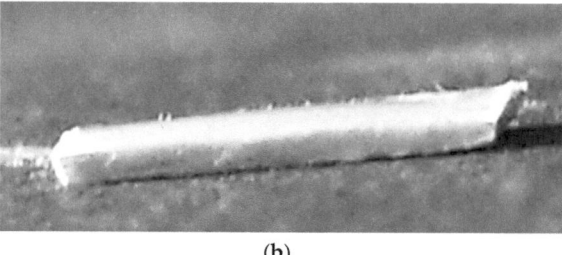
(b)

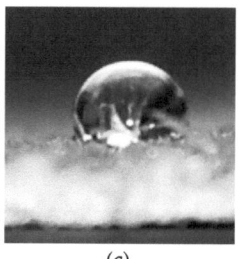
(c)

Figure 7. Exemplary water contact angles of the PS sphere coated pine wood sample before (**a**) and after O$_2$ plasma treatment (**b**), as well as on the uncoated pine wood (**c**).

These self-assembling colloidal films can be considered as one part of deposition process for self-cleaning coatings. Even though the results appear very promising, there are two issues that have to be overcome in order to use the coating in any application. Firstly, the film has to be fastened to gain mechanical stability. Secondly, the surface energy has to be varied e.g., via the attachment of surfactants, thus creating a lotus-effect film. Therefore, the overall coating process of such films can be subdivided into the following steps:

1. Deposition of the self-assembling units
2. Fixation of this colloidal film
3. Simultaneous plasma etching and functionalization
4. Attaching surfactant molecules

The colloidal deposition step (1) has successfully been carried out as demonstrated earlier in this section. The fixation step (2) however still has to be developed. The plasma-etching step (3) has successfully been demonstrated as well, where the results on the C 1s structure after the plasma treatment indicated a high amount of polar groups. These polar groups can then presumably be used as bonding sites for the attachment of surfactant molecules to adapt the surface energy corresponding to the film (step 4). This last step is crucial for the hydrophobic properties of the film and may even lead to an amphiphobization e.g., when using organofluorine compounds.

4. Conclusions

The microscopic results reveal a polycrystalline FCC structure of the PS sphere coating with typical domain sizes of 5–15 single spheres. The O$_2$ plasma treatment significantly shrank the PS spheres without any influence on the crystal-like structure. Furthermore, the surface of the PS spheres got chemically modified towards hydroxyl and carbonyl groups. This two-part, chemical and structural functionalization of the PS spheres induced very low, superhydrophilic contact angles. However, the coating did not become stabilized by the plasma treatment. Thus, some further treatment has to be applied in order to make the film applicable. A promising approach for the fixation of the film is the plasma polymerization of a further polymer matrix. Moreover, the chemical groups attached to the PS spheres by the O_2 plasma treatment can be utilized to attach functional groups in order to try and improve the chemical resistance or generate very high, superhydrophobic contact angles. These tasks will be addressed in prospective investigations.

Supplementary Materials: The following are available online at https://www.mdpi.com/2079-6412/11/2/114/s1, Figure S1: SEM images of larger areas of the PS sphere coated pine wood sample after O_2 plasma treatment.

Author Contributions: Conceptualization, S.D.; formal analysis, S.D.; investigation, S.D., J.M., R.G., and A.P.; resources, W.M.-F.; data curation, S.D.; writing—original draft preparation, S.D.; writing—review and editing, A.P., W.V., and W.M.-F.; visualization, S.D.; supervision, W.V. and W.M.-F.; project administration, W.M.-F.; funding acquisition, S.D., W.V., and W.M.-F. All authors have read and agreed to the published version of the manuscript.

Funding: This research was funded by the Deutsche Forschungsgemeinschaft (DFG), grant numbers MA 1893/18-1 and VI 359/9-1.

Data Availability Statement: All raw and analyzed data have been made available through Zenodo at https://doi.org/10.5281/zenodo.4290000.

Acknowledgments: We gratefully acknowledge the technical assistance of Dana Schulte genannt Berthold (AFM) and Silvia Löffelholz (LM, CLSM), the group of W. Daum (Institute of Energy Research and Physical Technologies, TU Clausthal) for providing the atomic force microscope, and the preparation of the pine wood substrates by the group of Professor H. Militz (Department of Wood Biology and Wood Products, Georg-August-Universität Göttingen), as well as the financial support by the Deutsche Forschungsgemeinschaft (DFG) under project numbers MA 1893/18-1 and VI 359/9-1.

Conflicts of Interest: The authors declare no conflict of interest. The funders had no role in the design of the study; in the collection, analyses, or interpretation of data; in the writing of the manuscript, or in the decision to publish the results.

References

1. Nagayama, K.; Dimitrov, A.S. Fabrication and Application of Particle-Crystalline Films. *ACS Symp. Ser.* **1996**, *648*, 468–489.
2. Dimitrov, A.S.; Nagayama, K. Continuous Convective Assembling of Fine Particles into Two-Dimensional Arrays on Solid Surfaces. *Langmuir* **1996**, *12*, 1303–1311. [CrossRef]
3. Sato, T.; Hasko, D.G.; Ahmed, H. Nanoscale colloidal particles: Monolayer organization and patterning. *J. Vac. Sci. Technol. B* **1997**, *15*, 45–48. [CrossRef]
4. Jiang, P.; Bertone, J.F.; Hwang, K.S.; Colvin, V.L. Single-Crystal Colloidal Multilayers of Controlled Thickness. *Chem. Mater.* **1999**, *11*, 2132–2140. [CrossRef]
5. Pillai, S.; Hemmersam, A.G.; Mukhopadhyay, R.; Meyer, R.L.; Moghimi, S.M.; Besenbacher, F.; Kingshott, P. Tunable 3D and 2D polystyrene nanoparticle assemblies using surface wettability, low volume fraction and surfactant effects. *Nanotechnology* **2009**, *20*, 025604. [CrossRef] [PubMed]
6. Masson, J.-F.; Murray-Méthot, M.-P.; Live, L.S. Nanohole arrays in chemical analysis: Manufacturing methods and applications. *Analyst* **2010**, *135*, 1483–1489. [CrossRef]
7. Singh, G.; Pillai, S.; Arpanaei, A.; Kingshott, P. Multicomponent colloidal crystals that are tunable over large areas. *Soft Matter* **2011**, *7*, 3290–3294. [CrossRef]
8. Asher, S.A.; Alexeev, V.L.; Goponenko, A.V.; Sharma, A.C.; Lednev, I.K.; Wilcox, C.S.; Finegold, D.N. Photonic Crystal Carbohydrate Sensors: Low Ionic Strength Sugar Sensing. *J. Am. Chem. Soc.* **2003**, *125*, 3322–3329. [CrossRef]
9. Holtz, J.H.; Asher, S.A. Polymerized colloidal crystal hydrogel films as intelligent chemical sensing materials. *Nature* **1997**, *389*, 829–832. [CrossRef]
10. Xia, Y.; Gates, B.; Yin, Y.; Lu, Y. Monodispersed Colloidal Spheres: Old Materials with New Applications. *Adv. Mater.* **2000**, *12*, 693–713. [CrossRef]
11. Bartlett, P.N.; Baumberg, J.J.; Birkin, P.R.; Ghanem, M.A.; Netti, M.C. Highly Ordered Macroporous Gold and Platinum Films Formed by Electrochemical Deposition through Templates Assembled from Submicron Diameter Monodisperse Polystyrene Spheres. *Chem. Mater.* **2002**, *14*, 2199–2208. [CrossRef]

12. Carstens, T.; Prowald, A.; El Abedin, S.Z.; Endres, F. Electrochemical synthesis of PEDOT and PPP macroporous films and nanowire architectures from ionic liquids. *J. Solid State Electrochem.* **2012**, *16*, 3479–3485. [CrossRef]
13. Al Zoubi, M.; Al-Salman, R.; Li, Y.; Endres, F. Highly Ordered 3D-Macroporous Poly(Para-Phenylene) Films Made by Electropolymerization of Benzene in an Ionic Liquid. *Z. Phys. Chem.* **2011**, *225*, 393–403. [CrossRef]
14. Nagayama, K. Two-dimensional self-assembly of colloids in thin liquid films. *Colloid. Surf. A* **1996**, *109*, 363–374. [CrossRef]
15. Gasparotto, L.H.S.; Prowald, A.; Borisenko, N.; El Abedin, S.Z.; Garsuch, A.; Endres, F. Electrochemical synthesis of macroporous aluminium films and their behavior towards lithium deposition/stripping. *J. Power Sour.* **2011**, *5*, 2879–2883. [CrossRef]
16. Bhawalkar, S.P.; Qian, J.; Heiber, M.C.; Jia, L. Development of a Colloidal Lithography Method for Patterning Nonplanar Surfaces. *Langmuir* **2010**, *26*, 16662–16666. [CrossRef]
17. Koch, K.; Bhushan, B.; Barthlott, W. Diversity of structure, morphology and wetting of plant surfaces. *Soft Matter* **2008**, *4*, 1943–1963. [CrossRef]
18. Solga, A.; Cerman, Z.; Striffler, B.F.; Spaeth, M.; Barthlott, W. The dream of staying clean: Lotus and biomimetic surfaces. *Bioinspir. Biomim.* **2007**, *2*, S126–S134. [CrossRef]
19. Nosonovsky, M. Multiscale Roughness and Stability of Superhydrophobic Biomimetic Interfaces. *Langmuir* **2007**, *23*, 3157–3161. [CrossRef]
20. Koch, K.; Bhushan, B.; Jung, Y.C.; Barthlott, W. Fabrication of artificial Lotus leaves and significance of hierarchical structure for superhydrophobicity and low adhesion. *Soft Matter* **2009**, *5*, 1386–1393. [CrossRef]
21. Tuteja, A.; Choi, W.; Ma, M.; Mabry, J.M.; Mazzella, S.A.; Rutledge, G.C.; McKinley, G.H.; Cohen, R.E. Designing Superoleophobic Surfaces. *Science* **2007**, *318*, 1618–1622. [CrossRef] [PubMed]
22. Deng, X.; Mammen, L.; Butt, H.-J.; Vollmer, D. Candle Soot as a Template for a Transparent Robust Superamphiphobic Coating. *Science* **2012**, *335*, 67–70. [CrossRef] [PubMed]
23. Allard, D.; Lange, B.; Fleischhaker, F.; Zentel, R.; Wulf, M. Opaline effect pigments by spray induced self-assembly on porous substrates. *Soft Mater.* **2005**, *3*, 121–131. [CrossRef]
24. Wang, C.; Lin, X.; Schäfer, C.G.; Hirsemann, S.; Ge, J. Spray Synthesis of Photonic Crystal Based Automotive Coatings with Bright and Angular-Dependent Structural Colors. *Adv. Function. Mater.* **2020**, *134*, 2008601. [CrossRef]
25. Lange, B.; Fleischhaker, F.; Zentel, R. Functional 3D photonic films from polymer beads. *Phys. Status Solidi A* **2007**, *204*, 3618–3635. [CrossRef]
26. Sprafke, A.N.; Schneevoigt, D.; Seidel, S.; Schweizer, S.L.; Wehrspohn, R.B. Automated spray coating process for the fabrication of large-area artificial opals on textured substrates. *Opt. Express* **2013**, *21*, A528. [CrossRef]
27. Wei, J.; Zhang, G.; Dong, J.; Wang, H.; Guo, Y.; Zhuo, X.; Li, Y. Facile, Scalable Spray-Coating of Stable Emulsion for Transparent Self-Cleaning Surface of Cellulose-Based Materials. *ACS Sustain. Chem. Eng.* **2018**, *6*, 11335–11344. [CrossRef]
28. Kim, J.B.; Lee, S.Y.; Lee, J.M.; Kim, S.-H. Designing Structural-Color Patterns Composed of Colloidal Arrays. *ACS Appl. Mater. Int.* **2019**, *11*, 14485–14509. [CrossRef]
29. Bermel, P.; Luo, C.; Zeng, L.; Kimerling, L.C.; Joannopoulos, J.D. Improving thin-film crystalline silicon solar cell efficiencies with photonic crystals. *Opt. Express* **2007**, *15*, 16986–17000. [CrossRef]
30. Üpping, J.; Salzer, R.; Otto, M.; Beckers, T.; Steidl, L.; Zentel, R.; Carius, R.; Wehrspohn, R.B. Transparent conductive oxide photonic crystals on textured substrates. *Photonic. Nanostruct.* **2011**, *9*, 31–34. [CrossRef]
31. Wang, Z.; Yan, Y.; Shen, X.; Sun, Q.; Jin, C. Candle soot nanoparticles decorated wood for efficient solar vapor generation. *Sustain. Energy Fuels* **2019**, *4*, 354–361. [CrossRef]
32. Vogel, N.; Retsch, M.; Fustin, C.-A.; del Campo, A.; Jonas, U. Advances in Colloidal Assembly: The Design of Structure and Hierarchy in Two and Three Dimensions. *Chem. Rev.* **2015**, *115*, 6265–6311. [CrossRef] [PubMed]
33. Edwards, E.; Wang, D.; Moehwald, H. Hierarchical Organization of Colloidal Particles: From Colloidal Crystallization to Supraparticle Chemistry. *Macromol. Chem. Phys.* **2007**, *208*, 439–445. [CrossRef]
34. Seo, K.; Kim, M.; Kim, D.H. Candle-based process for creating a stable superhydrophobic surface. *Carbon* **2014**, *68*, 583–596. [CrossRef]
35. Esmeryan, K.D.; Castano, C.E.; Bressler, A.H.; Abolghasemibizaki, M.; Mohammadi, R. Rapid synthesis of inherently robust and stable superhydrophobic carbon soot coatings. *Appl. Surf. Sci.* **2016**, *369*, 341–347. [CrossRef]
36. Chang, H.; Tu, K.; Wang, X.; Liu, J. Fabrication of mechanically durable superhydrophobic wood surfaces using polydimethylsiloxane and silica nanoparticles. *RSC Adv.* **2015**, *5*, 30647–30653. [CrossRef]
37. Lin, X.; Park, S.; Choi, D.; Heo, J.; Hong, J. Mechanically Durable Superhydrophobic PDMS-Candle Soot Composite Coatings with High Biocompatibility. *J. Ind. Eng. Chem.* **2019**, *74*, 79–85. [CrossRef]
38. Tu, K.; Wang, X.; Kong, L.; Chang, H.; Liu, J. Fabrication of robust, damage-tolerant superhydrophobic coatings on naturally micro-grooved wood surfaces. *RSC Adv.* **2016**, *6*, 701–707. [CrossRef]
39. Tu, K.; Kong, L.; Wang, X.; Liu, J. Semitransparent, durable superhydrophobic polydimethylsiloxane/SiO2 nanocomposite coatings on varnished wood. *Holzforschung* **2016**, *70*, 1039–1045. [CrossRef]
40. Wang, J.; Lu, Y.; Chu, Q.; Ma, C.; Cai, L.; Shen, Z.; Chen, H. Facile Construction of Superhydrophobic Surfaces by Coating Fluoroalkylsilane/Silica Composite on a Modified Hierarchical Structure of Wood. *Polymers* **2020**, *12*, 813. [CrossRef]
41. Tu, K.; Wang, X.; Kong, L.; Guan, H. Facile preparation of mechanically durable, self-healing and multifunctional superhydrophobic surfaces on solid wood. *Mater. Des.* **2018**, *140*, 30–36. [CrossRef]

42. Zhou, H.; Wang, H.; Niu, H.; Zhao, Y.; Xu, Z.; Lin, T. A Waterborne Coating System for Preparing Robust, Self-healing, Superamphiphobic Surfaces. *Adv. Funct. Mater.* **2017**, *27*, 1604261. [CrossRef]
43. Wu, X.; Liu, M.; Zhong, X.; Liu, G.; Wyman, I.; Wang, Z.; Wang, J. Smooth Water-Based Antismudge Coatings for Various Substrates. *ACS Sustain. Chem. Eng.* **2017**, *5*, 2605–2613. [CrossRef]
44. Bao, W.; Jia, Z.; Cai, L.; Liang, D.; Li, J. Fabrication of a superamphiphobic surface on the bamboo substrate. *Eur. J. Wood Wood Prod.* **2018**, *76*, 1595–1603. [CrossRef]
45. Bao, W.; Zhang, M.; Jia, Z.; Jiao, Y.; Cai, L.; Liang, D.; Li, J. Cu thin films on wood surface for robust superhydrophobicity by magnetron sputtering treatment with perfluorocarboxylic acid. *Eur. J. Wood Prod.* **2019**, *77*, 115–123. [CrossRef]
46. Wang, J.; Hu, J.; Wen, Y.; Song, Y.; Jiang, L. Hydrogen-Bonding-Driven Wettability Change of Colloidal Crystal Films: From Superhydrophobicity to Superhydrophilicity. *Chem. Mater.* **2006**, *18*, 4984–4986. [CrossRef]
47. Drelich, J.; Chibowksi, E.; Meng, D.D.; Terpilowski, K. Hydrophilic and superhydrophilic surfaces and materials. *Soft Matter* **2011**, *7*, 9804–9828. [CrossRef]
48. Svec, F.; Levkin, P.A.; Frechet, J.M.J. Superhydrophobic and Superhydrophilic Materials, Surfaces and Methods. EP2283067A2, 9 May 2008.
49. Son, J.; Kundu, S.; Verma, L.K.; Sakhuja, M.; Danner, A.J.; Bhatia, C.S.; Yang, H. A practical superhydrophilic self cleaning and antireflective surface for outdoor photovoltaic applications. *Sol. Energ. Mat. Sol. C.* **2012**, *98*, 46–51. [CrossRef]
50. Mantanis, G.I.; Young, R.A. Wetting of wood. *Wood Sci. Technol.* **1997**, *31*, 339. [CrossRef]
51. De Meijer, M. A review of interfacial aspects in wood coatings: Wetting, surface energy, substrate penetration and adhesion. In Proceedings of the COST E18 Final Seminar, Paris, France, 26–27 May 2005.
52. Petrič, M.; Knehtl, B.; Krause, A.; Militz, H.; Pavlič, M.; Pétrissans, M.; Rapp, A.; Tomažič, M.; Welzbacher, C.; Gérardin, P. Wettability of waterborne coatings on chemically and thermally modified pine wood. *J. Coat. Technol. Res.* **2007**, *4*, 203–206. [CrossRef]
53. Stehr, M.; Gardner, D.J.; Wålinder, M.E.P. Dynamic Wettability of Different Machined Wood Surfaces. *J. Adhes.* **2001**, *76*, 185–200. [CrossRef]
54. Davalos, J.F.; Qiao, P.; Trimble, B.S. Fiber-Reinforced Composite and Wood Bonded Interfaces: Part 1. Durability and Shear Strength. *J. Composit. Technol. Res.* **2000**, *22*, 224–231.
55. Wood, J.R.; Bader, M.G. Void control for polymer-matrix composites (1): Theoretical and experimental methods for determining the growth and collapse of gas bubbles. *Compos. Manuf.* **2014**, *5*, 139–147. [CrossRef]
56. NanoScan Lab—Scienta Omicron. Available online: https://scientaomicron.com/en/system-solutions/electron-spectroscopy/NanoScan-Lab (accessed on 23 December 2020).
57. ISO 4287:1997. *Geometrical Product Specifications (GPS)—Surface Texture: Profile Method—Terms, Definitions and Surface Texture Parameters*; International Organisation for Standardization: Geneva, Switzerland, 1997.
58. Frerichs, M.; Voigts, F.; Maus-Friedrichs, W. Fundamental processes of aluminium corrosion studied under ultra high vacuum conditions. *Appl. Surf. Sci.* **2006**, *253*, 950–958. [CrossRef]
59. Masendorf, R.; Dahle, S.; Wegewitz, L.; Korte, S.; Lilienkamp, G.; Voigts, F.; Maus-Friedrichs, W. On the origin of fatigue corrosion cracking in Al7075. *Materialwiss. Werkst.* **2013**, *44*, 311–318. [CrossRef]
60. Scofield, J.H. Hartree-Slater Subshell Photoionization Cross-Sections at 1254 and 1487 eV. *J. Electron Spectrosc. Relat. Phenom.* **1976**, *8*, 129–137. [CrossRef]
61. Powell, C.; Jablonski, A. Progress in quantitative surface analysis by X-ray photoelectron spectroscopy: Current status and perspectives. *J. Electron. Spectrosc. Relat. Phenom.* **2010**, *178*, 331–346. [CrossRef]
62. Reilman, R.F.; Msezane, A.; Manson, S.T. Relative intensities in photoelectron spectroscopy of atoms and molecules. *J. Electron. Spectrosc. Relat. Phenom.* **1976**, *8*, 389–394. [CrossRef]
63. Jablonski, A. Database of correction parameters for the elastic scattering effects in XPS. *Surf. Interface Anal.* **1995**, *23*, 29–37. [CrossRef]
64. National Institute of Standards and Technology Electron Inelastic-Mean-Free-Path Database 1.2. Available online: http://www.nist.gov/srd/nist71.cfm (accessed on 29 February 2012).
65. Willert, A.; Prowald, A.; El Abedin, S.Z.; Höfft, O.; Endres, F. Electrodeposition of Lithium in Polystyrene Sphere Opal Structures on Copper from an Ionic Liquid. *Aust. J. Chem.* **2012**, *65*, 1507–1512. [CrossRef]
66. Ferrand, P.; Minty, M.J.; Egen, M.; Ahopelto, J.; Zentel, R.; Romanov, S.G.; Torres, C.M.S. Micromoulding of three-dimensional photonic crystals on silicon substrates. *Nanotechnology* **2003**, *14*, 323–326. [CrossRef]
67. Wegewitz, L.; Dahle, S.; Höfft, O.; Voigts, F.; Viöl, W.; Endres, F.; Maus-Friedrichs, W. Plasma-oxidation of Ge(100) surfaces using a dielectric barrier discharge investigated by MIES, UPS and XPS. *J. Appl. Phys.* **2011**, *110*, 033302. [CrossRef]
68. Kogelschatz, U. Dielectric-Barrier Discharges: Their History, Discharge Physics, and Industrial Applications. *Plasma Chem. Plasma Process.* **2003**, *23*, 1–46. [CrossRef]
69. Dahle, S.; Hirschberg, J.; Viöl, W.; Maus-Friedrichs, W. Gas purification by the plasma-oxidation of a rotating sacrificial electrode. *Plasma Sources Sci. Technol.* **2015**, *24*, 035021. [CrossRef]
70. Stalder, A.F.; Melchior, T.; Müller, M.; Sage, D.; Blu, T.; Unser, M. Low-bond axisymmetric drop shape analysis for surface tension and contact angle measurements of sessile drops. *Colloid. Surf. A* **2010**, *364*, 72–81. [CrossRef]

71. Dushkin, C.D.; Nagayama, K.; Miwa, T.; Kralchevsky, P.A. Colored multilayers from transparent submicrometer spheres. *Langmuir* **1993**, *9*, 3695–3701. [CrossRef]
72. Wegewitz, L.; Prowald, A.; Meuthen, J.; Dahle, S.; Höfft, O.; Endres, F.; Maus-Friedrichs, W. Plasma chemical and chemical functionalization of polystyrene colloidal systems. *Phys. Chem. Chem. Phys.* **2014**, *16*, 18261–18267. [CrossRef]
73. Marlow, F.; Muldarisnur, M.; Sharifi, P.; Brinkmann, R.; Mendive, C. Opals: Status and prospects. *Angew. Chem. Int. Ed.* **2009**, *48*, 6212–6233. [CrossRef]
74. Beamson, G.; Briggs, D. *High Resolution XPS of Organic Polymers*; John Wiley & Sons Ltd.: Chichester, UK, 1992.
75. Lin, W.; Klein, J. Control of surface forces through hydrated boundary layers. *Curr. Opin. Colloid Interface Sci.* **2019**, *44*, 94–106. [CrossRef]

Interactions of Coating and Wood Flooring Surface System Properties

Matjaž Pavlič *, Marko Petrič and Jure Žigon

Department of Wood Science and Technology, Biotechnical Faculty, University of Ljubljana, Jamnikarjeva ulica 101, 1000 Ljubljana, Slovenia; marko.petric@bf.uni-lj.si (M.P.); jure.zigon@bf.uni-lj.si (J.Ž.)
* Correspondence: matjaz.pavlic@bf.uni-lj.si; Tel.: +386-1-320-3621

Abstract: Parquet flooring is one of the most common types of flooring, the surface of which can be covered with various coatings. To avoid possible damage to the parquet during use, it is necessary to test the surfaces before installation according to various non-standard and standard protocols. The present study provides an overview of the interactions between the properties of selected waterborne coatings (solids content, hardness, resistance to cracking, tensile strength) and the properties of oak wood flooring surfaces (dry film thickness, coating adhesion, resistance to scratching, impact, abrasion and cold liquids). The tests conducted showed that the performance of the surface systems was highly dependent on the coating formulations, as they were either one- or two-component systems. Although no major differences in surface resistance to cold liquids were found, there was a correlation between coating thickness, hardness and tensile strength. The harder coatings had higher tensile strengths and lower elongations. The coatings with higher tensile strength and better hardness achieved better adhesion properties. The coatings that exhibited ductile behavior showed the worst scratch resistance. A statistically significant relationship was found between the higher resistance of the flooring systems to impact stress and the improved abrasion resistance. The obtained results provide potential end users of surface coatings with valuable information on the quality that can be expected in wood flooring.

Keywords: coating; wood; flooring; abrasion; scratching; adhesion; cold liquids; hardness; impact; cracking

1. Introduction

Wood parquets of different types, structure and appearance are among the most common flooring elements in residential buildings, municipal buildings, public buildings and sports halls. Frequent exposure of floor surfaces to various influences (exposure to various liquids, detergents and other various staining or aggressive agents, scratching of the film or even greater loads leading to deeper plastic deformations, various impacts, abrasion of the coating film due to higher exposure of a particular place, etc.) in our daily life makes the flooring one of the most exposed elements in the interior. In order to maintain the functionality and appearance of wooden floorings, different types of surface finishes based on natural substances (oils and waxes) have been used throughout history [1]. The development of synthetic polymers based on phenolic, alkyd, epoxy or polyurethane resins in the last century has promoted the use of different film-forming coatings (lacquers or varnishes), mainly because of their better durability compared to natural-based finishes such as oils and waxes [2].

Nowadays, the parquet in the form of engineered wood is one of the most common type of flooring. Engineered wood flooring (EWF) consists of the so-called core layer, which is usually made of softwood species or low-density hardwood species, and the top layer. The wood species used for the top layer must meet various technical criteria, such as esthetics, sustainable harvesting, mechanical properties of the species and/or availability [3]. In recent years, oak (*Quercus petraea* (Matt.) Liebl. and *Quercus robur* L.) is the most popular wood species used for the top layer in European parquet manufacturing [4,5]. Both layers together influence the performance of EWF. However, besides the influence of the physical composition (panel dimensions, the number of layers, their orientation and thickness, wood species used) on the mechanical properties of EWF, the layer of the coating system also has a great influence on the distortion reduction in EWF parquet, as well as on the other crucial properties of the parquet flooring, such as thickness and width swelling, water absorption, dimensional changes, sound absorption, surface durability and color changes [5–7]. Nevertheless, the finishing of a wood floor is perhaps one of the most critical but rewarding steps [8]. Surface finishes are applied to wood for two main reasons: to protect the wood surface during the use of a product and to achieve a specific visual appearance, including color and gloss [9]. The overall performance of the surface system depends on the quality and preparation of the wood substrate, the type of coating material, and good compatibility between the substrate and the coating [10–14]. However, the final quality of wood flooring surfaces is closely related to the finishing process, including the application of the coating and the curing process [15]. Only when all three criteria are met, the surface system helps the flooring to withstand the various influences and damages [16,17]. These properties, which are of paramount importance for interior wood flooring, include, for example, deformability, resistance to impact, various liquids, friction, scratching and abrasion [18,19], all of which can be determined according to various international standards [20].

The properties of coating materials of any kind depend, among other things, particularly on the chemical composition and thickness of the coating film. The type of binder, solvent, and other substances such as pigments and fillers included in the coating formulation determine both the mechanical and physical properties of the coating and its interactions with a particular type of wood substrate [13,21]. Some of the most popular surface coatings for wood floors are acrylates and polyurethanes. Nowadays, many different coating manufacturers offer a variety of acrylate resins- and polyurethane resin-based waterborne coatings for different applications in the market. Generally, these coatings have a lower volatile organic compound content, while their films have excellent properties such as adhesion to substrates, flexibility, durability, compressive and tensile strength, scratch and abrasion resistance, and stability during weathering [22]. In recent years, various nanomaterials have been widely used to improve the properties of waterborne polyurethane coating formulations because they have a large effect despite their small size [23]. The nanoadditives in wood flooring coatings improve their resistance properties [24,25], but they can also strongly influence other physical and mechanical properties of free films [26,27].

All these presented facts lead to the general conclusion and hypothesis that the properties of coated EWF parquet strongly depend on the interactions in the so-called surface system formed by the chosen coating system and the wood substrate. The motivation for the present study was to investigate the properties of selected EWFs coated with different commercial coating systems. These coating systems differed in terms of coating formulation (i.e., binder types) and in the number of coatings layers applied. The performance of the created surface systems of the parquet was evaluated using various non-standard methods and methods, described in international standards. The investigation of the interactions between the properties of the coatings and wood flooring surface systems was additionally supported by statistical analyses.

2. Materials and Methods

2.1. Preparation of Wood Floring Surface Systems

As a substrate 3-layer Engineered Wood Flooring with a 3 mm thick top layer of European oak (*Quercus* sp.). The middle (profiling) and bottom layers were made of spruce (*Picea abies* (L.) Karst.). Five boards measuring 180 mm × 2200 mm × 12 mm ($w \times l \times t$), with an oak top layer without visual defects and similar texture appearance were taken from the stock. After conditioning for three weeks at (23 ± 2) °C and a RH of (50 ± 5) %, all top surfaces of the boards were sanded with sandpaper (120 g). After sanding, each board was sawn into six pieces with a length of (360 ± 5) mm. In this way, five sets of six consecutive samples were prepared, one for each finish.

Six commercial waterborne coating systems were selected for the finish (Table 1): two one-component, acrylate-polyurethane-based systems (System 1: 1C-1 and System 2: 1C-2) and four two-component, polyurethane-based systems (System 3: 2C-1, System 4: 2C-1+, System 5: 2C-2, and System 6: 2C-3). Systems 3 and 4 were from the same manufacturer. The difference was in the number of final coatings, System 4 had one more. System 6 was the only system that used a polyurethane-based primer. The coating systems were formed according to the instructions of the coating manufacturers (primer with an application rate of 120 g/m^2, the first layers of topcoat with an application rate of 120 g/m^2 and all others with an application rate of 100 g/m^2). The intermediate drying time between each coating was 24 h. All coatings were applied with a hand roller. After the application of the first layer of the topcoat, all surfaces were sanded again with sandpaper (150 g). For reasons of trademark protection, the trade names of the systems are not mentioned.

Table 1. Structure of the used waterborne coating systems.

System Label	Used Coatings
1C-1	One-component acrylate-based primer First coat of a one-component acrylate-polyurethane-based finish Second coat of a one-component acrylate-polyurethane-based finish
1C-2	One-component acrylate-based primer First coat of a one-component acrylate-polyurethane-based finish Second coat of a one-component acrylate-polyurethane-based finish
2C-1	One-component acrylate-based primer First coat of a two-component polyurethane-based finish Second coat of a two-component polyurethane-based finish
2C-1+	One-component acrylate-based primer First coat of a two-component polyurethane-based finish Second coat of a two-component polyurethane-based finish Third coat of a two-component polyurethane-based finish
2C-2	One-component acrylate-based primer First coat of a two-component polyurethane-based finish Second coat of a two-component polyurethane-based finish
2C-3	One-component polyurethane-based primer First coat of a two-component polyurethane-based finish Second coat of a two-component polyurethane-based finish

All properties of finished floorings were determined after drying and conditioning at (23 ± 2) °C and a RH of (50 ± 5) % for 21 days after application of the last coat.

The coating properties (solids content, hardness, resistance to cracking, tensile strength and attenuated total reflection Fourier transform infrared (ATR-FTIR) spectroscopy) were determined only for the top coats used, which accounted for the majority of the dry film thickness in the present study of wood flooring surface systems.

2.2. Determination of Solids Content

The solids content was determined following EN ISO 3251:2019 [28]. A total of (1 ± 0.1 g) of the coating was applied in a specified Petri dish and dried at a temperature of 125 °C for 60 min. The solids content was calculated from the initial coating mass and the mass after drying (*SC* in %). The result represents an average value from three measurements.

2.3. Determination of Coating Hardness

The pendulum-damped method using the König pendulum tester (EN ISO 1522:2007 [29], Pendulum Damping Tester Model 299/300, ERICHSEN GmbH & Co. KG, Hemer, Germany) was used to determine coating hardness. The coatings were applied to a glass plate using a manual quadruple film applicator at an application gap height of 360 µm and a speed of (3 ± 0.5) cm/s. After drying and conditioning at (23 ± 2) °C and a relative air humidity (RH) of (50 ± 5) % for 21 days, the coating hardness was measured (three samples per coating, five measurements per sample). The hardness value corresponded to the damping time of the pendulum swinging on the coating surface from 6° to 3°, with respect to the normal axis, measured with an electronic counter. The longer the damping time, the harder the coating.

2.4. Determination of Coating Resistance to Cracking

The resistance of the coating to cracking was determined using a bending test (EN ISO 1519:2012 [30]). The coatings were applied to a glossy photographic paper (grammage 200 g/m^2) using a manual quadruple film applicator (application gap height 360 µm, speed (3 ± 0.5) cm/s, Model 360, ERICHSEN GmbH & Co. KG, Hemer, Germany). After drying and conditioning at (23 ± 2) °C and an RH of (50 ± 5) % for 21 days, three coated paper strips of 30 mm width were prepared for each coating. These paper strips were then clamped in the mandrel tester (Erichsen 266, Erichsen GmbH, Hemer, Germany) with the coated side facing away from the mandrel. The bending process was performed over a period of from 1 to 2 s, bending the coated paper strip for 180°. By changing the mandrel with a smaller diameter, the process was repeated on a new part of the strip. The result was the average maximum diameter at which the coating cracked.

2.5. Tensile Strength Test

Prepared free dry coating films, as described in Section 2.4., were cut into tensile strength specimens of size 100 mm × 20 mm (eight replicates per coating type). The length of the specimens was parallel to the direction in which the coating films were applied. The film thickness and width were then measured and the specimens were then mounted in gauges of the Zwick Z005 universal testing machine (Zwick/Roell, Ulm, Germany), separated by 50 mm. The tensile strength and elongation of the coating free films were measured at the actuator speed of 25 mm/min. From the detected strength–elongation curves, the mean values were reported.

2.6. Attenuated Total Reflection Fourier Transform Infrared (ATR-FTIR) Spectroscopy

The ATR-FTIR spectrometer Perkin Elmer Spectrum Two (PerkinElmer Inc., Waltham, MA, USA), with an LiTaO$_3$ detector type, was used to determine the chemical bonds in the used top coatings. Spectra were recorded at three spots (16 scans per spot) of the surface of each coated flooring type, in a wavelength range from 600 to 4000 cm^{-1} at a resolution of 0.5 cm^{-1}. Relevant absorption bands were then interpreted using the corresponding software (Spectrum V.10.5.3, PerkinElmer Inc.).

2.7. Determination of the Dry Film Thickness

Dry film thicknesses were measured using the microscopic method according to EN ISO 2808:2007 [31]. Cross-sections of the finished floorings were made and viewed under the Olympus SZH stereo microscope (Olympus, Tokyo, Japan) at 120× magnification. The result was the representative (average) value of the dry film thickness measured at 10 points.

2.8. Determination of the Coating Adhesion

Coating adhesion was determined using a pull-off test described in EN ISO 4624:2016 [32]. Ten dollies were bonded to each coated surface using an epoxy adhesive (UHU plus endfest 300, UHU, Bühl, Germany). After 24 h, the cured epoxy adhesive and coating were cut to the substrate around the perimeter of each dolly and the testing was performed using a PosiTest AT adhesion tester (DeFelsco Corporation, Ogdensburg, NY, USA). A tensile stress was applied at an increasing rate of no more then 1 MPa/s until the dolly was pulled off. The breaking strength (in MPa) and the visually perceptible nature of the fracture were recorded.

2.9. Assessment of the Surface Resistance to Scratching

The scratch resistance of the surfaces was evaluated following to the standard EN 15186:2012 [33] using a circular method. The sample was clamped in the fixture and a diamond tip (radius of 90 µm) was applied to the specimen surface with a fixed force. During the single rotation of the sample, the tip caused a scratch on the sample surface. As a result, the minimum force (average of three measurements) that caused a visible scratch or mark on the sample surface in at least six of the eight observed areas in the enclosed template was recorded.

In addition, by applying a force of 1 N at the tip a circular mark with diameter of 50 mm) was induced on the sample surface of each surface system. Analysis of the deformation response with morphological examinations of the residual patterns was performed using the LEXT OLS5000 confocal laser scanning microscope (CLSM) (Olympus, Tokyo, Japan). Four different spots of 640 µm² were selected on each sample, where 5-line cross-sectional profiles perpendicular to the scratch were obtained. The measured average surface area of the scratch and the average depth of the scratch allowed the calculation of the volume of the scratch mark caused, as follows

$$V = \frac{A \times l \times d}{2} \qquad (1)$$

where V is the volume of the scratch (µm³), A is the measured surface area of the scratch (µm²), l is the length of the analyzed scratch (640 µm), and d is the average depth of the scratch (µm).

2.10. Assessment of Surface Resistance to Impact

A modified approach from the standard ISO 4211-4:1988 [34] was used to assess the impact resistance. A cylindrical steel weight with a mass of 500 g was dropped from a given height through a vertically mounted guide onto a 14 mm diameter steel ball positioned on the test surface. Six impacts were made on each flooring surface from each of the following drop heights: 10, 25, 50, and 100 mm. After the impacts, the surfaces were carefully examined with a magnifying glass (10× magnification) under direct light. For easier identification of surface cracks, we stained the impact sites with a whiteboard marker and wiped them off after a few seconds. The impact marks were then evaluated according to the descriptive numerical evaluation code from 5 to 1. If there were no changes after the impact, the assigned rating was 5. If there were no cracks in the coating film but the impact mark was visible, rating was 4. The rating of 3 showed cracks in the coating film. If more cracks were present, the rating was lower. As a result, the minimum height at which the coating film cracked was recorded (rating 3).

2.11. Assessment of the Surface Resistance to Abrasion

The abrasion resistance was determined according to the standard method EN 15185:2012 [35]. From the coated EWF boards, three samples of 100 mm × 100 mm were sawn, diagonally divided into four quadrants and individually placed in the Taber® rotary platform abrasion tester (Taber Industries, North Tonawanda, NY, USA). A rubber wheel with Taber® S-42 abrasive paper was placed on the surface of the sample. The load on each wheel was 500 g and the specimen started to rotate. The freely rotating grinding wheels caused a mark in the form of a circle on the surface of sample. As a result, the number of revolutions (average of the three measurements) required to abrade the coating film onto the substrate in all four quadrants was determined, which is called the Initial Point of wear (*IP*).

2.12. Assessment of Surface Resistance to Cold Liquids

The resistance of the surface to cold liquids was determined according to the method EN 12720:2009+A1:2014 [36]. Soft filter paper disks with a diameter of 25 mm (grammage 450 g/m^2) were immersed for 30 s in the selected liquids (CEN/TS 16209:2012 [37]): acetic acid (10%), acetone, ammonia (10%), citric acid (10%), cleaning agent, coffee, ethanol (48%), mustard, oil (paraffin), red wine (Merlot, 13%), water and sweat (alkaline solution). After the immersion time, each soaked disk was placed on the surface of the sample. The disk was immediately covered with a Petri dish with a straight glass rim (40 mm diameter, 25 mm high). The surfaces were than subjected to the liquids for a period of 10 s, 2 min, 10 min, 1 h, 6 h, 16 h, and 24 h, respectively. After the exposure period, the Petri dishes and paper disks were removed and the remainder of the liquids was carefully wiped off with a soft paper towel. After from 16 to 24 h, the test surfaces were cleaned, examined for damage and rated according to a numerical evaluation code from 5 to 1 (5—no change, 4—minor change, visible only when the light source was reflected, and so on to 1—severe change). The evaluation was performed in the laboratory light environment and in a standardized viewing cabinet. As a result, the minimum exposure time was recorded for each liquid, at which point the exposed surface was given a grade 4 or 5.

2.13. Statistical Analysis

The correlation between the properties was determined by calculating the Pearson correlation coefficient r_{xy}, which is the magnitude of the linear relationship between the variables x and y, measured on the same test item. It is defined as the sum of all the products of the standard deviations of both values with respect to the degrees of freedom or as the ratio between the covariance and the product of both standard deviations. In this way, the computer software STATGRAPHICS Plus for Windows 4.0 (Statgraphics Technologies, Inc., The Plains, VA, USA) was used. To find correlations between several properties, a correlation matrix was created in which the relationship between two properties is indicated by three values: the Pearson correlation coefficient, the number of values examined, and the *p*-value. In order to be sure that there is a statistically significant correlation at a 95% confidence level, the *p*-value must be less than 0.05. The values of the correlation coefficients range from −1 to +1. The higher the absolute value of the correlation coefficient, the greater the linear relationship between the two properties under study. If the value of the correlation coefficient is 0, it means that there is no relationship between the properties.

3. Results and Discussion

3.1. ATR-FTIR Spectra

The characteristic peaks in the recorded ATR-FTIR spectra (Figure 1) are associated with the chemical properties of the applied topcoat on the particular surface system [38]. Polyurethane typically consists of isocyanate-capped macromolecular chains containing urethane groups in their backbones, usually with a 1–15 wt.% free isocyanate concentration [39]. All the analyzed coatings exhibited some common peaks, typical of urethane-based polymers, but also some differences in the recorded spectra. Only 1C-1 showed a high peak at 700 cm^{-1}, which is related to the C–H bending of the styrene [40]. The sharp peak at 760 cm^{-1} is due to the C–N bond vibration, indicating the presence of urethane bonds. The peaks at 960 and 1160 cm^{-1} are attributed to the C–N–C stretch vibration in the isocyanurate ring [22]. The stretching vibration at 1190 cm^{-1} for C–O shows the reaction of hydroxyl (OH) groups present in wood and atmospheric humidity with the coating by hydrogen bonds or by ester formation [15]. The peaks at 1250, 1420 and 1490 cm^{-1} indicate the stretching of C–O–C, C–H and C–O, and –NH$_2$ vibrations in C–NH and (CO)–NH urethane groups in the polymer chains [26,41,42]. The lower intensity of the band at 1420 cm^{-1} of 1C-1 and 1C-2 indicates the polymerization of acrylate double bonds (CH$_2$=CH twisting vibration) [41]. The band at 1680 cm^{-1} shows the C=O stretching vibrations in unsaturated acids [22] and NH–(C=O)–NH of polyurea. This band was not detected in the spectra of 1C-1 and 1C-2. The sharp stretching region around 1740 cm^{-1} signalizes the free C=O stretching vibration of urethane [41,42] and shows the presence of an acrylate group in all types of coatings [22]. The same peak also clearly indicates the formation of urethane in cured coatings [26]. The NCO absorption band at 2275 cm^{-1} is not seen, indicating that the –NCO group has reacted completely with OH of water solvent [26,41,42]. The band at 2870 cm^{-1} is attributed to C–H$_2$ stretching [41,42]. All spectra show C=C–O, –NH, –OH and C=O functional groups, and a broad absorption band corresponding to the hydrogen bond between the –NH group and the C=O group can be observed at 3310–3500 cm^{-1} [22,26,41]. The intensity of this absorption band is particularly pronounced for 2C-2 and 2C-3.

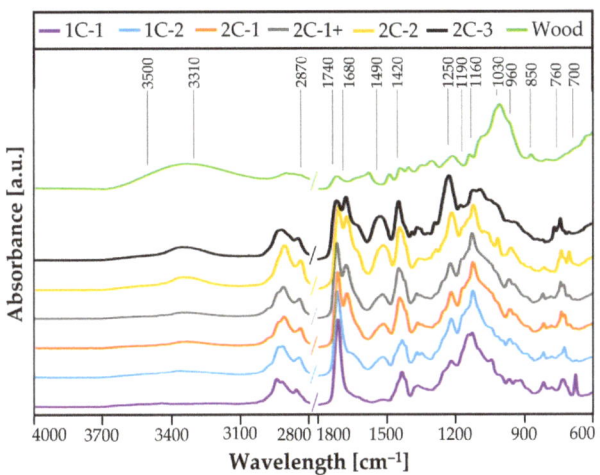

Figure 1. ATR-FTIR spectra in the wavelength regions from 4000 to 2800 cm^{-1} and from 1800 to 600 cm^{-1}.

3.2. Solids Content and Coating Hardness

The *SC* and coating hardness were the properties of the wet and dry topcoats that showed the least variation within the measurements (Table 2). Moreover, the difference between the coatings at *SC* was not so large, but it seems that two-component coatings (2C-1, 2C-2, 2C-3) tended to have higher values. On the other hand, the hardness of the two-component coatings (except) 2C-3 was significantly higher than the hardness of the one-component coatings.

Table 2. Solids content (*SC*) content and coating hardness.

Property	Coating (Average Values and STD)				
	1C-1	1C-2	2C-1	2C-2	2C-3
SC (%)	30.2 ± 0.6	30.2 ± 1.6	35.1 ± 0.8	38.2 ± 0.4	31.6 ± 0.6
Hardness (s)	43.0 ± 0.6	53.0 ± 1.0	80.0 ± 1.2	58.6 ± 0.1	47.1 ± 1.8

In general, the low *SC* coatings and the two-component polyurethane coatings tend to have greater film hardness [42,43].

3.3. Coating Resistance to Cracking and Tensile Strength

After performing the bending test on coated paper strips, we found that all the top coatings tested were highly resistant to cracking, as no cracks were found even at the smallest diameter (2 mm). This means that the coatings still showed elastic behavior 21 days after application and no difference could be found between them using this method. The situation was quite different when the tensile properties were determined in the tensile strength test (Table 3).

Table 3. Tensile strength and elongation of coating films.

Maximum Value	Coating (Average Values and STD)				
	1C-1	1C-2	2C-1	2C-2	2C-3
Tensile strength (MPa)	12.4 ± 1.7	14.3 ± 2.0	30.0 ± 6.6	20.0 ± 3.7	17.0 ± 2.7
Elongation (mm)	58.6 ± 16.2	20.4 ± 7.3	5.2 ± 1.3	38.6 ± 15.8	32.3 ± 13.8

The relation between standard force and elongation defines the ductile or brittle behavior of the coating films. The coating type 2C-1 exhibited the highest tensile strength (30.0 ± 6.6 MPa), but the same coating showed the lowest elongation (5.2 ± 1.3 mm), which together indicate the brittleness of the coating. The reason for this behavior could be the presence of strong C=O bonds in two-component coatings. On the other hand, the largest elongation (58.6 ± 16.2 mm) was observed for coating type 1C-1, but the maximum tensile strength was the lowest (12.4 ± 1.7 MPa), which emphasizes the ductility of the material.

3.4. Dry Film Thickness and Coating Adhesion

Table 4 shows the determined dry film thicknesses and measured coating films' adhesion strengths. The film thickness is mainly related to the amount of coating applied, SC in the coating and the porosity of the substrates [44]. On the other hand, the amount of coating applied can significantly affect the adhesion of coating films and the hardness of surface systems [15]. Indeed, the thinnest films were found on samples coated with one-component coatings, while the thickest films were found on samples coated with two-component coatings. The variability in film thickness is mostly related to the porosity of the wood.

Table 4. Dry film thickness and coating adhesion.

Property	Flooring Surface System (Average Values and STD)					
	1C-1	1C-2	2C-1	2C-1+	2C-2	2C-3
Dry film thick. (μm)	79 ± 1.9	80 ± 1.8	108 ± 2.1	131 ± 2.9	110 ± 2.2	82 ± 1.9
Adhesion (MPa)	3.51 ± 0.40	3.80 ± 0.26	4.53 ± 0.37	4.51 ± 0.81	3.24 ± 0.51	3.80 ± 0.53

In all cases of adhesion measurements, an adhesive type of failure was found, signaling that the adhesion strength of the coating film to the wood was higher than the cohesion strength of the wood substrate. The highest adhesion was found for 2C-1 and 2C-1+ surface systems, which reached values of about 4.5 MPa. In addition to the systems coated with one-component acrylate-polyurethane-based systems, which had adhesion strengths of 3.51 MPa (1C-1) and 3.80 MPa (1C-2), systems 2C-2 (3.24 MPa) and 2C-3 (3.80 MPa) also achieved strength values below 4 MPa, despite the two-component formulation. In general, the adhesion of the coating system to wood substrates is mainly conditioned by the adhesion between the primer and the substrate. Therefore, the determined adhesion strengths could not only be correlated with the chemical properties of the topcoats determined by ATR-FTIR spectroscopy, but these correlations seemed to be more complex and were probably related to the chemical properties of the whole coating system.

3.5. Surface Resistance to Scratching

The resistance of the surface systems to scratching was evaluated by the minimum force that caused a visible scratch (standard method) and by measuring the volume of the scratch mark caused by the applied force on the diamond tip of 1 N (non-standard method). The results of the first method were obtained visually, while the results of the other method were gained volumetrically (Table 5). According to the first method, there were no major differences in scratch resistance between the surface systems. In contrast, the volumes of scratches caused by the same force on the tip (1 N) differed more significantly. By far the largest scratch volume was measured on surface system 1C-1, followed by the scratch volume on 2C-3, which was 73% smaller than on 1C-1. Correlating this result with the chemical composition results, the reason for the significantly larger scratch volume on surface system 1C-1 is the absence of NH_2 and C=O groups in the polymer chains. The smallest scratch volume was measured on the 2C-1+ surface system, which is probably the consequence of the additional application of the topcoat. Considering the variability of the measured scratch volumes, the volumes of the scratches measured on all other surface systems were quite similar.

Table 5. Surface resistance to scratching.

Scratching	Flooring Surface System					
	1C-1	1C-2	2C-1	2C-1+	2C-2	2C-3
Scratch force (N)	0.6	0.8	0.7	0.7	0.7	0.6
Scratch volume at 1 N (μm^3)	253,000 ± 53,120	39,000 ± 11,456	34,000 ± 9625	22,000 ± 5148	41,000 ± 15,742	68,000 ± 12,756

3.6. Surface Resistance to Impact and Abrasion

Surface system 1C-1 had both the worst impact resistance and the worst abrasion resistance. Here, the lowest height at which the coating film cracked was determined to be only 10 mm, while the *IP* was reached after 110 revolutions (Table 6). Surface systems coated with a two-component coating system exhibited better impact resistance and better resistance to abrasion. The samples coated with 2C-3 showed the best impact and abrasion resistance (drop height of 200 mm, *IP* reached after 263 revolutions). It is worth showing the comparison of the abrasion resistance of the 2C-1 and 2C-1+ systems, where the additional topcoat on 2C-1+ contributed to the 28% higher number of revolutions required to reach the *IP*. The better impact and abrasion resistance of the two-component coating systems was associated with the presence of C–O–C, C=O and C–NH bonds in the coating polymers.

Table 6. Surface resistance to impact and abrasion.

Resistance Property	Flooring Surface System					
	1C-1	1C-2	2C-1	2C-1+	2C-2	2C-3
Impact (height in mm)	10	50	50	100	100	200
Abrasion (*IP* revolution)	110	117	177	227	170	263

3.7. Surface Resistance to Cold Liquids

Table 7 expresses the surface resistance to cold liquids and indicates the exposure time at which grade 4 or 5 was reached. In addition, the cells that have a grade worse than the maximum among all the surfaces tested are shaded in gray. From this, it can be seen that there is not much difference in surface resistance between one- and two-component surface systems. Among the one-component systems system 1C-1 (3 gray shaded cells) can be considered slightly less resistant to cold liquids. Among the two-component systems, system 2C-2 (also 3 gray shaded cells) was found to be slightly less resistant to cold liquids. Systems 2C-1 and 2C-1+ have the same classification, as expected, since the surface resistance to cold liquids depends mainly on the properties of the topcoat used, which was the same for these two systems. The resistance of the surface systems to cold liquids did not seem to be related to their chemical properties.

Table 7. Surface resistance to cold liquids.

Liquid	Resistance to Cold Liquids (Exposure Time at Which Grade 4 or 5 was Achieved)					
	1C-1	1C-2	2C-1	2C-1+	2C-2	2C-3
Acetic acid (10%) *	1 h	1 h	1 h	1 h	1 h	1 h
Acetone	10 s	10 min	10 min	10 min	10 min	10 s
Ammonia (10%) *	2 min	2 min	2 min	2 min	2 min	2 min
Citric acid (10%) *	16 h	16 h	16 h	16 h	16 h	16 h
Cleansing agent	16 h	16 h	16 h	16 h	16 h	16 h
Coffee	1 h	1 h	6 h	6 h	1 h	6 h
Ethanol (48%) *	6 h	6 h	1 h	1 h	6 h	1 h
Mustard	16 h	16 h	16 h	16 h	6 h	16 h
Paraffin oil	24 h	24 h	24 h	24 h	24 h	24 h
Red wine (merlot, 13%)	10 min	10 min	10 min	10 min	10 min	16 h
Water	24 h	24 h	24 h	24 h	24 h	24 h
Perspiration, basic	1 h	1 h	1 h	1 h	1 h	1 h

* (m/m) aqueous solution.

3.8. Statistical Analysis

In order to find the correlation between the numerically evaluated properties of the coatings (SC, coating hardness, tensile strength and elongation) and the flooring surface systems (dry film thickness, coating adhesion, surface resistance to scratching—scratch force and volume at 1 N, surface resistance to impact and abrasion), a correlation matrix was constructed (Table 8), as described in Section 2.13. The correlation matrix was used as the basis for analyzing the data from Tables 2–6. This matrix shows the interconnectedness of the properties under consideration. The cross boxes of statistically significantly related variables (p-value less than 0.05) are shaded in green. It must be emphasized again that, in this analysis, only a linear relationship was found between the properties, based on the studied values of the results of coatings and flooring surface systems together.

The correlation matrix (Table 8) confirms some already suspected relationships. As expected, SC has no influence on the properties of the surface system; on the other hand, there is a correlation between SC and the dry film thickness. Even though the correlation is not statistically significant, it shows that a higher SC leads to a higher dry film thickness. However, as expected, the hardness of the coating, tensile strength and elongation are highly correlated. The harder the coating, the higher the tensile strengths achieved and the shorter the elongation before breaking. Of course, these three properties are also related to coating adhesion. A higher coating adhesion of flooring surface systems was achieved when coatings with higher tensile strength and hardness and lower flexibility (shorter elongation before breaking) were used. Surprisingly, a statistically significant relationship between coating hardness, tensile strength and dry film thickness can also be observed. Another expected relationship also proves to be statistically significant. The scratch volume at 1 N was higher for surface systems with more flexible (shorter elongation before breaking) coatings. The last statistically significant relationship in the matrix shows that flooring systems that are more resistant to impact stress are also more resistant to abrasion.

Table 8. Correlation matrix of coatings and flooring surface systems properties.

Properties	SC	Hardness	Tensile Strength	Elong.	Dry Film Thick	Adhesion	Scratch Force	Scratch Volume	Impact	Abrasion
1-11 SC	–	0.6158 * (6) ** 0.1931 ***	0.6561 (6) 0.1570	−0.3209 (6) 0.5351	0.8041 (6) 0.0538	0.0809 (6) 0.8789	0.1475 (6) 0.7803	−0.5366 (6) 0.2724	0.1712 (6) 0.7457	0.3113 (6) 0.5482
1-11 Hardness	0.6158 (6) 0.1931	–	0.9743 (6) 0.0010	−0.8627 (6) 0.0270	0.8694 (6) 0.0245	0.8101 (6) 0.0507	0.3811 (6) 0.4560	−0.6358 (6) 0.1748	−0.0721 (6) 0.8920	0.3006 (6) 0.5627
1-11 Tensile strength	0.6561 (6) 0.1570	0.9743 (6) 0.0010	–	−0.8168 (6) 0.0473	0.8809 (6) 0.0204	0.8032 (6) 0.0543	0.1902 (6) 0.7182	−0.6111 (6) 0.1974	0.0883 (6) 0.8678	0.4719 (6) 0.3447
1-11 Elong.	−0.3209 (6) 0.5351	−0.8627 (6) 0.0270	−0.8168 (6) 0.0473	–	−0.6163 (6) 0.1926	−0.8608 (6) 0.0277	−0.5604 (6) 0.2474	0.8197 (6) 0.0458	−0.1343 (6) 0.7997	−0.3908 (6) 0.4437
1-11 Dry film thickness	0.8041 (6) 0.0538	0.8694 (6) 0.0245	0.8809 (6) 0.0204	−0.6163 (6) 0.1926	–	0.5242 (6) 0.2858	0.2160 (6) 0.6811	−0.5537 (6) 0.2543	0.0768 (6) 0.8850	0.3971 (6) 0.4357
1-11 Adhesion	0.0809 (6) 0.8789	0.8101 (6) 0.0507	0.8032 (6) 0.0543	−0.8608 (6) 0.0277	0.5242 (6) 0.2858	–	0.1966 (6) 0.7089	−0.4370 (6) 0.3862	−0.0090 (6) 0.9866	0.3723 (6) 0.4674
1-11 Scratch force	0.1475 (6) 0.7803	0.3811 (6) 0.4560	0.1902 (6) 0.7182	−0.5604 (6) 0.2474	0.2160 (6) 0.6811	0.1966 (6) 0.7089	–	−0.6219 (6) 0.1874	−0.3021 (6) 0.5606	−0.3480 (6) 0.4990
1-11 Scratch volume	−0.5366 (6) 0.2724	−0.6358 (6) 0.1748	−0.6111 (6) 0.1974	0.8197 (6) 0.0458	−0.5537 (6) 0.2543	−0.4370 (6) 0.3862	−0.6219 (6) 0.1874	–	−0.4402 (6) 0.3824	−0.4841 (6) 0.3306
1-11 Impact	0.1712 (6) 0.7457	−0.0721 (6) 0.8920	0.0883 (6) 0.8678	−0.1343 (6) 0.7997	0.0768 (6) 0.8850	−0.0090 (6) 0.9866	−0.3021 (6) 0.5606	−0.4402 (6) 0.3824	–	0.8917 (6) 0.0170
1-11 Abrasion	0.3113 (6) 0.5482	0.3006 (6) 0.5627	0.4719 (6) 0.3447	−0.3908 (6) 0.4437	0.3971 (6) 0.4357	0.3723 (6) 0.4674	−0.3480 (6) 0.4990	−0.4841 (6) 0.3306	0.8917 (6) 0.0170	–

* correlation coefficient. ** number of samples. *** p-value (value below 0.05 shows a statistically significant correlation at 95% confidence level, cross box shaded with green).

4. Conclusions

Studies on EWF parquet coated with different coating systems showed that the chemical formulation of the coatings and the number of layers applied have a great influence on the performance of the flooring surfaces. The ATR-FTIR spectra revealed the differences in the chemical composition of the coating systems, including the presence of certain chemical bonds and functional groups. The largest difference was found in the presence of the C=O group, which was only present in the two-component systems. The two-component coatings had a higher SC and greater hardness than one-component coatings. The tensile properties of the coating films differed between formulations; the films exhibited either brittle or ductile behavior. Three of the four two-component coating systems formed thicker films on wood than one-component systems. In addition, only two of the two-component systems achieved an adhesion strength greater than 4 MPa. By far the worst scratch, impact, and abrasion resistance was found for coating system 1C-1. Resistance to cold liquids showed no major differences between the resistance of the one- and two-component surface systems. In conclusion, the statistical analysis showed some correlations between the properties of the coating systems and the properties of the coated flooring surfaces. The presented results give a good overview of the interactions between the properties of the chosen coating type and the expected quality of wood flooring surfaces.

Author Contributions: Conceptualization, M.P. (Matjaž Pavlič) and J.Ž.; methodology, M.P. (Matjaž Pavlič) and J.Ž.; formal analysis, M.P. (Matjaž Pavlič) and J.Ž.; investigation, M.P. (Matjaž Pavlič) and J.Ž.; resources, M.P. (Matjaž Pavlič) and J.Ž.; data curation, M.P. (Matjaž Pavlič) and J.Ž.; writing—original draft preparation, M.P. (Matjaž Pavlič) and J.Ž.; writing—review and editing, M.P. (Matjaž Pavlič), J.Ž., and M.P.(Marko Petrič); visualization, M.P. (Matjaž Pavlič) and J.Ž.; supervision, M.P. (Matjaž Pavlič), J.Ž., and M.P. (Marko Petrič). All authors have read and agreed to the published version of the manuscript.

Funding: The authors acknowledge the financial support of the Slovenian Research Agency (research core funding No. P4-0015).

Data Availability Statement: Data is contained within the article or available on reasonable request.

Conflicts of Interest: The authors declare no conflict of interest.

References

1. Williams, R.S. Finishing of Wood. In *Wood Handbook—Wood as an Engineering Material*; General technical report FPL–GTR–113; U.S. Department of Agriculture, Forest Service, Forest Products Laboratory: Madison, WI, USA, 1999; 463p. [CrossRef]
2. Arminger, B.; Jaxel, J.; Bacher, M.; Gindl-Altmutter, W.; Hansmann, C. On the drying behavior of natural oils used for solid wood finishing. *Prog. Org. Coat.* **2020**, *148*, 105831. [CrossRef]
3. Grubii, V.; Johansson, J. Performance of multi-layered wood flooring elements produced with sliced and sawn lamellas. *Pro Ligno* **2019**, *15*, 166–172.
4. FEP, European Federation of the Parquet Industry. *The European Parquet Industries in 2008. A Challenging Year for the European Parquet Sector*; European Federation of the Parquet Industry: Brussels, Belgium, 2009; pp. 1–5.
5. Németh, R.; Molnárné Posch, P.; Molnár, S.; Bak, M. Performance evaluation of strip parquet flooring panels after long-term, in-service exposure. *Drewno* **2014**, *57*, 119–134. [CrossRef]
6. Belleville, B.; Blanchet, P.; Cloutier, A.; Deteix, J. Wood–adhesive interface characterization and modeling in engineered wood flooring. *Wood Fiber Sci.* **2008**, *40*, 484–494.
7. Blanchet, P.; Beauregard, R.; Cloutier, A.; Gendron, G.; Lefebvre, M. Evaluation of various engineered wood flooring constructions. *For. Prod. J.* **2003**, *53*, 30–37.
8. Cassens, D.L.; Feist, W.C. *Finishing and Maintaining Wood Floors*; North Central Regional Extension Publication-Michigan State University, Cooperative Extension Service: Lansing, MI, USA, 1989.
9. Jafarian, H.; Demers, C.M.H.; Blanchet, P.; Laundry, V. Effects of interior wood finishes on the lighting ambiance and materiality of architectural spaces. *Indoor Built Environ.* **2018**, *27*, 786–804. [CrossRef]
10. Nemli, G.; Örs, Y.; Kalaycıoğlu, H. The choosing of suitable decorative surface coating material types for interior end use applications of particleboard. *Constr. Build. Mater.* **2005**, *19*, 307–312. [CrossRef]
11. Rolleri, A.; Roffael, E. Influence of the surface roughness of particleboards and their performance towards coating. *Maderas-Cienc. Tecnol.* **2011**, *12*, 143–148. [CrossRef]
12. Bardak, S.; Sarı, B.; Nemli, G.; Kırcı, H.; Baharoğlu, M. The effect of décor paper properties and adhesive type on some properties of particleboard. *Int. J. Adhes. Adhes.* **2011**, *31*, 412–415. [CrossRef]
13. Keskin, H.; Tekin, A. Abrasion resistances of cellulosic, synthetic, polyurethane, waterborne and acid-hardening varnishes used woods. *Constr. Build. Mater.* **2011**, *25*, 638–643. [CrossRef]
14. Veigel, S.; Grüll, G.; Pinkl, S.; Obersriebnig, M.; Müller, U.; Gindl-Altmutter, W. Improving the mechanical resistance of waterborne wood coatings by adding cellulose nanofibers. *React. Funct. Polym.* **2014**, *85*, 214–220. [CrossRef]
15. Wang, J.; Wu, H.; Liu, R.; Long, L.; Xu, J.; Chen, M.; Qiu, H. Preparation of a fast water-based UV cured polyurethane-acrylate wood coating and the effect of coating amount on the surface properties of oak (*Quercus alba* L.). *Polymers* **2019**, *11*, 1414. [CrossRef]
16. Scrinzi, E.; Rossi, S.; Deflorian, F.; Zanella, C. Evaluation of aesthetic durability of waterborne polyurethane coatings applied on wood for interior applications. *Prog. Org. Coat.* **2011**, *72*, 81–87. [CrossRef]
17. Sell, J.; Feist, W.C. Role of density in the erosion of wood during weathering. *For. Prod. J.* **1986**, *36*, 57–60.
18. Gurleyen, L.; Ayata, U.; Esteves, B.; Cakicier, N. Effects of heat treatment on the adhesion strength, pendulum hardness, surface roughness, color and glossiness of scots pine laminated parquet with two different types of UV varnish application. *Maderas-Cienc Tecnol.* **2017**, *19*, 213–224. [CrossRef]
19. Brischke, C.; Ziegeler, N.; Bollmus, S. Abrasion resistance of thermally and chemically modified timber. *Drv. Ind.* **2019**, *70*, 71–76. [CrossRef]
20. International Organization for Standardization. *ISO 17959:2014 General Requirements for Solid Wood Flooring*; International Organization for Standardization: Geneva, Switzerland, 2014.
21. Podgorski, L.; De Meijer, M.; Lanvin, J.-D. Influence of coating formulation on its mechanical properties and cracking resistance. *Coatings* **2017**, *7*, 163. [CrossRef]
22. Xu, J.; Jiang, Y.; Zhang, T.; Dai, Y.; Yang, D.; Qiu, F.; Yu, Z.; Yang, P. Synthesis of UV-curing waterborne polyurethane-acrylate coating and its photopolymerization kinetics using FT-IR and photo-DSC methods. *Prog. Org. Coat.* **2018**, *122*, 10–18. [CrossRef]

23. Kong, L.; Xu, D.; He, Z.; Wang, F.; Gui, S.; Fan, J.; Pan, X.; Dai, X.; Dong, X.; Liu, B.; et al. Nanocellulose-Reinforced polyurethane for waterborne wood coating. *Molecules* **2019**, *24*, 3151. [CrossRef]
24. Utgof, S.S.; Ignatovich, L.V.; Romanova, A.M. Application of nanoadditives for wear resistance improvement of parquet protective and decorative coatings. In *Proceedings of BSTU, Chemistry, Organic Substances Technology and Biotechnology*; BSTU: Minsk, Belarus, 2012; Volume 4, pp. 100–102.
25. Sow, C.; Riedl, B.; Blanchet, P. UV-waterborne polyurethane-acrylate nanocomposite coatings containing alumina and silica nanoparticles for wood: Mechanical, optical, and thermal properties assessment. *J. Coat. Technol. Res.* **2011**, *8*, 211–221. [CrossRef]
26. Landry, V.; Blanchet, P.; Riedl, B. Mechanical and optical properties of clay-based nanocomposites coatings for wood flooring. *Prog. Org. Coat.* **2010**, *67*, 381–388. [CrossRef]
27. Roessler, A.; Schottenberger, H. Antistatic coatings for wood-floorings by imidazolium salt-basedionic liquids. *Prog. Org. Coat.* **2014**, *77*, 579–582. [CrossRef]
28. European Committee for Standardization. *EN ISO 3251:2019 Paints, Varnishes and Plastics—Determination of Non-Volatile-Matter Content (ISO 3251:2019)*; European Committee for Standardization: Brussels, Belgium, 2019.
29. European Committee for Standardization. *EN ISO 1522:2007 Paints and Varnishes—Pendulum Damping Test (ISO 1522:2007)*; European Committee for Standardization: Brussels, Belgium, 2007.
30. European Committee for Standardization. *EN ISO 1519:2012 Paints and Varnishes—Bend Test (Cylindrical Mandrel) (ISO 1519:2011)*; European Committee for Standardization: Brussels, Belgium, 2012.
31. European Committee for Standardization. *EN ISO 2808:2007 Paints and Varnishes—Determination of Film Thickness (ISO 2808:2007)*; European Committee for Standardization: Brussels, Belgium, 2007.
32. European Committee for Standardization. *EN ISO 4624:2016 Paints and Varnishes—Pull-Off Test for Adhesion (ISO 4624:2016)*; European Committee for Standardization: Brussels, Belgium, 2016.
33. European Committee for Standardization. *EN 15186:2012 Furniture—Assessment of the Surface Resistance to Scratching*; European Committee for Standardization: Brussels, Belgium, 2012.
34. International Organization for Standardization. *ISO 4211-4:1988 Furniture—Tests for Surfaces—Part 4: Assessment of Resistance to Impact*; International Organization for Standardization: Geneva, Switzerland, 1988.
35. European Committee for Standardization. *EN 15185:2012 Furniture—Assessment of the Surface Resistance to Abrasion*; European Committee for Standardization: Brussels, Belgium, 2012.
36. European Committee for Standardization. *EN 12720:2009+A1:2014 Furniture—Assessment of Surface Resistance to Cold Liquids*; European Committee for Standardization: Brussels, Belgium, 2014.
37. European Committee for Standardization. *CEN/TS 16209:2012 Furniture—Classification for Properties for Furniture Surfaces*; European Committee for Standardization: Brussels, Belgium, 2012.
38. Beveridge, A.; Fung, T.; Macdougall, D. Use of infrared spectroscopy for the characterization of paint fragments. In *Forensic Examination of Glass and Paint: Analysis and Interpretation*; Caddy, B., Ed.; CRC Press: Boca Raton, FL, USA, 2001.
39. Lu, R.; Wan, Y.-Y.; Honda, T.; Ishimura, T.; Kamiya, Y.; Miyakoshi, T. Design and characterization of modified urethane lacquer coating. *Prog. Org. Coat.* **2006**, *57*, 215–222. [CrossRef]
40. Defeyt, C.; Langenbacher, J.; Rivenc, R. Polyurethane coatings used in twentieth century outdoor painted sculptures. Part I: Comparative study of various systems by means of ATR-FTIR spectroscopy. *Herit. Sci.* **2017**, *5*, 11. [CrossRef]
41. Masson, F.; Decker, C.; Jaworek, T.; Schwalm, R. UV-radiation curing of waterbased urethane–acrylate coatings. *Prog. Org. Coat.* **2000**, *39*, 115–126. [CrossRef]
42. Huang, S.; Xiao, J.; Zhu, Y.; Qu, J. Synthesis and properties of spray-applied high solid content two component polyurethane coatings based on polycaprolactone polyols. *Prog. Org. Coat.* **2017**, *106*, 60–68. [CrossRef]
43. Kaboorani, A.; Auclair, N.; Riedl, B.; Landry, V. Mechanical properties of UV-cured cellulose nanocrystal (CNC) nanocomposite coating for wood furniture. *Prog. Org. Coat.* **2017**, *104*, 91–96. [CrossRef]
44. Bessières, J.; Maurin, V.; George, B.; Molina, S.; Masson, E.; Merlin, A. Wood-coating layer studies by X-ray imaging. *Wood Sci. Technol.* **2013**, *47*, 853–867. [CrossRef]

Article

Enhancing Thermally Modified Wood Stability against Discoloration

Dace Cirule [1,*], Errj Sansonetti [1], Ingeborga Andersone [1], Edgars Kuka [1,2] and Bruno Andersons [1]

[1] Latvian State Institute of Wood Chemistry, Dzerbenes 27, LV-1006 Riga, Latvia; harrysansonetti@hotmail.com (E.S.); i.andersone@edi.lv (I.A.); edgars.kuka@edi.lv (E.K.); bruno.andersons@edi.lv (B.A.)
[2] Faculty of Material Science and Applied Chemistry, Riga Technical University, Paula Valdena 3, LV-1048 Riga, Latvia
* Correspondence: dace.cirule@edi.lv

Abstract: Thermal modification of wood has gained its niche in the production of materials that are mainly used for outdoor applications, where the stability of aesthetic appearances is very important. In the present research, spectral sensitivity to discoloration of thermally modified (TM) aspen wood was assessed and, based on these results, the possibility to delay discoloration due to weathering by non-film forming coating containing transparent iron oxides in the formulation was studied. The effect of including organic light stabilizers (UVA and HALS) in coatings as well as pretreatment with lignin stabilizer (HALS) was evaluated. Artificial and outdoor weathering was used for testing the efficiency of different coating formulations on TM wood discoloration. For color measurements and discoloration assessment, the CIELAB color model was used. Significant differences between the spectral sensitivity of unmodified and TM wood was observed by implying that different strategies could be effective for their photostabilization. From the studied concepts, the inclusion of the transparent red iron oxide into the base formulation of the non-film forming coating was found to be the most effective approach for enhancing TM wood photostability against discoloration due to weathering.

Keywords: thermally modified wood; weathering; discoloration; photosensitivity; transparent iron oxides; non-film forming coatings

1. Introduction

Thermal treatment has been found to be a feasible modification method for increasing wood hydrophobicity, biological durability, and dimensional stability [1–3]. In addition, such modification is considered to be more environmentally friendly compared to chemical modification or preservative treatment because no chemicals are used. Furthermore, thermal modification does not entail any obstacles for easy disposal of wood after its service life. Apart from the enhanced functionality, during thermal modification, the appearance of wood is essentially altered by forming a lighter or darker brown color, which for some species is observed to be more homogenous compared to the color of unmodified wood [4,5]. Furthermore, the characteristic brown color imparted to wood by thermal modification is often regarded as an extra or even one of the main benefits as it is highly favored by customers [6–8].

Among the variety of possible TM wood applications, the dominating areas are decking, claddings, façades, garden accessories, and similar outdoor applications [2]. Given this aspect of TM wood application, the stability of the aesthetic appearance is of great importance for its successful competition with other materials in the area [2]. The surface of wood used outdoors is degraded by weathering, which is a set of complex phenomena mostly including complementary, synergetic, and antagonistic effects of solar irradiation, water, fungi, and atmosphere [9,10]. Although wood functionality usually is not deteriorated due to weathering, which is known to exclusively be a superficial phenomenon, it exerts disastrous effect on the aesthetic value of wood [10,11]. Some researches claim improved weathering resistance of wood due to thermal treatment [12,13]. Nevertheless, discoloration of TM wood during outdoor exposure has been observed to be quite a rapid process [13–15].

Among different wood protection strategies against weathering, the most widespread approach is the use of coatings due to economical, application simplicity, as well as custom practice considerations [10]. The use of transparent coatings is the most common practice for wood protection from discoloration with simultaneous retention of its natural color and texture [16]. However, due to the high photosensitivity of wood, photoprotective additives such as UV absorbers (UVA), which convert the absorbed UV radiation into heat, and hindered amine light stabilizers (HALS), which inhibit autooxidation by scavenging free radicals, are routinely included in coating formulations to improve their performance. In addition, it has been found that the inclusion of such additives into formulation does not reduce adhesion strength between the coating and wood, which is an important characteristic for providing good performance of the coating [17]. The combination of organic UVA and HALS has proved to exert a synergistic effect at inhibiting both coating and wood photodegradation [18–20]. Light stabilizers of both types are often included in a single commercial additive for wood coatings. A variety of UVA chemical compositions and product forms differing in efficiency, spectral coverage, and resistance to weathering has been developed to improve the performance of coatings [21,22]. Promising results have also been reached when a concept about the wood photostabilization by pre-treatment with HALS inhibiting lignin photooxidation was implemented [23]. However, organic UVA may be decomposed by solar irradiation, resulting in the loss of their efficiency [24]. Recently a number of studies have focused on different aspects of incorporation of inorganic nanoparticles into transparent coating formulations to enhance performance of wood coatings as they have inherent high photostability and, what is of high importance, they are non-toxic [15,23]. The most widely used UV-absorbing inorganic additives are zinc oxide (ZnO) and titanium dioxide (TiO_2) due to their high absorption of radiation mainly in the UV range, thus providing good retention of the natural wood color after application of coating [15,20,21]. Another easily available inorganic UVA are iron oxides. However, considering transparent coatings for pale woods, iron oxides are recognized as inappropriate because of their typical strong absorption not only in UV but also in wide spectral region of visible light, resulting in inevitable alteration of the wood original appearance [23,25].

Although wood is substantially transformed during thermal modification, in a number of studies, it has been observed that the coating formulations intended for the use on unmodified wood suits as well for application on TM wood as a substrate [26–28]. Nonetheless, the increased hydrophobic character of heat-treated wood and the changes of the polarity of surface energy of wood could cause problems with the adhesion of waterborne coatings [29].

Concerning TM wood protection against photodegradation, different approaches have been studied. Introduction of visible light non-absorbing inorganic and organic UVA as well as HALS into coating formulations has been found to be ineffective for TM wood protection from discoloration [30–32]. A promising result was achieved when TM wood was treated with TiO_2 sol [33]. However, due to the complexity of the treatment process, the proposed approach does not seem to be an appropriate solution for commercial purposes.

The focus of the present research was the investigation of the possibility to improve TM wood resistance to color change. TM wood spectral sensitivity to photodiscoloration was evaluated since such data are important for the designing of effective protective coatings but hardly any information is reported in the literature. Such knowledge provides the possibility to tailor coating formulation matching TM wood peculiarities. Basing on these results, the inclusion of transparent iron oxides into the non-film forming coating formulation was tested. The use of iron oxides could be a simple, effective, and economical way to reduce TM wood photodiscoloration during outdoor exposure without substantially altering its characteristic appearance. In addition, the effect of wood pretreatment with the lignin stabilizer and the inclusion of a light stabilizer containing organic UVA and HALS into coating formulation were evaluated. European aspen (*Populus tremula* L.) was used for the experiment as it is a fast-growing species with inadequate commercial value and thermal modification could be considered a potential way to increase its market share.

2. Materials and Methods

2.1. Wood Material

Aspen (*Populus tremula* L.) wood was used for the study. Defect-free boards measuring 1000 mm × 100 mm × 25 mm were subjected to thermal modification in a WTT (Wood Treatment Technology, Brande, Denmark) experimental wood modification device, in a water vapor medium under elevated pressure (0.6 MPa) at 170 °C, for 1 h at the peak temperature. Before preparing the specimens, the boards of unmodified and TM wood were conditioned for a month in an atmosphere of 20 °C and 65% relative humidity. The specimens measuring 150 mm × 70 mm × 15 mm for exposure to solar radiation and to artificial UV weathering and the specimens measuring 300 mm × 90 mm × 20 mm for outdoor exposure on weathering racks were cut from the boards. For all experiments, six replicate specimens from different boards were prepared for each treatment. The surfaces of all specimens were furnished by planing.

2.2. Color and Reflectance Spectra Measurement

A portable spectrophotometer CM-2500d Konica Minolta, Tokyo, Japan (standard illuminant D65, d/8° measuring geometry, 10° standard observer, measuring area Ø 8 mm) was used for measurements of wood reflectance spectra and color measurements. Reflectance spectra were recorded for the wavelengths ranging from 360 to 740 nm, using a scanning interval of 10 nm. Color was expressed according to the CIELAB color model (Commission Internationale de l'Eclairage 1976) as the color parameters L^*, a^*, b^*. Measurements were performed at five surface points and average values were calculated for representation of each specimen color and reflectance spectra. The color change (ΔE_{ab}) was calculated from the differences between the initial and resulting values of color parameters ΔL^*, Δa^*, Δb^* according to the equation:

$$\Delta E_{ab} = \sqrt{(\Delta L)^2 + (\Delta a)^2 + (\Delta b)^2} \qquad (1)$$

For evaluation of color change, the measurements were always performed at the same spots on the tested surface of the specimen.

2.3. Coating Formulations and Finishing of Specimens

The base coating formulation was solvent-borne alkyd prepared in a laboratory by using long oil and medium oil alkyd resins provided by a local paint factory. The only additive was cobalt siccative and the solid content of the base formulation was 20%. Pigmented coatings were prepared by adding to the base formulation equal amounts of yellow, red, or a mixture (1:1) of both iron oxides with pigment concentration in the dry coating film being 16%. Sicoflush® L Red 2817 and Sicoflush® L Yellow 1916 (from the BASF, Ludwigshafen, Germany) pigment pastes for solvent-based coatings containing transparent pigments were used. The coatings were brush applied on the specimens in two coats for the artificial weathering experiment and two and three coats for outdoor weathering. In preliminary experiments with the base coating formulation, it was assessed that the spread rate of 160 g/m^2 is the upper limit for uniform coverage without formation of a visibly detectable film. The spread rate was assessed by weighing specimens before and after finishing. The average spread rate was 116 g/m^2 for two coats and 136 g/m^2 for three coats. For evaluating the effect of light stabilizer containing organic UVA and HALS on prevention TM wood from discoloration due to UV irradiation, 2% (based on the dry weight of the alkyd resins) of a commercial light stabilizer Tinuvin 5060 (from the BASF, Ludwigshafen, Germany) was added into the pigmented formulation (containing the mixture of both iron oxides) and into the base non-pigmented formulation. The specimens were then conditioned at RH 65% and 20 °C for two weeks to assure proper curing of the coating. For the experiment in which usefulness of TM wood pre-treatment with the lignin stabilizer (HALS) was evaluated, before exposure to UV irradiation, unmodified and TM wood specimens were pre-treated with lignin photostabilizing product "Lignostab 1198 L" of BASF production by spraying it on the specimen surface.

2.4. Weathering Experiments

Two outdoor experiments were carried out during the present study. They both were located in Riga (56°58′ N 24°11′ E). To assess spectral sensitivity of TM wood, the effect of incident light spectral composition on unmodified and TM wood discoloration was studied. The specimens of unmodified and TM wood specimens were exposed to full solar radiation as well as to solar radiation transmitted through glass filters of different absorption spectra. The experiment was performed by exposure of specimens to solar irradiation outdoors during summer between May and August, when the incident solar radiation has the highest intensity. Specimens were exposed outdoors only on sunny hours, when the intensity of the total UV radiation (290 to 390 nm) was above 10 W/m^2. During exposure, the average UV radiation intensity was 23 W/m^2 and the average solar radiation intensity was 590 W/m^2. UV light meter (Lutron UV-340, Taipei City, Taiwan) and solar power meter (Amprobe SOLAR-100, Glottertal, Germany) were applied for the measurements of radiation intensity. The total exposure time was 100 h. The specimens were kept in the dark while they were not exposed outdoors.

During the second outdoor experiment, the specimens with pigmented coatings were exposed on weathering racks inclined at 45° and facing south. The experiment was carried out from January to November; the total exposure time was 220 days.

Artificial weathering was performed by exposing the specimens to UV irradiation in an accelerating weathering chamber QUV (Q-Lab, Westlake, OH, USA). UVA-340 type fluorescent lamps, which simulate the UV portion (295–360 nm) of the solar radiation, were used with irradiation intensity of 0.89 W/m^2 at 340 nm.

3. Results and Discussion

3.1. TM Wood Color and Its Photosensitivity

The color of pale wood species indicates a poor amount of chromophores in their chemical composition as only a small portion of the incident visible light is absorbed while the majority is reflected. On the contrary, one of the consequences of wood subjection to high temperature treatment is the changed composition of its chromophores, imparting wood the characteristic brown color (Figure 1), which is a result of intense visible light absorption.

It has been found that the main chemical transformations processes in wood during thermal treatment imparting wood the typical brown color are condensation and oxidation reactions, leading to an extensive formation of conjugated structures [34,35]. It is suggested that the formation of quinoid compounds due to degradation and oxidation of lignin and aromatic extractives could be involved in wood color changes upon thermal treatment [35]. In addition, most of the chromophoric substances are of high molecular mass and/or degree of cross-linking [7]. However, precise composition and formation processes of the chromophores of TM wood is still unknown. From the reflectance spectra (Figure 2), it can be seen that due to thermal modification, wood reflectance decreases in the whole range of the visible light with more pronounced changes in the wavelength range above 500 nm, indicating substantially increased absorption in this region. It implies that colored quinoid with the typical absorption at the shorter wavelength region of visible light are not the dominant cause of wood color changes due to thermal treatment.

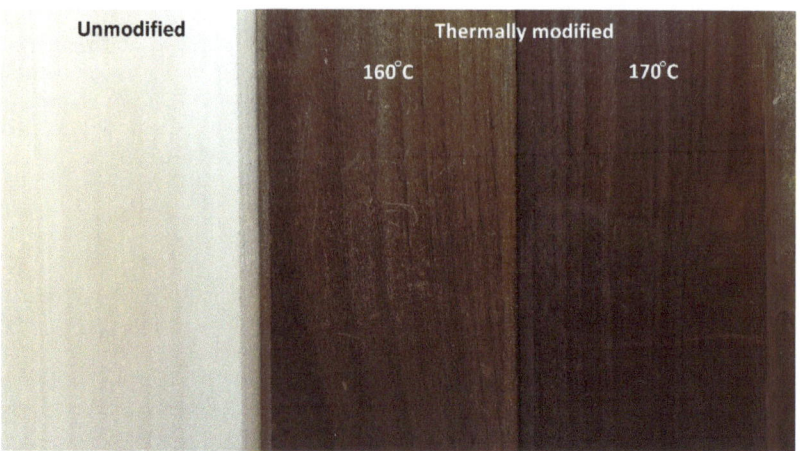

Figure 1. Unmodified and thermally modified aspen wood.

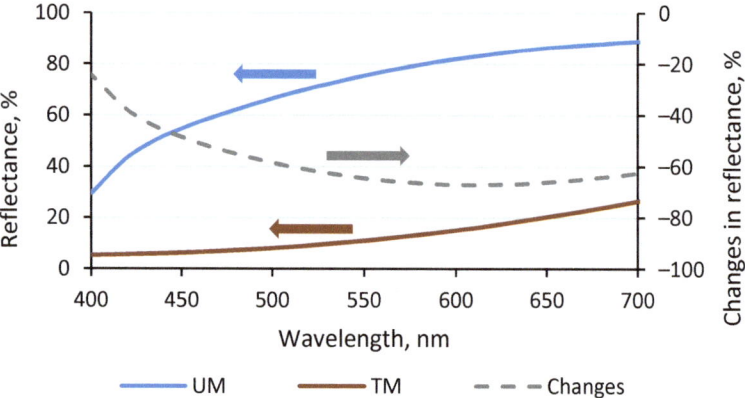

Figure 2. Reflectance spectra of unmodified (UM) and thermally modified (TM) wood.

Although the absorption of radiation of definite wavelengths is mandatory for wood photo-discoloration, the absorption of light does not necessarily lead to chemical transformations. Moreover, only the chemical changes affecting chromophores result in discoloration [36]. Therefore, information about spectral sensitivity of the material is important to develop coating formulations providing effective protection, especially when transparent coatings are preferable for the maintenance of wood's natural appearance. For the identification of thermally modified wood, spectral sensitivity and evaluating the changes caused by the thermal treatment, both unmodified and thermally modified wood specimens were exposed to direct solar radiation and to solar radiation filtered through glass filters transmitting definite wavelength ranges of the solar spectrum. The results of the experiment (Figure 3) clearly demonstrate differences in spectral sensitivity regarding discoloration of the two tested wood types.

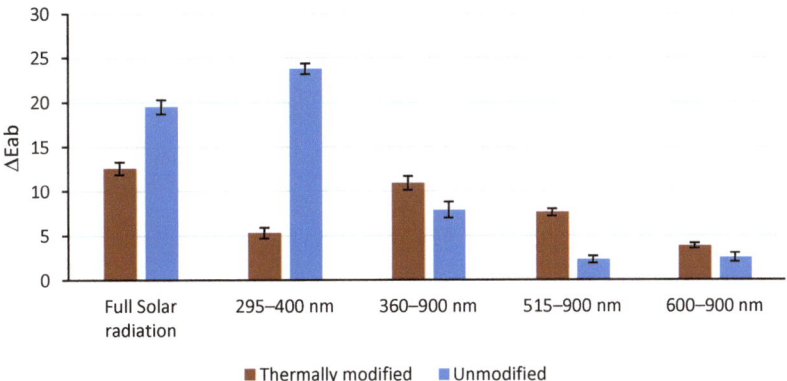

Figure 3. Discoloration ΔE_{ab} of unmodified and thermally modified wood caused by wood exposure to full solar radiation and to solar radiation through glass filters transmitting definite wavelength ranges.

Substantially greater discoloration of the unmodified wood subjected to full solar spectrum compared to TM wood implies that in general the wood transformations during thermal treatment impart to wood a higher color stability. In addition, much greater improvement of resistance to discoloration due to the wood thermal treatment can be observed when specimens covered by filters transmitting solely UV radiation (295–400 nm) are compared. It is easy to see that TM wood exposure to the UV portion of the solar radiation, which caused even greater discoloration of unmodified wood than radiation of full solar spectrum, resulted in much smaller color changes compared with those entailed by the full solar radiation. Reduced discoloration caused by UV irradiation due to wood thermal modification has been observed in a number of researches [37–41]. The phenomena of enhanced resistance to photodiscoloration of TM wood is associated mainly with chemical transformations of lignin, which is well recognized to be the key wood component responsible for its discoloration [37,38,42,43]. However, in the literature, contradicting results have also been reported showing greater discoloration for TM wood than for unmodified wood caused by UV irradiation during artificial weathering [44–46]. Such discrepancies in results could be caused by differences in wood species, thermal treatment processes, as well as design of experiments used in the studies.

However, the exposure of TM wood to the visible light portion of the solar radiation (360–900 nm) resulted in substantial discoloration, while only minor color changes were detected for unmodified wood. In addition, essentially greater discoloration was observed for TM wood specimens covered with filters transmitting radiation in the range of 515-900 nm compared with the specimens covered with UV light transmitting filters. Considerable changes in color were detected even for the TM wood specimens exposed to the radiation of wavelengths longer than 600 nm. Due to the complicated chemical structure of wood, precise mechanisms of wood photochemical transformations have not been entirely established [10,43]. However, the main reactions of photochemical transformations are similar for common organic materials and start with the formation of free radicals, which is initiated by the absorption of photons of adequate energy for cleavage off a radical [47]. The results of wood discoloration implies that TM wood contains chromophores, the transformation of which can be caused by less energetic photons of visible light in addition to UV radiation. These results suggest that shielding from considerable portion of the visible light could be an effective protection strategy to prevent TM wood discoloration and contribute to retention of its natural appearance, especially when the envisaged application includes exposure to solar radiation. Apart from differences in spectral sensitivity, there are cardinal differences in the discoloration pattern of unmodified and TM wood. For unmodified wood, photodiscoloration shows up as a darkening and is pertained to transformations of lignin with formation of chromophores such as quinones and quinine-like structures [23,36,38]. In contrary, TM wood color becomes lighter, implying a reduction in chromophoric structures that absorb visible light. Obviously, radically different processes are involved in transformation of chromophores for unmodified and TM wood.

3.2. Effect of Lignin Stabilizer

It has been found that, similar to unmodified wood, lignin is the most light-sensitive wood component also in TM wood [39,45,46] Therefore, an experiment was carried out by employing an approach that has proved its suitability in case of unmodified wood [23]. The approach includes wood pretreatment prior to coating application with a lignin stabilizer, which is a special HALS designed for trapping the radicals forming at the wood surface. The results proved that pretreatment with the lignin stabilizer can reduce rate and amount of unmodified wood discoloration caused by exposure to UV irradiation (Figure 4).

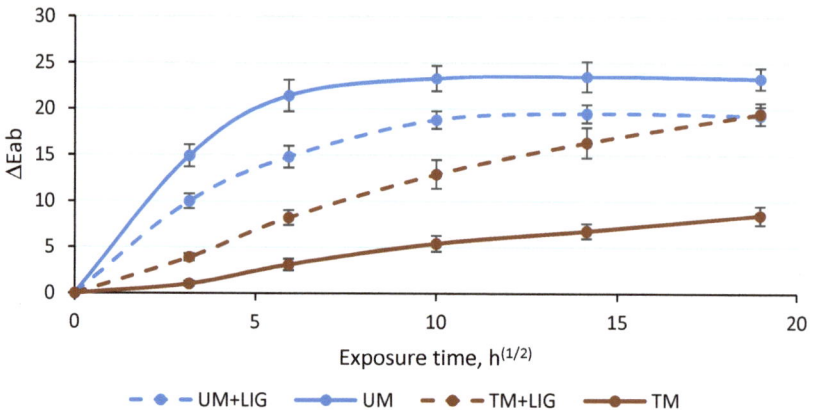

Figure 4. UV radiation caused discoloration of unmodified (UM) and thermally modified (TM) wood with and without pretreatment with wood lignin stabilizing agent (LIG).

An adverse effect of pretreatment with lignin stabilizer was detected for TM wood when more than twice as large discoloration was observed for the pretreated specimens compared to those without pretreatment. From these results, it is clearly evident that the application of lignin stabilizer did not improve TM wood resistance to UV-caused discoloration. The adverse effect of the lignin stabilizer on discoloration of unmodified and TM wood could be related to the differences of the two wood types regarding chemical structures of the components involved in the processes of free radicals formation and wood discoloration. It has been proved by measurements of EPR (electron paramagnetic resonance) spectroscopy that during thermal treatment, stable free radicals are formed in wood, which may also take part in photodiscoloration processes of TM wood [38,48]. Miklečić et al. [32] also found that pre-treatment of TM wood with lignin stabilizer of HALS type did not prove efficient and even caused increase of discoloration. However, Kocaefe and Saha [49] have reported better TM wood color protection effective use of coatings modified by adding a HALS type lignin stabilizer and bark extracts compared to that of organic UVA and HALS for softwood (jack pine), while similar discoloration was observed for the tested TM hardwoods (aspen and birch) for both additive combinations. These results imply that there is no consensus on the effect of HALS type lignin stabilizer on the discoloration of TM wood.

3.3. Effect of Incorporation of Pigments into Coating on TM Wood Color

As mentioned above, when maintenance of the natural appearance of wood is desirable, iron oxides are usually avoided in transparent wood coatings because of the color they impart to the coating due to the characteristic absorption in visible light range. However, this property could be beneficial in the case of TM wood for which considerable discoloration was caused by exposure to visible light (Figure 3), implying the necessity to screen wood from this part of solar radiation. In addition, due to the inherent brown color of TM wood, application of colored iron oxides should not lead to excessive transformations of the TM wood original color, which itself absorbs visible light significantly. For evaluation of the effect of transparent iron oxide pigments on TM wood appearance, reflectance spectra of wood coated with unpigmented and pigmented formulations were compared (Figure 5).

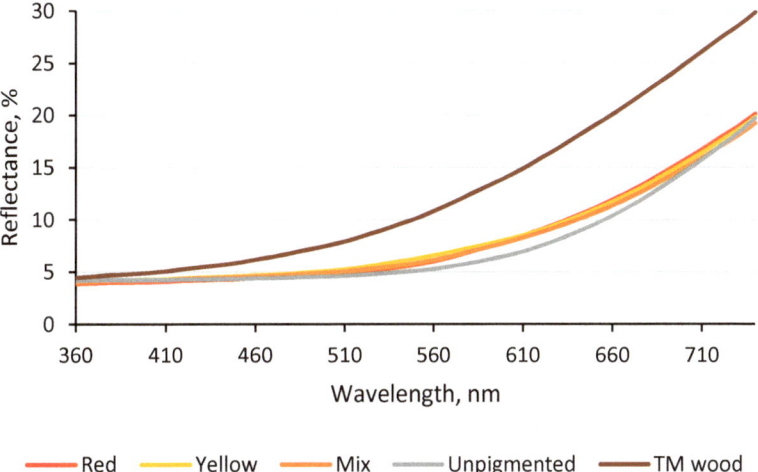

Figure 5. Reflectance spectra of uncoated thermally modified (TM) wood and thermally modified wood coated with unpigmented and with yellow, red, and a mixture (mix) of red and yellow (1:1) iron oxides pigmented transparent coatings.

The reflectance spectra demonstrate that the application of all the tested coating formulations reduced the amount of the reflected radiation of the TM wood surface in the whole range of visible light with greater reduction recorded for the longer wavelengths. However, the differences between the effect of unpigmented formulation and those containing transparent iron oxides were statistically insignificant. In addition, the pigmented coatings imparted quite similar appearance regardless of the applied pigments (red, yellow or a mixture). These results indicate that the use of iron oxide pigments in coating formulations does not cause extra changes in TM wood color, apart from those imparted by the coating base matrix.

3.4. Effect of Incorporation of a Light Stabilizer Additive into Pigmented Coating Formulation

Iron oxide pigments absorb certain radiation in both UV and visible light range, depending on the oxide type. The effect of addition of organic UVA into pigmented coating formulation was tested by exposure specimens to UV irradiation in an artificial weathering chamber for 1500 h with regular evaluation of surface discoloration. The results showed that the addition of organic UVA to the coating formulation provided only slight delay of discoloration in the very beginning (Figure 6).

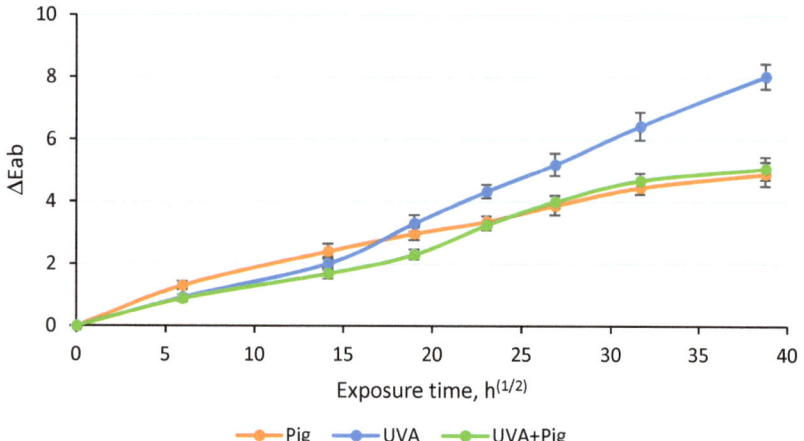

Figure 6. Discoloration of thermally modified wood with coatings containing different additives (Pig—iron oxide pigments; UVA—UV absorber) during exposure to UV irradiation.

For the specimens finished with both types of coatings with the UVA additive, discoloration during the first 100 h (in the chart the unit of exposure time is a square root of an hour) was quite similar and slightly smaller than color changes detected for specimens coated with the pigmented formulation without UVA additive, implying a positive effect of the additive in delaying discoloration of TM wood. On the other hand, the further exposure resulted in almost equal discoloration of specimens with pigmented coatings, regardless of extra UVA addition or not, while the specimens treated with the unpigmented formulation that contained UVA discolored substantially more. It indicates that UVA additive in the transparent formulation containing iron oxide pigments exerts certain positive effect against TM wood discoloration only for the initial period but is ineffective for long-term protection. Therefore, there is no ground for use of UVA in such coatings when prevention of TM wood discoloration is considered. This finding is consistent with a research reporting that addition of organic UV absorbers to dark pigmented systems does not improve their performance in protection of unmodified wood from discoloration [23]. However, the long-term efficiency differs depending on the UVA chemistry [22]. The UVA included in the light stabilizer used in the experiments was of the BTZ (2-(2-hydroxyphenyl)-benzotriazile) class, which has been found to provide less long-lasting protection than those of HPT (2-hydroxy-phenyl-s-triazine) [22]. For more well-grounded inference, studies focusing on other UVA types should be performed. In addition, the potential importance of the UVA additive in elimination of coating microfilm degradation should be investigated as unimpaired coating film may be crucial for proper wood protection in outdoor environments, where, apart from solar radiation, other factors contribute to weathering with moisture exhibiting the most serious effect.

3.5. Effect of Red and Yellow Iron Oxide Pigmenst in Coating Formulation on TM Wood Discoloration during Natural Weathering

In the artificial weathering chamber, only the potential effect of UV radiation was simulated without considering the contribution of visible light which, as it has previously been identified (Figure 3), has a strong implication or even dominance in the total TM wood discoloration under exposure to solar radiation. Yellow and red iron oxides have similar UV absorption capacity but significantly differ in visible light absorption with the absorbance significantly red-shifted for the red iron oxide [50]. Mixture of both iron oxides as the pigment additive was used in the coating formulation for the artificial weathering test with UV irradiation. Taking into account the differences in visible light absorption spectra of yellow and red iron oxides, formulations containing only one iron oxide or mixture of both were used for the outdoor test to evaluate the protective potency depending on the kind of iron oxide. Two sets of specimens differing in number of coats were prepared for exposure to outdoor weathering. The results of specimen discoloration caused by 220 days outdoor exposure are presented in Figure 7. The results of discoloration clearly show that the red iron oxide has a higher capacity for prevention of TM wood discoloration.

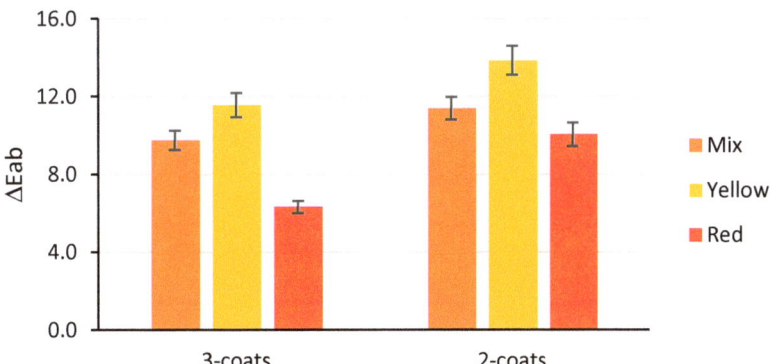

Figure 7. Discoloration of thermally modified wood exposed outdoors for 220 days depending on the iron oxide pigments used in the coating formulation.

As expected, less color changes were detected for all the tested formulations when three coats were applied. However for both sets of specimens, a similar trend was observed regarding the protective force of each formulation. The results of outdoor weathering clearly show that the coating formulation with red iron oxide prevented TM wood from discoloration more effectively than the other two formulations. This effect was especially pronounced in the case of three coats when almost twice less discoloration was detected for specimens with coating containing red iron oxide than for those with yellow iron oxide. Higher efficiency of the red iron oxide in comparison with the yellow one considering protection from discoloration was observed also in a case of unmodified pine sapwood [25]. The authors propose that the observed phenomena is due to the masking effect of the red iron. However, when TM wood is considered, a more possible explanation for red iron oxide's better performance in reducing discoloration is the red-shifted absorbance, resulting in less solar radiation of wavelengths that cause TM wood discoloration reaching wood.

The addition of nanoparticles as well as pigment concentration may affect other coating properties [51]. Therefore, further research is needed to optimize coating formulations by assessing the most eligible pigment concentrations, taking into account all aspects of coating performance. In addition, systematic investigation into the compatibility and interaction of iron oxides with different kinds of coating systems regarding binders, solvents, and other additives should be carried out.

4. Conclusions

The study has shown that TM wood considerably differs from unmodified wood regarding wood spectral sensitivity. For unmodified wood, UV radiation is the dominant cause of photodiscoloration. In contrary, the wide wavelength range of visible light was found to cause significant discoloration of TM wood. This finding suggests that a cardinally different strategy should be applied for providing effective protection of TM wood from the main strategy used for unmodified wood, which mostly focuses on protecting wood surface from the UV radiation. In addition, the results showed that pre-treatment of wood with a lignin stabilizer designed for unmodified wood is not effective for the protection of TM wood from discoloration.

Inclusion of iron oxides in transparent coating formulation does not radically alter the appearance of the finished TM wood. The obtained results suggest that including of iron oxides and preferentially the red one into transparent coating formulations can provide effective TM wood protection from photodiscoloration. On the other hand, addition of the light stabilizer containing organic UV absorber and HALS to the pigmented coating formulation does not improve the color retention of TM wood.

Author Contributions: Conceptualization, D.C. and I.A.; investigation, E.S. and E.K.; data curation, E.S.; writing—original draft preparation, D.C.; writing—review and editing, I.A. and E.K.; supervision, B.A.; funding acquisition, B.A. All authors have read and agreed to the published version of the manuscript.

Funding: This research was funded by the European Regional Development Fund project: "Innovative wood and its processing materials with upgraded service properties", Nr.2010/0324/2DP/2.1.1.0/APIA/VIAA/057.

Institutional Review Board Statement: Not applicable.

Informed Consent Statement: Not applicable.

Data Availability Statement: The data presented in this study are available on request from the corresponding author.

Conflicts of Interest: The authors declare no conflict of interest.

References

1. Hill, C.A.; Hill, S. *Wood Modification: Chemical, Thermal, and Other Processes*; John Wiley and Sons Ltd.: West Sussex, UK, 2006; pp. 99–127.
2. Esteves, B.M.; Pereira, H.M. Wood modification by heat treatment: A review. *BioResuorces* **2009**, *4*, 370–404. [CrossRef]
3. Militz, H.; Altgen, M. Processes and properties of thermally modified wood manufactured in Europe. *Deterior. Prot. Sustain. Biomater. ACS Symp. Ser.* **2014**, *1158*, 269–285. [CrossRef]
4. Schnabel, T.; Zimmer, B.; Petutschnigg, A.J.; Schönberg, S. An approach to classify thermally modified hardwoods by color. *For. Prod. J.* **2007**, *57*, 105–110.
5. Cirule, D.; Kuka, E. Effect of thermal modification on wood colour. *Res. Rural Dev.* **2015**, *2*, 87–92.
6. Brischke, C.; Welzbacher, C.R.; Brandt, K.; Rapp, A.O. Quality control of thermally modified timber: Interrelationship between heat treatment intensities and CIE L*a*b* color data on homogenized wood samples. *Holzforschung* **2007**, *61*, 19–22. [CrossRef]
7. Esteves, B.; Marques, A.V.; Domingos, I.; Pereira, H. Heat-induced colour changes of pine (*Pinus pinaster*) and eucalypt (*Eucalyptus globulus*) wood. *Wood Sci. Technol.* **2008**, *42*, 369–384. [CrossRef]
8. Tuong, V.M.; Jian, L. Effect of heat treatment on the change in color and dimensional stability of acacia hybrid wood. *BioResources* **2010**, *5*, 1257–1267.
9. Hon, D.N.S.; Minemura, N. Color and Discoloration. In *Wood and Cellulosic Chemistry*, 2nd ed.; Hon, D.N.S., Shiraishi, N., Eds.; Marcel Dekker: New York, NY, USA, 2000; pp. 385–442.
10. Cogulet, A.; Blanchet, P.; Landry, V. The multifactorial aspect of wood weathering: A review based on holistic approach of wood degradation protected by clear coating. *BioResources* **2018**, *13*, 2116–2138. [CrossRef]
11. Kropat, M.; Hubbe, M.A.; Laleicke, F. Natural, accelerated, and simulated weathering of wood: A review. *Bioresources* **2020**, *15*, 9998–10062.
12. Nuopponen, M.; Wikberg, H.; Vuorinen, T.; Maunu, S.L.; Jämsä, S.; Viitaniemi, P. Heat-treated softwood exposed to weathering. *J. Appl. Polym. Sci.* **2004**, *91*, 2128–2134. [CrossRef]
13. Yildiz, S.; Yildiz, U.C.; Tomak, E.D. The effects of natural weathering on the properties of heat-treated alder wood. *BioResources* **2011**, *6*, 2504–2521. [CrossRef]
14. Metsä-Kortelainen, S.; Paajanen, L.; Viitanen, H. Durability of thermally modified Norway spruce and Scots pine in above-ground conditions. *Wood Mater. Sci. Eng.* **2011**, *6*, 163–169. [CrossRef]
15. Auclair, N.; Riedl, B.; Blanchard, V.; Blanchet, P. Improvement of photoprotection of wood coatings by using inorganic nanoparticles as ultraviolet absorbers. *For. Prod. J.* **2011**, *61*, 20–27. [CrossRef]
16. Evans, P.D.; Haase, J.G.; Shakri, A.; Seman, B.M.; Kiguchi, M. The search for durable exterior clear coatings for wood. *Coatings* **2015**, *5*, 830–864. [CrossRef]
17. Reinprecht, L.; Tiňo, R.; Šomšák, M. Impact of fungicides, plasma, UV-additives and weathering on the adhesion strength of acrylic and alkyd coatings to the Norway spruce wood. *Coatings* **2020**, *10*, 1111. [CrossRef]
18. Evans, P.D.; Chowdhury, M.J.; Mathews, B.; Schmalzl, K.; Ayer, S.; Kiguchi, M.; Kataoka, Y. Weathering and surface protection of wood. In *Handbook of Environmental Degradation of Materials*; Myer, K., Ed.; William Andrew Publishing: Norwich, NY, USA, 2005; pp. 277–297.
19. Shenoy, M.A.; Marathe, Y.D. Studies on synergistic effect of UV absorbers and hindered amine light stabilisers. *Pigment Resin Technol.* **2007**, *36*, 83–89. [CrossRef]

20. Forsthuber, B.; Grüll, G. The effects of HALS in the prevention of photo-degradation of acrylic clear topcoats and wooden surfaces. *Polym. Degrad. Stab.* **2010**, *95*, 746–755. [CrossRef]
21. Aloui, F.; Ahajji, A.; Irmouli, Y.; George, B.; Charrier, B.; Merlin, A. Inorganic UV absorbers for the photostabilisation of wood-clearcoating systems: Comparison with organic UV absorbers. *Appl. Surf. Sci.* **2007**, *253*, 3737–3745. [CrossRef]
22. Schaller, C.; Rogez, D.; Braig, A. Organic vs inorganic light stabilizers for waterborne clear coats: A fair comparison. *J. Coat. Technol. Res.* **2012**, *9*, 433–441. [CrossRef]
23. Schaller, C.; Rogez, D. New approaches in wood coating stabilization. *J. Coat. Technol. Res.* **2007**, *4*, 401–409. [CrossRef]
24. Blanchard, V.; Blanchet, P. Color stability for wood products during use: Effects of inorganic nanoparticles. *BioResources* **2011**, *6*, 1219–1229.
25. Schauwecker, C.F.; McDonald, A.G.; Preston, A.F.; Morrell, J.J. Use of iron oxides to influence the weathering characteristics of wood surfaces: A systematic survey of particle size, crystal shape and concentration. *Eur. J. Wood Prod.* **2014**, *72*, 669–680. [CrossRef]
26. Jämsä, S.; Ahola, P.; Viitaniemi, P. Long-term natural weathering of coated Thermo Wood. *Pigment Resin Technol.* **2000**, *29*, 68–74. [CrossRef]
27. Altgen, M.; Militz, H. Thermally modified Scots pine and Norway spruce wood as substrate for coating systems. *Coat. Technol. Res.* **2017**, *14*, 531–541. [CrossRef]
28. Nejad, M.; Dadbin, M.; Cooper, P. Coating Performance on exterior oil-heat treated wood. *Coatings* **2019**, *9*, 225. [CrossRef]
29. Jirouš-Rajković, V.; Miklečić, J. Heat-treated wood as a substrate for coatings, weathering of heat-treated wood, and coating performance on heat-treated wood. *Adv. Mater. Sci. Eng.* **2019**, *2019*, 8621486. [CrossRef]
30. Saha, S.; Kocaefe, D.; Boluk, Y.; Pichette, A. Enhancing exterior durability of jack pine by photo-stabilization of acrylic polyurethane coating using bark extract. Part 1: Effect of UV on color change and ATR-FT-IR analysis. *Prog. Org. Coat.* **2011**, *70*, 376–382. [CrossRef]
31. Saha, S.; Kocaefe, D.; Sarkar, D.K.; Boluk, Y.; Pichette, A. Effect of TiO_2-containing nano-coatings on the color protection of heat-treated jack pine. *J. Coat. Technol. Res.* **2011**, *8*, 183–190. [CrossRef]
32. Miklečić, J.; Turkulin, H.; Jirouš-Rajković, V. Weathering performance of surface of thermally modified wood finished with nanoparticles-modified waterborne polyacrylate coatings. *Appl. Surf. Sci.* **2017**, *408*, 103–109. [CrossRef]
33. Shen, H.; Zhang, S.; Cao, J.; Jiang, J.; Wang, W. Improving anti-weathering performance of thermally modified wood by TiO_2 sol or/and paraffin emulsion. *Constr. Build. Mater.* **2018**, *169*, 372–378. [CrossRef]
34. Chen, Y.; Fan, Y.; Gao, J.; Stark, N.M. The effect of heat treatment on the chemical and color change of black locust (*Robinia pseudoacacia*) wood flour. *BioResources* **2012**, *7*, 1157–1170. [CrossRef]
35. Yao, C.; Yongming, F.; Jianmin, G.; Houkun, L. Coloring characteristics of in situ lignin during heat treatment. *Wood Sci. Technol.* **2012**, *46*, 33–40. [CrossRef]
36. Tolvaj, L.; Faix, O. Artificial ageing of wood monitored by DRIFT spectroscopy and CIE L*a*b* color measurements. *Holzforschung* **1995**, *49*, 397–404. [CrossRef]
37. Ayadi, N.; Lejeune, F.; Charrier, F.; Charrier, B.; Merlin, A. Color stability of heat-treated wood during artificial weathering. *Holz Roh Werkst* **2003**, *61*, 221–226. [CrossRef]
38. Deka, M.; Humar, M.; Rep, G.; Kričej, B.; Šentjurc, M.; Petrič, M. Effects of UV light irradiation on colour stability of thermally modified, copper ethanolamine treated and non-modified wood: EPR and DRIFT spectroscopic studies. *Wood Sci. Technol.* **2008**, *42*, 5–20. [CrossRef]
39. Miklečić, J.; Jirouš-Rajković, V.; Antonović, A.; Španić, N. Discolouration of thermally modified wood during simulated indoor sunlight exposure. *BioResources* **2011**, *6*, 434–446. [CrossRef]
40. Tolvaj, L.; Nemeth, R.; Pasztory, Z.; Bejo, L.; Takats, P. Colour stability of thermally modified wood during short-term photodegradation. *BioResources* **2014**, *9*, 6644–6651. [CrossRef]
41. Li, X.; Li, T.; Li, G.; Lu, Q.; Qin, S.; Li, J. Effect of UV light irradiation on color changes in thermally modified rubber wood based on FTIR. *BioResources* **2020**, *15*, 5179–5197.
42. George, B.; Suttie, E.; Merlin, A.; Deglise, X. Photodegradation and photostabilisation of wood—The state of the art. *Polym. Degrad. Stab.* **2005**, *88*, 268–274. [CrossRef]
43. Evans, P.D. Weathering of wood and wood composites. In *Handbook of Wood Chemistry and Wood Composites*, 2nd ed.; Rowell, R.M., Ed.; CRC Press: Boca Raton, FL, USA, 2013; pp. 151–216.
44. Huang, X.; Kocaefe, D.; Kocaefe, Y.; Boluk, Y.; Pichette, A. A spectrocolorimetric and chemical study on color modification of heat-treated wood during artificial weathering. *Appl. Surf. Sci.* **2012**, *258*, 5360–5369. [CrossRef]
45. Huang, X.; Kocaefe, D.; Kocaefe, Y.; Boluk, Y.; Pichette, A. Study of the degradation agents of heat-treated jack pine (*Pinus banksiana*) under artificial sunlight irradiation. *Polym. Degrad. Stab.* **2012**, *97*, 1197–1214. [CrossRef]
46. Srinivas, K.; Pandey, K.K. Photodegradation of thermally modified wood. *J. Photochem. Photobiol. B* **2012**, *117*, 140–145. [CrossRef]
47. Zayat, M.; Garcia-Parejo, P.; Levy, D. Preventing UV-light damage of light sensitive materials using a highly protective UV-absorbing coating. *Chem. Soc. Rev.* **2007**, *36*, 1270–1281. [CrossRef]
48. Sivonen, H.; Maunu, S.L.; Sundholm, F.; Jämsä, S.; Viitaniemi, P. Magnetic resonance studies of thermally modified wood. *Holzforschung* **2002**, *56*, 648–654. [CrossRef]

49. Kocaefe, D.; Saha, S. Comparison of the protection effectiveness of acrylic polyurethane coatings containing bark extracts on three heat-treated North American wood species: Surface degradation. *Appl. Surf. Sci.* **2012**, *258*, 5283–5290. [CrossRef]
50. Hayashi, K. Practical issue of nanosized colorant particles. In *Nanoparticle Technology Handbook*, 3rd ed.; Naito, M., Yokoyama, T., Hosokawa, K., Nogi, K., Eds.; Elsevier: Amsterdam, The Netherlands, 2018; pp. 607–612.
51. Feist, W.C. Role of pigment concentration in the weathering of semitransparent stains. *For. Prod. J.* **1988**, *38*, 41–44.

Article

Study of the Adhesion of Silicate-Based Coating Formulations on a Wood Substrate

Arnaud Maxime Cheumani Yona [1,2,*], Jure Žigon [1], Sebastian Dahle [1] and Marko Petrič [1]

[1] Department of Wood Science and Technology, Biotechnical Faculty, University of Ljubljana, Jamnikarjeva ulica 101, SI-1000 Ljubljana, Slovenia; Jure.Zigon@bf.uni-lj.si (J.Ž.); Sebastian.Dahle@bf.uni-lj.si (S.D.); Marko.Petric@bf.uni-lj.si (M.P.)
[2] Macromolecular Research Team, Faculty of Science, University of Yaoundé 1, P.O. 812 Yaoundé, Cameroon
* Correspondence: ArnaudMaximeCheumani.Yona@bf.uni-lj.si

Abstract: Silicate coatings are environmentally friendly inorganic-based products that have long been used for mineral substrates and protection of steel against corrosion. The development and acceptance of these coatings in the wood sector require some adjustments in formulations or special preparation of the surface to be coated to obtain durable finishes. In this work, the adhesion of various silicate-based formulations to a beech wood substrate (*Fagus sylvatica* L.), was assessed with the main objective to study relevant parameters and potential improvements. Adhesion strength was determined by pull-off and cross-cut tests. Other coating properties such as scratch, impact, and water resistance were also determined. Surface roughness and interface were analyzed using confocal laser scanning microscopy (CLSM) and scanning electron microscopy (SEM), and coating curing was studied by attenuated total reflection-infrared spectroscopy (ATR FTIR). The results showed that adhesion was highly dependent on formulation, penetration of the coatings into wood, and mechanical anchoring. Increasing the content of solid particles in the coating formulations or adding a polyol (glycerol, xylose), which probably acted as a coalescent, considerably decreased the adhesion strength, probably by blocking penetration into the wood by forming aggregates. Adhesion was improved by pre-mineralization of the surface, and substitution of a part of the potassium silicate binder with potassium methyl siliconate reduced the formation of cracks caused by dimensional instability of the wood.

Keywords: coating; silicate; water glass; wood; potassium methyl siliconate; adhesion

Citation: Cheumani Yona, A.M.; Žigon, J.; Dahle, S.; Petrič, M. Study of the Adhesion of Silicate-Based Coating Formulations on a Wood Substrate. *Coatings* **2021**, *11*, 61. https://doi.org/10.3390/coatings11010061

Received: 21 December 2020
Accepted: 1 January 2021
Published: 7 January 2021

Publisher's Note: MDPI stays neutral with regard to jurisdictional claims in published maps and institutional affiliations.

Copyright: © 2021 by the authors. Licensee MDPI, Basel, Switzerland. This article is an open access article distributed under the terms and conditions of the Creative Commons Attribution (CC BY) license (https://creativecommons.org/licenses/by/4.0/).

1. Introduction

Wood coatings are used extensively as domestic and industrial finishes for the decoration and surface protection of wood and other wood-based materials. To perform durably, these wood finishing products specifically require good adhesion, flexibility, water resistance, and resistance to other damage-causing parameters (photodegradation, scratching, and impact). The global wood coatings market is currently dominated by organic-based materials. However, there is a growing interest in the development of purely inorganic or inorganic-rich alternative solutions stimulated by new requirements and regulations such as the reduction of volatile organic compounds emissions or the demand for ultraviolet-resistant and fire-retardant products. A large number of research works have been published in recent years on surface coatings of wood by depositing thin layers of various inorganic nanoparticles (e.g., TiO_2, SiO_2, Al_2O_3, ZnO, CuO) or even metallic particles (e.g., Ag, Zn, Cu, Al) using techniques such as sol–gel impregnation, or physical and chemical vapor deposition [1–4]. Silicate-based coating systems are also potential inorganic alternatives that have the added advantage of being formulated for application by any end user with a minimum of experience and for wood renovation.

Silicate coatings consist of water glass (e.g., potassium water glass) or silica sol (e.g., ethyl silicate) or mixtures thereof as the main binder, calcium carbonate, zinc oxide, zinc powder, talc, fine sand extender, and mineral pigments (e.g., TiO_2, Fe_2O_3) [5–7]. Organic additives such as aqueous dispersions of polymers (e.g., styrene-acrylic, styrene-butadiene, polystyrene), thickeners (e.g., hydroxyethyl cellulose), dispersants, and stabilizers can be added to some extent. Silicate coatings are established for concrete, masonry and stone, brick, steel or aluminum and have appeared to be one of the most resilient paints in exterior use. Silicate coatings cure through chemical reactions, which involve formation of strong chemical bonds with inorganic species at the surface of mineral substrates and promote adhesion of the coating to these substrates [5,8]. The higher dimensional stability of mineral substrates compared to wood reduces the stresses exerted on the coatings and favors durability despite the lower flexibility inherent to inorganic materials. Composite primers made of alkyd resins or drying oils modified with mineral particles have been proposed by coating companies (e.g., Keimfarben®, Beeck®) for durable wood silicate coatings. However, scientific literature about their applications for wood and wood products is sparse. To the best of our knowledge, there are no reports in the literature on fundamental studies of the adhesion mechanisms and potential performance improvements of such coatings for wood surfaces. Previous research in this field focused on the fire performances [9,10]. Kumar and co-authors [9] observed that vermiculite-sodium silicate composites coated on wood formed a solid foamy layer and released water molecules that improved the flame retardancy, when exposed to fire. Silicate paints containing hydromagnesite provided the first group of fire resistance for a wooden surface [10]. The authors presented a coating formulation with relatively good adhesion strength values on wood (5–6 MPa), but the research data and optimization work leading to these results were not published (to the best of our knowledge) [10]. Silicate-based geopolymer cements have also been tested for wood coatings, but they are slightly different systems [11–13].

The focus of this work was to fundamentally study the adhesion of silicate coating formulations on wood. Beech wood (*Fagus sylvatica* L.) was used as a substrate. The influence of organic additives such as glycerol, D-(+)-xylose, sucrose and dextrin as well as of the surface pretreatment of the wood are reported. The properties of the fresh liquid formulations were determined by rheological measurements. Adhesion was studied by pull-off (standard ISO 4624-2016 [14]) and cross-cut tests (EN ISO 2409 (1997) [15]), scratch resistance by pencil hardness (standard EN ISO 1518-1 (2000) [16]) and impact resistance according to the standard ISO 4211-4 (1995) [17]. The interface between the coatings and the wood was studied using scanning electron microscopy combined with energy dispersive analysis of X-rays (SEM-EDAX) and the surface roughness was investigated using confocal laser scanning microscopy (CLSM). The resistance of the coating to liquid water was determined and the chemical changes were studied by ATR-FTIR analyses.

2. Materials and Methods

2.1. Materials

European beech wood (*Fagus sylvatica* L.) was used in this study. The dimensions of the plates were approximately 380, 80 and 10 mm for longitudinal, tangential and radial directions, respectively. Potassium hydroxide flakes (90%), lithium hydroxide (98%), silica gel (porosity 60 Å, particle size 63–200 μm) and methyltrimethoxysilane (MTMS 98%) were used to prepare potassium water glass, lithium water glass, and potassium methyl siliconate. The components were mixed with demineralized water to obtain silicate or siliconate solutions with different modulus (molar ratio $SiO_2/(K,Li)_2O$ for water glasses or MTMS/KOH for the siliconate) and solid content of 35%. The prepared water glasses were kept for at least 24 h for maturation before use in formulation of coatings. A precipitated calcium carbonate and a calcium carbonate sample (≤ 50 μm particle size) were used as fine and coarse powders, respectively, in a 50:50 mass ratio. Glycerol (98%), sucrose ($\geq 99.5\%$), D-xylose ($\geq 98\%$), titanium (IV) oxide (99.5%, 21 nm mean particle size) and silicon antifoam (30% in water, emulsion) were used as additives. These chemicals were all purchased from Sigma Aldrich Chemie GmbH, Steinheim, Germany, except the precipitated calcium carbonate (Sigma Aldrich, Darmstadt, Germany), lithium hydroxide (Acrōs organics,

Geel, Belgium) and glycerol (Honeywell, Seelze, Germany). Domemul SA 9263 (styrene-acrylic emulsion, non-volatile matter (39–41%, pH 8–8.5, viscosity at 23 °C 20–350 mPa s) was provided by Helios TBLUS (Količevo, Slovenia). Zinc oxide ($\geq$99%) was obtained from Fluka-Honeywell (Seelze, Germany), talcum (98%) and dextrin were purchased from Roth (Karlsruhe, Germany). The chemicals were all used as received without further purification steps.

2.2. Preparation of the Coatings

The different formulations developed in this study are shown in Table 1. Solid components, styrene-acrylic emulsion, demineralized water and eventually glycerol were weighed in a plastic cup (polypropylene), and premixed with a glass rod. Water glass and silicon antifoam were then added to the mixture and the whole was thoroughly mixed using an IKa® T25 digital ultra-Turrax® (Staufen, Germany) at 3200 rpm for 2 min and 5000 rpm for 3 min. Polyols (glycerol, xylose, dextrin, or sucrose) were added to the formulation as potential plasticizers and adhesion promoters to the wood. A part of potassium water glass was replaced with a potassium methyl siliconate water solution, an organosilicate solution prepared from methyltrimethoxysilane.

Table 1. Recipes for the preparation of different coating formulations.

Samples	Water Glass (g)	Acrylic Resin (g)	$CaCO_3$ (g)	ZnO (g)	Talc (g)	Gly (g)	Xyl (g)	Sucr (g)	Dex (g)	TiO_2 (g)	Water (g)	Silicon (g)
CF1 [a]	35	4	10	5	5	0	0	0	0	0	4	0.2
CF2 [a]	35	4	10	5	5	4	0	0	0	0	4	0.2
CF3 [a]	35	4	10	5	5	0	0	4	0	0	4	0.2
CF4 [a]	35	4	10	5	5	0	4	0	0	0	4	0.2
CF5 [a]	35	4	10	5	5	0	0	0	4	0	4	0.2
CF6 [a]	35	5	30	10	5	0	0	0	0	0	15	0.2
CF7 [a]	35	4	10	5	5	0	0	0	0	5	20	0.2
CF8 [b]	35	4	10	5	5	0	0	0	0	0	4	0.2
CF9 [c]	35	4	10	5	5	0	0	0	0	0	4	0.2
CF10 [d]	35	4	10	5	5	0	0	0	0	0	4	0.2
CF11 [e]	35	4	10	5	5	0	0	0	0	0	4	0.2
CF12 [f]	35	4	10	5	5	0	0	0	0	0	4	0.2

Potassium water glass: (a) module 3.2, (b) module 0.5, (c) module 1.5, (d) module 2.5, (e) lithium silicate module 3.2, and (f) potassium silicate module 3.2 + potassium methyl siliconate module 3.2 75:25 wt.%. Gly: glycerol, Xyl: D-(+)-xylose, Sucr: Sucrose and Dex: Dextrin.

2.3. Rheological Measurements

Rheological properties are important parameters that determine the application and penetration of a coating formulation into a substrate (wood). Rheological measurements of the fresh coating formulations were carried out with an ARES G2 rheometer (TA instruments, New Castle, DE, USA) using two plate parallel geometry. Both plates were 25 mm in diameter, and the gap was fixed between 0.9 and 1 mm. Flow ramp tests were performed at a shear rate from 0 to 1000 s^{-1} at a temperature of 25 °C controlled by an air flow. Three replicates were performed for each sample and the results are mean values of the replicates.

2.4. Application of Coatings on Wood Substrate

The coatings were applied to the longitudinal tangential surface. The surface was sanded with sandpaper 120-grit and dusted off with compressed air. The coatings were applied manually using a coating applicator with a 240 μm wet film thickness outflow at a movement speed of approximately 30 mm·s^{-1}. The coated substrates were stored under ambient conditions (temperature (23 ± 3) °C and (50 ± 5)% relative humidity) for two weeks for drying and curing prior to characterization. The coating layers dried within a few hours, but the curing and development of resistances of such reactive inorganic materials required more time for silicification and reactions of silicates with mineral additives.

2.5. Pretreatment of the Wood Samples

Pre-mineralized wood samples were also used as substrates. The pretreatment was carried out by a double impregnation process as proposed by many authors including a silicification step and a curing step to produce insoluble silica [18–20]. Beech wood samples were dipped in a 10 wt.% potassium silicate solutions (module 3.2) for 2 h and dried for 24 h under ambient conditions. They were then dipped again in a 5 wt.% boric acid or 5 wt.% sodium bicarbonate solution for 2 h. Finally, the samples were dried under ambient conditions for 24 h and further in an oven at 105 °C for 24 h. The treated samples were cooled down and kept under ambient conditions for one weak to allow moisture equilibrium before being used for coating experiments.

2.6. Characterization of the Samples

2.6.1. Pull-Off Adhesion Test of the Samples

Pull-off adhesion tests were carried out according to the standard ISO 4624-2016 [14]. Aluminum dollies (20 mm) were glued on the surface of the coatings using a 2-component polyurethane adhesive and allowed to cure for 24 h. The coating around the dollies was carefully cleaned down to the substrate to isolate the glued zone from the rest of the coating layer. The tensile stress (adhesion strength) applied to peel off the coating from wood surface was measured by using a Defelsko Positest® Adhesion tester (Defelsko instruments corporation, Ogdensburg, NY, USA). In general, two types of failure are distinguished: If less than 40% of the coating layer remained on the substrate, the failure is considered adhesive and the measured strength is representative to the adhesion of the coating with the substrate. Otherwise, the failure is regarded as a cohesive type (cohesion of the coating). Three replicates were performed for each sample and the results are mean values of the replicates.

2.6.2. Cross-Cut Analysis of the Samples

The adhesion of the coatings to the wood substrate was also assessed by cross-cut tests carried out according to EN ISO 2409 (1997) [15]. The coatings were cut approximately at 45° to the grain direction using a 2 mm normalized cutting tool. The two series of parallel cuts were crossed at an angle of 90° to obtain a pattern of squares. The sample was brushed and carefully examined using a lighter magnifier (2.5×) and rate based on the step classification given by the standard EN ISO 2409 (1997) [18]. Three replicates were performed for each sample.

2.6.3. Scratch Resistance of the Samples

The scratch resistance was determined according to the standard EN ISO 1518-1 (2000) [16]. The scratching needle with a hard hemispherical tip (tungsten carbide) of 1 mm diameter was drawn across the surface of the coated test specimen at a constant speed of 30–440 mm·s^{-1} perpendicular to the grain direction. Scratching was performed on different points of the coated wood specimens with increasing load on the scratch needle, until the coating cracked or the scratch was wider than 0.5 mm. The force level in N, which produced such a damage, is defined as a critical scratch, and this represents the scratch resistance.

2.6.4. Resistance to Impact of the Samples

The resistance to impact was determined using the standard method ISO 4211-4 (1995) [17]. A steel cylinder of (500 ± 5) g was placed at different heights for free fall onto a steel ball (diameter 14 mm) placed on the surface of the coating. After impact, the surface was examined with a magnifier (10×) and the diameter of the hole left at the surface of the coating was measured. The impact resistance of the coating was evaluated using numerical grades according to the standard. Three replicates were performed for each sample and drop height.

2.6.5. ATR-FTIR of the Samples

Attenuated total reflection-infrared spectroscopy (ATR FTIR) measurements of water glass, fresh silicate formulation and cured coating layers were performed using a Perkin Elmer Spectrum Two (PerkinElmer Inc., Waltham, MA, USA) ATR FTIR spectrometer, with a $LiTaO_3$ detector in the absorbance mode. The spectra were corrected for background noise and 16 scans per sample were collected at a wavelength from 400 to 4000 cm^{-1} at a resolution of 0.5 cm^{-1}.

2.6.6. SEM-EDAX of the Samples

The interface between the wood and the coating layer was examined with Scanning Electron Microscopy (SEM) and Energy-Dispersive Analysis of X-rays (EDAX). The micrographs were taken at a 20 kV voltage and a pressure of 50 Pa using a large field (LFD) detector in a FEI Quanta 250 SEM microscope (FEI Company, Hillsboro, OR, USA) at working distances between 7 and 11 mm under magnification by 300×. The interface and EDAX analyses were performed on fracture surfaces (obtained by splitting the coated wood samples in the longitudinal direction) to minimize the effect of polishing and particle dispersions of a cut-off saw.

2.6.7. Confocal Laser Scanning Microscopy Analyses of the Samples

The surface roughness of the coatings were studied using a Confocal Laser Scanning Microscope (CLSM) Olympus LEXT OLS5000 (Olympus Corporation, Tokyo, Japan) with the following objectives: MPLFLN5× (numerical aperture 0.15, working distance 20 mm). The microscope is equipped with a 405 nm violet laser, which enables a lateral resolution of down to 0.12 µm.

2.6.8. Water Absorption Measurements

The samples of 10 cm × 5.5 cm in size were cut from coated wood and used for water absorption tests. The tests were performed according to EN 927-5 [21] with modification (different sample sizes, no pre-conditioning). The uncoated surfaces were covered with a waterproof epoxy-based coating (Eplor HB, Color, Medvode, Slovenia). Each test specimen was exposed to demineralized water (300 mL) floating on the surface of the water with the investigated coated side facing down. The mass increase associated to water uptake was followed for 72 h. Two to three replicates were performed for each sample and the results are averages of the replicates.

3. Results and Discussion

3.1. Coating Performances

Pull-off adhesion, cross-cut, scratch and resistance to impact of the different coatings applied on beech wood are shown in Table 2.

Adhesion is an important coating parameter that defines the resistance of a coating to detachment from the substrate and is related to the quality of the coating, the physicochemical and mechanical interactions with the surface, and the degree of preparation of the surface before painting [22]. The failure mode after the pull-off test was mainly of the adhesive type, so the values in Table 1 actually reflect the adhesion strength of the coatings on wood. From the results, it can be seen that adhesion strength was significantly dependent on the formulation. Pull-off adhesion strengths comparable to the adhesion of some commercial organic-based coatings were obtained with CF1, CF7 and CF10–CF12. Adhesion strength values reported in the literature for wood coatings generally range from 2 to 5 MPa, but can reach up to 10 MPa for high performance coatings [23–25]. Based on observations of six polyurethane coatings, Oblak and co-authors [26] suggested a minimum acceptable adhesion value of 2.5 MPa for wood, and more than 5 MPa for excellent adhesion. It should be noted that this grade is applicable for organic-based coatings that are film-forming and flexible, and different values could be required for inorganic coatings. The cross-cut results are in line with the classes of adhesion obtained by pull-off tests.

The best result according to the EN ISO 2409 rate was 1, which corresponds to the cross-cut area with small flakes of coatings detached at the cutting intersections, as shown in Figure 1. An increase in rate is related to a decrease in adhesion and more detachment of the coatings as observed with CF4 (Rate 3).

Table 2. Adhesion strength, scratch resistance, and impact resistance of the coating formulations after two weeks of curing under ambient conditions (23 ± 3 °C and (50 ± 5)% relative humidity).

Coating Formulation	Pull-Off Adhesion Strength (MPa) [#]	Assessment of the Cross-Cut Test (2 mm)	Scratch Resistance (N)	Assessment of the Impact Test *	
				100 mm	400 mm
CF1	2.68 (0.33)	1	6	4 (4.5)	3 (6.3)
CF2	1.44 (0.40)	3	5	3.5 (5.2)	2.5 (6.7)
CF3	1.72 (0.08)	2	6	4 (4.5)	3 (6.3)
CF4	1.21 (0.11)	3	4	3 (5.8)	2.5 (6.8)
CF5	2.25 (0.08)	1	5	4 (4.2)	3 (6.6)
CF6	1.07 (0.13)	5	2	4 (5.8)	2.5 (7.0)
CF7	2.22 (0.06)	1	3	4 (3.5)	2 (6.3)
CF8	0.84 (0.15)	4	2	2.5 (4.5)	2 (6.1)
CF9	0.95 (0.08)	4	3	2.5 (4.8)	2 (7.0)
CF10	2.65 (0.46)	2	5	4 (4.0)	2.5 (7.1)
CF11	2.71 (0.37)	2	4	4 (4.8)	3 (6.3)
CF12	2.65 (0.22)	1	5	4 (4.6)	3 (6.5)

[#] Standard deviation in the parentheses. * 100 and 400 mm is the drop height and values in parentheses the diameter of the crater caused by the impact.

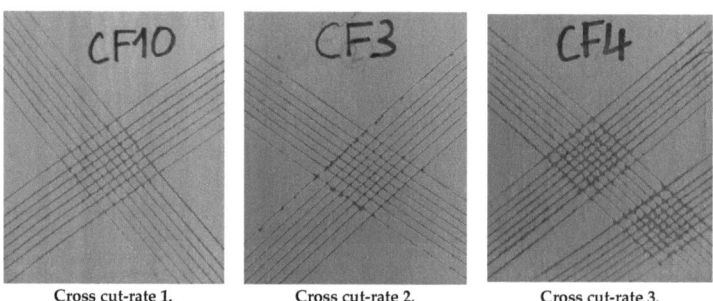

Cross cut-rate 1.   Cross cut-rate 2.   Cross cut-rate 3.

Figure 1. Cross-cut adhesion of CF3, CF4 and CF10.

Rate 1 is generally well accepted for commercial organic coatings. Relatively poor adhesion values were measured for the high solid content (CF6) formulation and for formulations with added glycerol, xylose and sucrose (pull-off adhesion lower than 2 MPa and cross-cut results of 2 to 5). The adhesion was higher with dextrin (CF5) than with other sugars and glycerol. The decrease of the modulus of water glass to 1.5 and 0.5 considerably reduced the adhesion of the coatings (CF8 and CF9). The scratch resistance and impact resistance measured with these formulations were also lower, suggesting that the failure was not only at the adhesion level but also on the cohesion and curing state of the coatings. The polymerization degree of silicates in water glass solution decreases with the modulus [5,8], it is probable that low molecular weight silicates are unable to form high strength compounds with the mineral additives (calcium carbonate, zinc oxide or talc). However, the reaction of water glass of low module (around to 0.5) with aluminosilicate compounds such as kaolin, metakaolin or activated fly ash has been demonstrated in the formation of high mechanical strength geopolymer cement binders [11–13]. CF12 showed that potassium water glass can be substituted with potassium methyl siliconate without a considerable loss of performance of the coatings.

The coatings showed acceptable resistance to scratch for some formulations. For comparison, the scratch resistances of cellulose nitrate lacquer, polyurethane, and various wood varnishes were reported between 1.5 and 4 N [27]. The coatings could withstand impact tests up to a drop height of 400 mm (corresponding to an impact energy of 2 J) without being severely destroyed. Impact holes with increasing diameter were observed, but no cracking and detachment of the coatings.

3.2. Adhesion Mechanisms

The adhesion was studied by interfacial observations. The interface between wood and coatings was analyzed using SEM coupled with EDAX. Four coating formulations were selected for these analyses based on their different level of adhesion strengths and the results are shown in Figure 2.

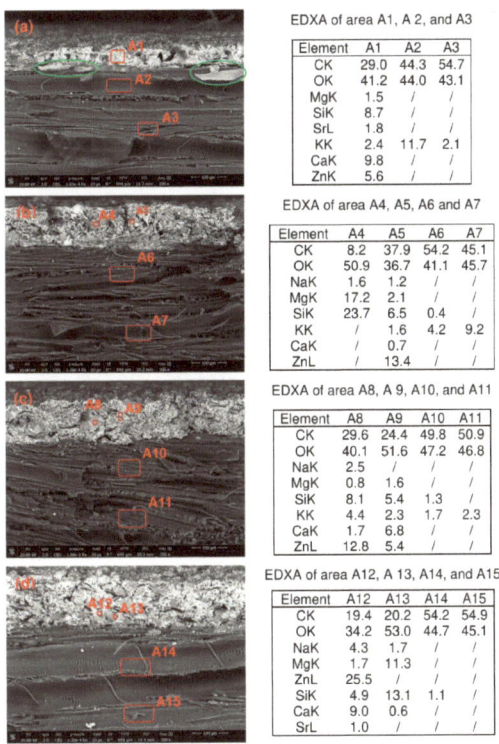

Figure 2. SEM-Energy-Dispersive Analysis of X-rays (EDAX) interfacial analyses of beech wood coated with (**a**) CF1 (**b**) CF2 (**c**) CF3 and (**d**) CF11.

The micrograph of CF1 clearly showed the penetration of the coating into the wood structure. The cured coatings (white area in the image of CF1 with green encirclements) can be observed in a few microns into the wood structure. The other white particles observed throughout the wood structure were attributed to the detachment and dispersion of the coating layers from the fracture zone during the preparation of the samples for analyses. The thickness of the coating layer also appeared to be less for CF1 than for other coatings (approx. 50 µm for CF1 and between 90 and 120 µm for CF2, CF3 and CF11).

The differences in the adhesion strength between the samples can be explained by the penetration of the coatings into the wood structure. Adhesion of a coating to a substrate occurs via two main mechanisms: mechanical anchoring and chemical interactions. Chemical interactions include adsorption and chemisorption with the formation of chemical bonds as well as electrostatic interactions at the interface between the substrate and the coating [22,28]. Mechanical adhesion results from penetration and hardening of the coating into the substrate. Wood is a porous material and therefore, mechanical adhesion generally plays an important role in the adhesion of paints and adhesives onto wood. Chemicals migrate into the wood structure by capillary permeation through open porosity and the phenomenon is modulated by parameters such as the size and shape of the chemical, or the nature of the solvent. It was reported that water-based coatings are prone to penetrate wood to a lesser extent and exhibited lower adhesion strengths than organic solvent-based ones [22,29]. The absorption of water or solvent from the coating by the wood surface changes the flow properties such as viscosity, solid content, and dispersion (particulate agglomerations). The penetration of the coating components into the wood was investigated using EDAX. Potassium was observed at depths down to around 200 μm, whereas silicon was detected only down to 50–100 μm. Potassium ions and probably hydroxide ions are the most water-transportable coating species due to their reduced geometric dimensions. Silicon is present in the coatings in various species ranging from simple silicate monomers to oligomer components with low mobility; a decrease of silicate modulus generally reduces the size of silicates species and hence increases their penetration into substrates. The contribution of penetration to adhesion was attributed to a few micrometers of hardened coatings at the interface between wood and coating layer. Alkaline metal silicates introduced into wood need additional presence of a curing agent (an acid or polyvalent metal cations such as Ca^{2+}, Zn^{2+}, Al^{3+}) to polymerize and form insoluble and mechanically resistant materials [30,31]. The usefulness of the penetration of silicates for adhesion is influenced by the penetration of other chemicals such as calcium carbonate or zinc oxide that could act as curing agents. The lower adhesion strengths measured with some coating formulations were attributed to lower ingress of such solid particles into the wood. It was observed that the application of water glass alone at the surface of wood before coating failed in increasing the adhesion, and even worse adhesion was measured in certain cases (results not shown).

Typical viscosities as a function of shear rate curves of potassium water glass, lithium water glass, and some coating formulations are shown in Figure 3.

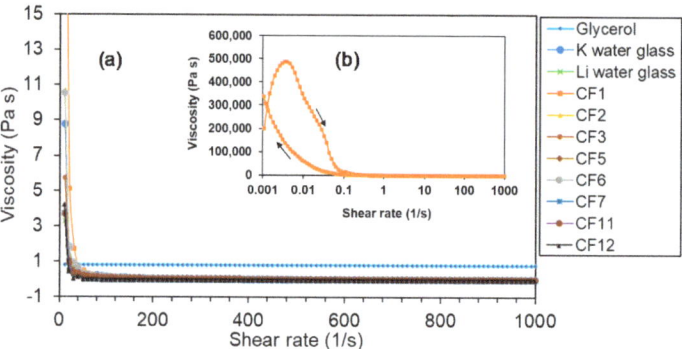

Figure 3. (a) Viscosity as a function of shear rate for selected formulations and (b) a typical flow behavior at lower shear rates (CF1).

The flow curves of water glasses and the coatings exhibited a shear thinning profile characteristic for suspensions, such as heterogeneous dispersion paints [32,33]. Only glycerol showed a Newtonian profile. In general, dispersed solid particles agglomerate and form a network that hinders random movements, the increase of the shear rate breaks the microstructure and thus decreases the viscosity, while the constraint against the particle flow decreases. The flow response of water glass is complex and can range from Newtonian to complex shear thickening. It changes with solid content, modulus, and the presence of impurities like metal salts [34–36]. A suspension-like model was proposed for sodium water glass in which monomeric and small oligomeric silicate species act as binders, whereas large oligomeric colloidal particles and small metallic cations act as effective rigid particles [35]. However, the microstructure was easily destroyed, resulting in a rapid drop of viscosity with an increase of the shear rate. The formulations showed a high viscosity at lower shear rate and a very low viscosity at high shear rate, between 0.016 Pa s and 0.25 Pa s at 100 to 200 s^{-1} estimated as the shear rate range generated by paint brushing on a substrate [37]. The flow curve measured by increasing (forward measurement) or decreasing (backward measurement) the shear rate for CF1 shown in Figure 3b, exhibited a typical flow behavior of the formulations at lower shear rates. The high viscosity suggests a steady equilibrium shear reversible gel-like structure at rest with weak interactions between particles. A detailed description of the rheological properties of the coatings at low shear rates and relation with properties such as leveling, sagging and sedimentation requires further extensive analyses, and that is beyond the scope of this research. The increase of the viscosity after coating application, the size and dispersion of particles and the settling properties could prevent penetration for some coatings. An increase of the solid content or addition of glycerol or sugar compounds are susceptible to cause the aggregation of particles. Glycerol is known as a coalescent additive. Low adhesion values were also measured when glycerol was substituted by the same amount of polyglycerol samples obtained by polymerization of glycerol catalyzed by lithium hydroxide at 240 °C for 6 h (results not shown). Large particles probably originating from aggregation can be seen in the micrographs (e.g., CF2, CF3 and CF11). However, For CF11, a more uniform interface was observed as compared to other coatings without signs of debonding or deposition of the coatings.

A relatively high amount of silicon was noticed in wood with CF3 (with sucrose additive) and was attributed to the ability of sugar to depolymerize and bond silicates in more soluble compounds. The formation of soluble organo-silicate complexes has been demonstrated in the literature for certain sugars including xylose [38]. The depolymerization of silicates by xylose could reduce or delay their contribution to the hardening process and explains the lower adhesion strength and performances measured with this sugar additive (CF4). The better adhesion with dextrin could be due to its contribution as a binding polymeric material able to adhere to some extent at the wood surface.

3.3. ATR-FTIR Analyses

ATR-FTIR analyses were performed on water glass, fresh coating formulations, and coatings after two weeks of curing. The spectra obtained with CF1 and CF12 are displayed in Figure 4. The spectra obtained with other samples were showing the same results (spectra not shown).

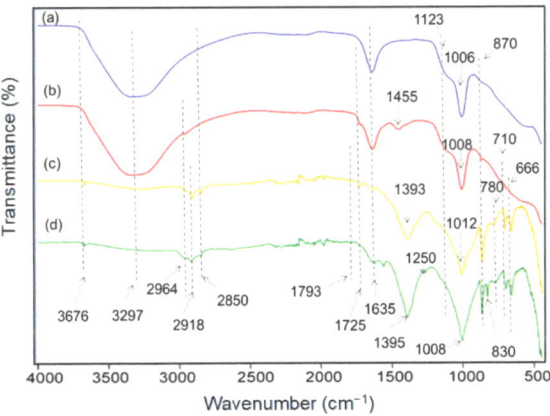

Figure 4. ATR-FTIR spectra of (**a**) potassium water glass (**b**) fresh liquid CF1 (**c**) hardened CF1 and (**d**) hardened CF12.

The vibration bands and corresponding functional groups are summarized in Table 3.

Table 3. FTIR spectra vibration bands assignments [39–45].

Bands	Assignments
3676 cm^{-1}	Stretching of isolated and mutually hydrogen bonded- Si–OH and Mg–OH
3297 cm^{-1}	Stretching O–H in water and silanol SiO–H hydrogen bonded- with water
2935, 2918, 2850 cm^{-1}	Asymmetric and symmetric stretching of C–H in methylene and methyl groups
1793, 1725 cm^{-1}	C=O carbonyls (CO_3^{2-} and carbonyl in acrylic resins
1635 cm^{-1}	Bending vibration of water (H–O–H)
1455, 1393 cm^{-1}	Asymmetric stretching of CO_3^{2-}
1120 cm^{-1}	Asymmetric Stretching of Si–O–Si in highly cross-liked zone, e.g., (Si–O)4 closed cage
1012, 1008, 1006 cm^{-1}	Asymmetric stretching of Si–O in short siloxane-chain
872 cm^{-1}	Out of plane bending (CO_3^{2-}), symmetric stretching Si–O
780 cm^{-1}	–CH_3 rocking and Si–C wagging stretching in Si–CH_3
710 cm^{-1}	In plane bending (CO_3^{2-})
666 cm^{-1}	Stretching of Si–O–Si in tetrahedral unit, stretching of Si–O–Mg

The band at 3673 cm^{-1} can be attributed to stretching of Si–OH hydroxyl of silanol groups of mutually H-bonded silicate compounds as well as to Mg–OH stretching in Talc. The broad band centered at 3297 cm^{-1} showed contribution of hydroxyl groups of adsorbed water and water in pores of the materials and stretching of Si–OH silanols H-bonded to water molecules. This band was very large for water glass and the fresh coating formulation because of their high-water content and overlapping with vibration bands of other functional groups (Si–OH and Mg–OH). The presence of water can also be seen from the bending vibration band at 1637 cm^{-1} occurring in molecular water. The bands at 2925, 2918 and 2850 cm^{-1} generally resulting from symmetric and asymmetric stretching of C–H were attributed to the presence of acrylic-styrene resins, glycerol, or others organic additives. Styrene-acrylic resins and carbonate groups (CO_3^{2-}) contributed to stretching bands at 1793 and 1725 cm^{-1} assigned to carbonyl (C=O) groups. The bands observed at 1393, 872 and 713 cm^{-1} were assigned to asymmetric stretching, out of plane bending and in-plane bending in carbonate compounds, respectively [42]. The domain between 1200 and 800 cm^{-1} displayed asymmetric and symmetric stretching of Si–O and the positions of bands revealed the degree of cross-linking of silicate species and association with metallic species (e.g., calcium ions) in metal silicates compounds. The main stretching band of Si–O–Si bonds appeared between 1090 and 1080 cm^{-1} in amorphous pure silica [40,46]. This vibration band is shifted towards lower wavenumbers around 1010 cm^{-1} or less for stretching of species such as $SiO(OH)_3^-$ (Q1) and $SiO_2(OH)_2^{2-}$ (Q2) groups as observed in the spectra of water glass and of fresh coatings. The stretching of Si–O in isolated SiO_4^{4-} (Q0) can be found down to approximately 850 cm^{-1} [39]. The combination of silicates with metal species such as calcium ions or zinc ions in calcium silicate or zinc silicate compounds reduced the interconnection between silicates and polymerization degree of the chains, while also shifting the vibrations bands of Si–O–Si towards lower wavenumbers. For example, the stretching of Si–O–Si was observed around 970 cm^{-1} in calcium silicates when the proportion of calcium ions incorporated in the structure increased [44]. The fact that the maximum absorption of the shoulder band in the Si–O region appeared at 1012–1008 cm^{-1} for the coatings suggested a curing mechanism involving preferably reaction of silicates with calcium, zinc or magnesium species from calcium carbonate, zinc oxide and talc additives instead of formation of silica networks. The results are in agreement with X-ray diffraction (XRD) analyses reported by Kazmina et al. [10] for similar systems. Curing of silicates can also occur naturally by reaction with carbon dioxide of the environment. Carbon dioxide diffuses inside the material and is dissolved in the pore solution and forms carbonic acid that neutralizes the alkalinity and leads to the occurrence of silicate polymerization. The dissolution of zinc oxide, calcium carbonate, or liberation of metal cations or ions into pore solutions and the reaction of silicates with surface metallic species yield to calcium silicates, zinc silicates and bonding of the particles.

3.4. Influence of Wood Mineralization

Mineralization was performed under mild conditions to allow mainly modification at the wood surface. The weight percent gain of wood determined as percentage of mass increase of the samples after the double impregnation processes were between 2% and 3% (values measured without leaching). The coating performances of formulation CF7 applied onto modified wood samples are shown in Table 4.

Table 4. Coating performances of CF7 on pre-mineralized wood samples (MW1 and MW2 wood mineralized with potassium water glass, and then boric acid or sodium bicarbonate, respectively).

Pre-Mineralized Wood	Pull-off Adhesion Strength (MPa) [#]	Assessment of the Cross-Cut Test	Scratch Resistance (N)	Assessment of the Impact Test (400 mm) *
MW1	3.8 (0.26)	1	6	3.5 (6.4)
MW2	2.6 (0.18)	1	5.5	3.5 (6.8)

[#] Standard deviation is in the parentheses. * 400 mm is the drop height and values in parentheses the diameter of the crater caused by impact.

The results showed an increase of the adhesion strength after mineralization of wood. A considerable increase was observed for wood mineralized with potassium water glass and cured with boric acid, from 2.2 MPa on untreated wood to 3.8 MPa on pretreated wood. The silicification of wood by a double treatment with boric acid curing has been reported as one of the best methods to produce insoluble silica in wood [14,15]. Chemical reactions between silicate species in the coatings and silica at the wood surface could contribute to an increase of the adhesion of coatings. Low adhesion was measured when potassium water glass was applied alone (result not shown) without a second curing treatment. The results showed that adhesion between silicate coatings and wood can be significantly improved by using pre-mineralized wood. These results are in agreement with commercial offers using mineralized alkyds or drying oils primers.

3.5. Water Resistance of the Samples

Liquid water absorptions expressed as the increase of mass per surface area of coating exposed to water (kg/m^2) are shown in Figure 5 for selected (based on adhesion strength values) coatings.

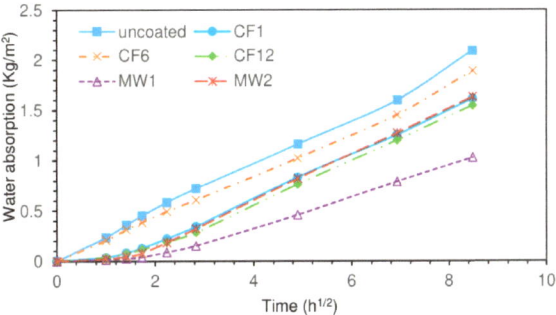

Figure 5. Water absorption of uncoated wood and coated wood (CF1, CF6, CF12, MW1 and MW2).

Water absorption was slightly reduced by most of the coatings. The lowest water uptake was measured for MW1. The samples (CF1, CF12, MW1 and MW2) showed low water absorption during the first two hours in water, but after the amount of water uptake increased considerably. Water absorption was not reduced by the addition of methyl siliconate. The results could be due to the method used for the application. The application was carried out by one single pass that led to craters and pores allowing for water transport as observed in confocal microscopy analyses of the coating surfaces (Figure 6).

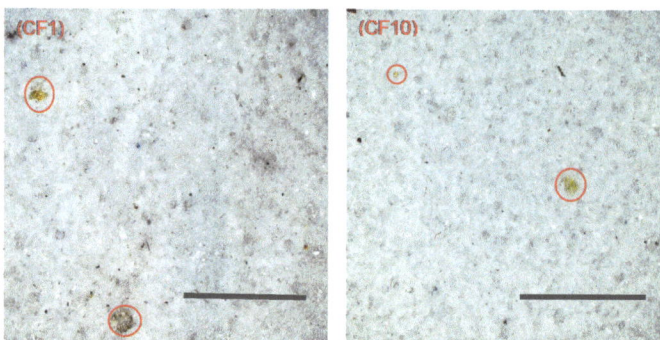

Figure 6. Confocal microscopy images of the surfaces of CF1 and CF10 (magnification 5×, scale bars represent 1 mm).

These coatings are not film-forming materials and a multi-layer application could be useful to reduce the formation of such open porosity. The dissolution of soluble components (e.g., potassium salts) could exacerbate the water permeability. It is important to note that cracks were observed at some coatings after 24 h in water, especially at coatings with low adhesion. Cracking and debonding increased while drying the samples, which resulted from the dimensional instability of wood. A picture of a coating sample with one of the highest surface damages (CF6) due to moisture (soaking in water for five days and indoor air drying for height weeks) is shown in Figure 7. The samples with higher adhesion strength values were less degraded by water as can be seen with CF1.

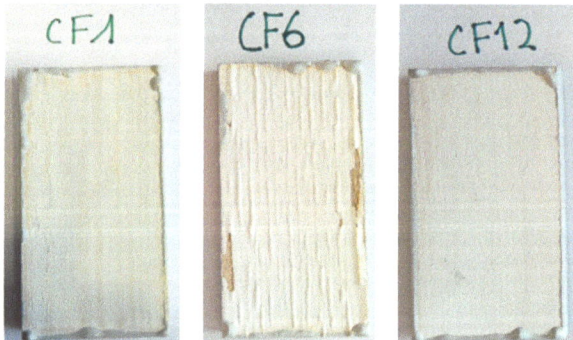

Figure 7. Photographs of CF1, CF6 and CF12 soaked in liquid water for 5 days and exposed for drying to ambient indoor conditions (temperature (23 ± 3) °C and (50 ± 5)% relative humidity) for eight (08) weeks.

No signs of cracks were noticed on the sample with methyl siliconate (CF12); only some discoloration appeared and was attributed to wood extractives. Silicate modified with organoalkoxysilanes showed an increase in flexibility as reported for the decrease of modulus of elasticity of organically-modified silicates on steel [47]. The increase of flexibility could explain the resistance of the coatings to the formation of cracks despite high water absorption. The surface of this coating was not chalking after residence in water as opposed to other coating formulations. Optimization of the formulations to the reduced water absorption is under progress (results will be shown elsewhere).

4. Conclusions

Various silicate-based formulations were prepared and applied on wood to study the coating performances (adhesion, scratch and impact resistance, and water resistance). The adhesion mechanism was found to be dependent on mechanical anchoring for untreated wood and related to penetration of the coatings into the wood. Poor adhesion strength (<2 MPa) to adhesion acceptable for coatings (2.5–2.70 MPa) were measured depending on the formulations. The surface pre-mineralization of wood increased the adhesion strength by increasing chemical bonding. An increase of the adhesion to 3.8 MPa was obtained for wood pretreated with a double impregnation of potassium silicate and boric acid solutions. An addition of polyols (glycerol, sucrose or D-xylose) was not effective in improving the coatings' properties. The coatings showed acceptable scratch and impact resistances, but the formulations must be improved to increase water resistance. Substitution of a part of potassium water glass by potassium methyl siliconate improved the resistance to crack formation without significant changes of the other performances of the coatings.

Author Contributions: Conceptualization and methodology, A.M.C.Y.; analyses, A.M.C.Y. and J.Ž.; writing—original draft, A.M.C.Y.; review and editing, A.M.C.Y., J.Ž., S.D., and M.P.; supervision, M.P.; funding acquisition A.M.C.Y. and M.P. All authors have read and agreed to the published version of the manuscript.

Funding: This research was funded by the Slovenian National Research Agency (ARRS) ("SilWoodCoat" N4-0117 and "Wood and lignocellulosic composites" P4-0015).

Data Availability Statement: The raw data presented in this study are available on request from the corresponding author.

Acknowledgments: The authors are grateful to Angela Balzano for her technical assistance in SEM-EDAX analyses.

Conflicts of Interest: The authors declare no conflict of interest.

References

1. Popescu, C.-M.; Pfriem, A. Treatments and modification to improve the reaction to fire of wood and wood based products—An overview. *Fire Mater.* **2020**, *44*, 100–111. [CrossRef]
2. Borges, C.C.; Denzin, T.G.H.; Moreira, C.T.; Duarte, P.J.; Junqueira, T.A. Nanoparticles-based wood preservatives: The next generation of wood protection? *CERNE* **2018**, *24*, 397–407. [CrossRef]
3. Bao, W.; Zhang, M.; Jia, Z.; Jiao, Y.; Cai, L.; Liang, D.; Li, J. Cu thin films on wood surface for robust superhydrophobicity by magnetron sputtering treatment with perfluorocarboxylic acid. *Eur. J. Wood Prod.* **2019**, *77*, 115–123. [CrossRef]
4. Köhler, R.; Sauerbier, P.; Ohms, G.; Viöl, W.; Militz, H. Wood protection through plasma powder deposition—An alternative coating process. *Forests* **2019**, *10*, 898. [CrossRef]
5. Parashar, G.; Bajpayee, M.; Kamani, P.K. Water-borne, non-toxic, high-performance inorganic silicate coatings. *Surf. Coat. Int. B* **2003**, *86*, 209–216. [CrossRef]
6. Pal, S.; Contaldi, V.; Licciulli, A.; Marzo, F. Self-cleaning mineral paint for application in architectural heritage. *Coatings* **2016**, *6*, 48. [CrossRef]
7. Loganina, V.I. Polymer silicate paints for interior decorating. *Contemp. Eng. Sci.* **2015**, *8*, 171–177. [CrossRef]
8. Fraci, A.T. *Waterborne Silicates in Coatings and Construction Chemicals in Inorganic Zinc Coatings, History, Chemistry, Properties, Applications and Alternatives*, 2nd ed.; Francis, R.A., Ed.; The Australasian Corrosion Association Inc.: Victoria, Australia, 2013; pp. 227–258.
9. Kumar, S.P.; Takamori, S.; Arakia, H.; Kuroda, S. Flame retardancy of clay–sodium silicate composite coatings on wood for construction purposes. *RSC Adv.* **2015**, *5*, 34109–34116. [CrossRef]
10. Kazmina, O.; Lebedeva, E.; Mitina, N.; Kuzmenko, A. Fire-proof silicate coatings with magnesium-containing fire retardant. *J. Coat. Technol. Res.* **2018**, *15*, 543–554. [CrossRef]
11. Ramasamy, S.; Hussin, K.; Al Bakri, M.M.A.; Ghazali, C.M.R.; Sandhu, A.V.; Binhussain, M.; Shahedan, N.F. Adhesiveness of kaolin based coating material on lumber wood. *Key Eng. Mater.* **2016**, *673*, 47–54. [CrossRef]
12. Ramasamy, S.; Al Bakri Abdullah, M.M.; Kamarudin, H.; Yue, H.; Jin, W. Improvement of kaolin based geopolymer coated wood substrates for use in NaOH molarity. *Mater. Sci. Forum* **2019**, *967*, 241–249. [CrossRef]

13. Shaikh, F.U.A.; Haque, S.; Sanjayan, J. Behavior of fly ash geopolymer as fire resistant coating for timber. *J. Sustain. Cem. Based Mater.* **2018**, *8*, 1–16. [CrossRef]
14. Standard ISO 4624. Paints and Varnishes-Pull Off Test for Adhesion. 2016. Available online: https://www.iso.org/standard/62351.html (accessed on 5 January 2021).
15. Standard EN ISO 2409. Paints and Varnishes-Cross Cut Test. 1997. Available online: https://www.iso.org/standard/76041.html (accessed on 5 January 2021).
16. Standard EN ISO 1518-1: 2000 Paint and Varnishes-Scratch Test. Available online: https://standards.iteh.ai/catalog/standards/cen/a2ff84bf-9c58-4b9e-a651-4c3899ae56fa/en-iso-1518-2000 (accessed on 5 January 2021).
17. Standard ISO 4211-4. Furniture-Tests for Surfaces-Assessment of Resistance to Impact. 1995. Available online: https://standards.iteh.ai/catalog/standards/sist/44c7b7b9-ec61-485e-8909-6e317e0e9f72/sist-iso-4211-4-1995 (accessed on 5 January 2021).
18. Chen, G.C. Treatment of wood with polysilicic acid derived from sodium silicate for fungal decay protection. *Wood Fiber Sci.* **2009**, *410*, 220–228.
19. Furuno, T.; Shimada, K.; Uehara, T.; Jodai, S. Combinations of wood and silicate II. Wood-mineral composites using water glass and reactants of barium chloride, boric acid, and borax, and their properties. *Mokuzai Gakkaishi* **1992**, *38*, 448–457.
20. Thougaard, L.; Hayden, J.P. Wood Preservation Method Using Sodium Silicate and Sodium Bicarbonate, WO 2014/101979 A2. Available online: https://patents.google.com/patent/WO2014101979A2/en (accessed on 5 January 2021).
21. EN 927-5:2006—Paints and Varnishes—Coating Materials and Coating Systems for Exterior Wood—Part 5: Assessment of the Liquid Water Permeability. Available online: https://www.en-standard.eu/csn-en-927-5-paints-and-varnishes-coating-materials-and-coating-systems-for-exterior-wood-part-5-assessment-of-the-liquid-water-permeability/ (accessed on 5 January 2021).
22. Meijer, M.D. *A review of interfacial aspects in wood coatings: Wetting, surface energy, substrate penetration and adhesion.* COST E18 Final Seminar; European Cooperation in Science and Technology: Brussels, Belgium, 2005.
23. Sonmez, A.; Budakci, M.; Bayram, M. Effect of wood moisture content on adhesion of varnish coatings. *Sci. Res. Essays* **2009**, *4*, 1432–1437.
24. Kesik, H.I.; Akyildiz, M.H. Effect of the heat treatment on the adhesion strength of water based wood varnishes. *Wood Res.* **2015**, *60*, 987–994.
25. Hazir, E.; Koc, K.H. Evaluation of wood surface coating performance using water based, solvent based and powder coating. *Maderas. Cienc. Tecnol.* **2019**, *21*, 467–480. [CrossRef]
26. Oblak, L.; Kričej, B.; Lipušček, I. The comparison of the coating systems according to the basis criteria. *Wood Res.* **2006**, *51*, 77–86.
27. Çakıcıer, N.; Korkut, S.; Korkut, D.S. Varnish layer hardness, scratch resistance, and glossiness of various wood species as affected by heat treatment. *BioResources* **2011**, *6*, 1648–1658.
28. Ahola, P. Adhesion between paints and wooden substrates: Effects of pre-treatments and weathering of wood. *Mater. Struct.* **1995**, *28*, 350–356. [CrossRef]
29. De Meijer, M.; Van de Velde, B.; Militz, H. Rheological approach to the capillary penetration of coating into wood. *J. Coat. Technol.* **2001**, *73*, 39. [CrossRef]
30. Furuno, T.; Uehara, T.; Jodai, S. Combinations of wood and silicate I. Impregnation by water glass and applications of aluminum sulfate and calcium chloride as reactants. *Mokuzai Gakkaishi* **1991**, *37*, 462–472.
31. Pereyra, A.M.; Giudice, C.A. Flame-retardant impregnants for woods based on alkaline silicates. *Fire Saf. J.* **2009**, *44*, 497–503. [CrossRef]
32. Arora, S.; Laha, A.; Majumdar, A.; Butola, B.S. Prediction of rheology of shear thickening fluids using phenomenological and artificial neural network models. *Korea Aust. Rheol. J.* **2017**, *29*, 185–193. [CrossRef]
33. Ge, J.; Tan, Z.; Li, W.; Zhang, H. The rheological properties of shear thickening fluid reinforced with SiC nanowires. *Results Phys.* **2017**, *7*, 3369–3372. [CrossRef]
34. Yang, Q.; Yang, X.H.; Wang, P.; Zhu, W.L.; Chen, X.Y. The viscosity properties of zinc-rich coatings from sodium silicate solution modified with aluminium chloride. *Pigment Resin Technol.* **2009**, *38*, 153–158. [CrossRef]
35. Yang, X.; Zhu, W.; Yang, Q. The viscosity properties of sodium silicate solutions. *J. Solut. Chem.* **2008**, *37*, 73–83. [CrossRef]
36. Yang, X.; Zhang, S. Characterizing and modeling the rheological performances of potassium silicate solutions. *J. Solution Chem.* **2016**, *45*, 1890–1901. [CrossRef]
37. Fischer, E.K. Rheological properties of commercial paints. *J. Colloid Sci.* **1950**, *5*, 271–281. [CrossRef]
38. Lambert, J.B.; Lu, G.; Singer, S.R.; Kolb, V.M. Silicate complexes of sugars in aqueous solution. *J. Am. Chem. Soc.* **2004**, *126*, 9611–9625. [CrossRef]
39. Bobrowski, A.; Stypuła, B.; Hutera, B.; Kmita, A.; Drożyński, D.; Starowicz, M. FTIR spectroscopy of water glass—The binder moulding modified by ZnO nanoparticles. *Metalurgija* **2012**, *51*, 477–480.
40. Khan, A.S.; Khalid, H.; Sarfraz, Z.; Khan, M.; Iqbal, J.; Muhammad, N.; Fareed, M.A.; Rehman, I.U. Vibrational spectroscopy of selective dental restorative materials. *Appl. Spectrosc. Rev.* **2016**. [CrossRef]
41. Al-Oweini, R.; El-Rassy, H. Synthesis and characterization by FTIR spectroscopy of silica aerogels prepared using several Si(OR)$_4$ and R00Si(OR')$_3$ precursors. *J. Mol. Struct.* **2009**, *919*, 140–145. [CrossRef]
42. Cai, G.-B.; Chen, S.-F.; Liu, L.; Jiang, J.; Yao, H.-B.; Xu, A.-W.; Yu, S.-H. 1,3-Diamino-2-hydroxypropane-N, N, N′, N′-tetraacetic acid stabilized amorphous calcium carbonate: Nucleation, transformation and crystal growth. *Cryst. Eng. Comm.* **2010**, *12*, 234–241. [CrossRef]

43. Ferrage, E.; Martin, F.; Petit, S.; Pejo-Soucaille, S.; Micoud, P.; Fourty, G.; Ferret, J.; Salvi, S.; De Parseval, P.; Fortune, J.P. Evaluation of talc morphology using FTIR and H/D substitution. *Clay Miner.* **2003**, *38*, 141–150. [CrossRef]
44. Giraudo, N.; Bergdolt, S.; Wohlgemuth, J.; Welle, A.; Schuhmann, R.; Königer, F.; Thissen, P. Calcium Silicate Phases Explained by High-Temperature-Resistant Phosphate Probe Molecules. *Langmuir* **2016**, *32*, 13577–13584. [CrossRef]
45. Maddalena, R.; Hall, C.; Hamilton, A. Effect of silica particle size on the formation of calcium silicate hydrate [C-S-H] using thermal analysis. *Thermochim. Acta* **2019**, *672*, 142–149. [CrossRef]
46. Li, K.-M.; Jiang, J.-G.; Tian, S.-C.; Chen, X.-J.; Yan, F. Influence of silica types on synthesis and performance of amine−silica hybrid materials used for CO_2 capture. *J. Phys. Chem. C* **2014**, *118*, 2454–2462. [CrossRef]
47. Latella, B.A.; Ignat, M.; Barbe, C.J.; Cassidy, D.J.; Bartlett, J.R. Adhesion behaviour of organically-modified silicate coatings on stainless steel. *J. Sol Gel Sci. Technol.* **2003**, *26*, 765–770. [CrossRef]

Article

Fire Characteristics of Selected Tropical Woods without and with Fire Retardant

Linda Makovicka Osvaldova [1,*], **Patricia Kadlicova** [1] **and Jozef Rychly** [2]

1. Department of Fire Engineering, Faculty of Security Engineering, University of Zilina, 010 26 Zilina, Slovakia; patriciakadlicova@gmail.com
2. Polymer Institute, Slovak Academy of Sciences, 845 41 Bratislava, Slovakia; jozef.rychly@savba.sk
* Correspondence: linda.makovicka@fbi.uniza.sk

Received: 29 April 2020; Accepted: 27 May 2020; Published: 29 May 2020

Abstract: The flammability of tropical woods and the effect of a selected fire protection coating were evaluated using a cone calorimeter at a cone radiancy of 35 kW/m^2. Three samples were from the South American continent (Cumaru, Garapa, Ipe), and two were from the Asian continent (Kempas and Merbau). Samples were treated with commercial fire retardant (FR) containing ferrous phosphate as an essential component. The untreated samples were used as reference materials that were of particular interest concerning their flammability. It was shown that there is unambiguous correlation between the effective heat of combustion (EHC) and total oxygen consumed (TOC) related to mass lost during burning for both the untreated and treated samples. In the case of Cumaru and Garapa, there exists an inverse relation between the amount of smoke and carbon residue. The decisive effect on the time of ignition was performed by the initial mass of the sample. This is valid for the spruce and the Cumaru, Ipe, and Kempas, both treated and untreated with retardant, while Garapa and Merbau were found to decline. According to the lower maximum average rate of heat emission (MARHE) parameter, a lower flammability was observed for the treated samples of wood, except for Garapa wood. Fire-retardant treated Garapa and Merbau also have a significantly lower time to ignition than untreated ones.

Keywords: tropical wood; flame retardant; cone calorimetry testing; heat release rate; total smoke production; amount of carbon residue; total oxygen consumed

1. Introduction

By the end of the 1970s, tropical and subtropical tree species started to be imported, processed, and sold in Central Europe. This became a growing trend, due to their positive physical and mechanical characteristics and innovative color finish in comparison with the native trees growing in a different climate zone. Exotic tree species, because of their good stability and lifespan, are used primarily as terraces, balconies, swimming pool flooring, etc., that is, in places with increased humidity and probability of sudden weather changes [1,2]. These tree species are used both in their natural and modified form commonly known as Thermwood [3].

Wood is an organic material, and therefore there is a wide range of substances that can be aimed at the protection of it from any undesirable impacts of the environment. From the fire-fighting perspective, many types of fire retardants have been developed and manufactured, putting stress on their efficiency, aging, and reaction to fire classification [4–6]. It is necessary to point out that lots of attention has been paid to the fire technical properties of common woods (flammable material), such as spruce, oak, beech, etc., and other materials, such as textiles and plastics [7–10]. Until now, less attention has been given to tropical woods, as they represent more complex systems [1,11,12]. The morphology, the different content of hemicelluloses, cellulose, and lignin, as well as the characteristics and amount of inorganic and low molar mass organic substances may exert a certain influence on their fire behavior.

Three types of trees species that are chosen for our experiments originate in America (Cumaru, Garapa, Ipe) and two tree species in Asia (Kempas and Merbau). The samples come from the Slovak supplier DLH SLOVAKIA s.r.o. (Bratislava, Slovakia) in the form of boards without any surface treatment and have the same dimensions. Laboratory tests are carried out in a cone calorimeter with the aim to evaluate their flammability by comparing the series of parameters of the respective tropical wood as well as with those of the samples treated by coating with fire retardant containing ferrous phosphate. In the particular case of the selected tropical woods, Reference [11] deals with the burning of Merbau wood; however, the initial content of water was unexpectedly high, and thus the effective heat of combustion (EHC) measured at 50 kW/m^2 was very low (4 MJ/kg) when compared with the EHC of common woods (14–18 MJ/kg).

2. Materials and Methods

2.1. Materials

Figure 1 shows the cross-section and microscopic structure (type of microscope: TESCAN SEM Solutions) of the selected tree species—three of them originate in America: (a) Cumaru, (b) Garapa, (c) Ipe, and the remaining two in Asia, (d) Kempas, and (e) Merbau.

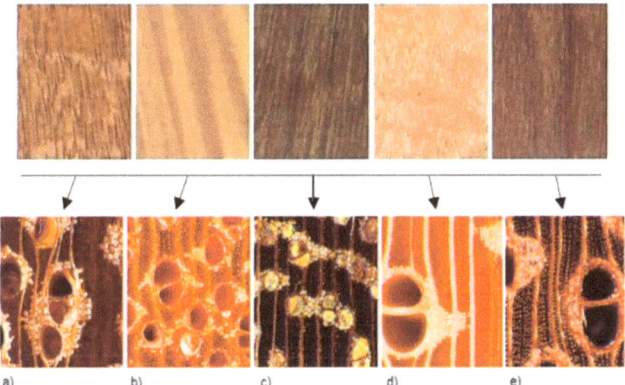

Figure 1. Microscopic structure (cross-section) of tropical tree species from listed as (**a**) Cumaru, (**b**) Garapa, (**c**) Ipe, and the two remaining in Asia, (**d**) Kempas, (**e**) Merbau.

Based on [13,14], Kempas and Merbau have the most distinct vessels, followed by Cumaru, Garapa and Ipe, in descending order. Ipe wood contains traces of lapachol, a substance having a protective function. Each tree type has single as well as double vessels. Garapa wood even has triple vessels. These trees also contain distinct medullary rays. Aside from the vessels and the rays, the pictures show the wood fiber of the particular tree species.

Cumaru (*Dipteryx spp.*) is a tree species with yellowish brown to reddish brown heartwood, clearly demarcated yellowish white sapwood, and with spiral, demarcated, and interlocked fibers of a darker color. Red-brown is the reference color. Sapwood is 2 to 3 cm thick [13,15].

Garapa (*Apuleia leiocarpa*) is a tree species with a yellowish beige to yellowish brown heartwood, yellowish white sapwood, and with straight and just a few spiral but clearly demarcated and interlocked fibers. Sapwood is 5 to 11 cm thick. Orange yellow is the reference color. As the wood ages, the yellow color changes to light brown [13,16].

Ipe (*Tabebuia spp.*) is a tree species with olive green to dark brown heartwood, yellowish white sapwood, and thin spiral and interlocked fibers. Traces of lapachol—which is liquid protecting the wood—can be found in the pores. Reference color is brown, and sapwood is 3 to 9 cm thick [13,17].

Kempas (*Koompassia malaccencis*) is a wood of roseate and gradually darkening reddish brown heartwood, light yellow to white sapwood, straight and just a few spiral interlocked fibers as well as frequent concentric layers of phloem. Reference color is red-brown, and sapwood is 2 to 5 cm thick [13,18].

Merbau (*Intsia bijuga*) is a tree species having reddish brown to dark brown heartwood, light yellow sapwood, and straight and only a few spiral interlocked fibers (so do Garapa and Kempas). If exposed to light without any surface protection for a longer period of time, the wood changes its color by becoming dark red or dark brown in color [13,19].

Samples of $100 \times 100 \times 20$ mm ± 1 mm were used during the test. Four untreated samples and one sample coated with fire retardant (FR) were selected for each tree type described above. Chemical composition of the FR is in the Table 1.

Table 1. Chemical composition of FR.

Name
$FeHPO_4$
Citric acid
Polyoxyethylene stearyl alcohol
Water

The coating was applied twice using a paint roller with FR (minimum 300 g/m^2) on the surface of the sample. After drying at different intervals, there was no visible measurable coating layer on the surface of the samples, and therefore we assumed that the majority of it was absorbed by the material.

2.2. Methods

2.2.1. Test Equipment

The device is used to study the fire behavior of small-size samples made out of various materials in solid form. The peak release rate (PkHRR) is a key factor needed to assess the fire characteristics. The heat release rate from a cone calorimeter was obtained by monitoring the oxygen consumption based on the difference in the cone entrance and in a cone extraction pipe of air flows in a cone calorimeter system. It is assumed that an average of 13.1 MJ of energy is released per one kilogram of consumed O_2 [20]. The heat flow was set at 35 kW/m^2. Other readouts were as follows: ignition time, total burning time, weight loss, effective heat consumption (EHC), O_2 consumption, smoke release rate, etc. No CO detection was available. The device (the product of Fire Testing Laboratories Ltd., West Sussex, UK) consists of a number of parts, as seen in Figure 2.

Figure 3 shows the test procedure in a cone calorimeter. It starts by preheating the samples using a radiant heat source (heating coil), followed by the ignition of the combustible gaseous components evaporating from the sample. The fire starts to spread, and the charred layers are observed.

Figure 2. Cone calorimeter.

Figure 3. Samples tested in the cone calorimeter.

2.2.2. Measurement Procedure

The sample of the given dimensions was mounted into the steel frame in a horizontal position so that the only part of the sample exposed to heat is its surface, not its edge. In order to prevent the material from peeling and the chemical components from dripping, the sample was wrapped up into an aluminum foil from the bottom and the sides. The radiator, which has a shape of a truncated cone is placed above the sample (radiant heat source). It represents a constant radiation source that the sample was exposed to. The temperature was regulated using 3 thermocouples and a thermostat. Weight measurements were carried out using a load cell of a tensometer with a readability of 0.01 g. A 10 kV spark generator equipped with a safety shut-off mechanism is used to ignite the vapors of the test specimens. There is a steady flow of air of 24 l/h and a constant temperature of 770 °C [21].

In the tests, the sample is ignited using a spark. The timer is activated the moment the sample ignites and is turned off the moment the sample goes out spontaneously. Smoke and combustion gases from the sample are absorbed by the suction hood with a tube made out of stainless steel equipped with a fan of adjustable flow rate (0 to 50 g/s), a measuring probe for the products of combustion, a thermocouple, a pressure transmitter, and a parameter analyzer of O_2 (0–25%). The smoke release rate is determined using a laser with photodiodes. The heat release rate is calibrated using methane 99.5% [21].

3. Results

Attention has been given to a proper selection of samples with respect to their humidity and density. The samples were climatized under standard conditions (22 °C and 55% of air humidity) for 7 days before and after the application of the retardant. The content of water in the samples was controlled gravimetrically to a constant weight achieved by conditioning. The samples were selected from the cut boards (see Table 2) so that differences not higher than ± 5% kg/m^3 were allowed.

Table 2. The density of samples.

The Sample	Density ^{1} (kg/m^3)	Density ^{2} (kg/m^3)
Cumaru	1200	1070
Garapa	1050	790
Ipe	1300	1050
Kempas	1050	880

^{1} The density of fresh wood and density; ^{2} the density at 12% of water content.

The amount of retardant absorbed is given in Table 3. As the properties of tropical woods differ due to different densities and thus porosities, the penetrability of the coating into the mass of the sample was rather different, as was the amount of retardant coated.

Table 3. The acceptance of wet retardant in grams.

The Sample	Cumaru	Garapa	Ipe	Kempas	Merbau
The acceptance (g)	2.4	2.5	6.5	8.4	2.6

The results are documented in the following section either in a tabular form (see Table 4) or graphically (see Figures 4–6). Figure 4 is a representation of four parallel HRR (heat release rate)–time runs for Merbau samples. One can see that repeatability of the respective experiments is quite satisfactory. Typical HRR–time runs involve a sharp increase of the heat released just after ignition. At that time, the flame is spread along the surface, and in its close proximity below, the flame then penetrates through the layer of carbonaceous residue where high molar mass products of the degradation are pushed into the bottom of the carbonaceous layer by the heat front and accumulate there. In the final stage, there appears a second maximum, which characterizes the burning of the accumulated products, followed by the extinction of the flame. The lines HRR–time are in fact a first derivative of mass loss runs, which may also be seen in Figure 4. It is of interest that the retardation effect on the Merbau samples performed by the FR is seen particularly at the final stages of burning. This is more or less distinct for all retarded samples. The prolongation of the time of burning is the most significant for Cumaru and Garapa. Garapa and Merbau samples showed a significant reduction in time to ignition (90 vs. 65 s, and 91 vs. 54 s). From the set of respective parallel runs, the average curves were determined, which may be seen in Figure 5 (Cumaru—C, Garapa—G, Ipe—I, Kempas—K, Merbau—M), as well as the standard deviations concerning the time of ignition, the time of burning, the amount of carbon residue in percentage, the peak of HRR, and MARHE (Table 4). The graphical results representing HRR curves for the samples treated with FR are in Figure 6.

Table 4. Parameters read from cone calorimetry measurements of untreated samples C (Cumaru), G (Garapa), K (Kempas), M (Merbau), and I (Ipe), and samples treated with FR.

Sample	Parameters								
	m (g)	m_{lost} (g)	Time to Ignition/Time of Burning (s)	EHC (MJ/kg)	Peak HRR (kW/m^2)	TOC (g)	TSR (m^2/m^2)	MARHE (kW/m^2)	% of Carbon Residue
C_{av}	163.3	121.5	45 ± 4/1067 ± 30	15.6	457.2 ± 42.3	115.9	662.2	176.5 ± 5.1	25.5 ± 0.3
C_{fr}	173.2	114.4	51/1514	13.4	264.4	93.5	189.0	98.1	33.9
G_{av}	180.0	118.5	90 ± 10/1439 ± 49	13.6	275.0 ± 28.2	98.4	205.7	107.4 ± 6.3	34.2 ± 0.5
G_{fr}	189.8	166.0	65/1731	15.9	342.5	161.0	923.0	148.6	24.5
K_{av}	139.9	112.7	37 ± 2/1200 ± 34	17.1	307.8 ± 9.8	115.8	385.8	155.6 ± 7.4	20.6 ± 1.7
K_{fr}	138.7	105.7	42/1237	17.5	268.1	112.8	426.8	145.0	23.8
M_{av}	253.3	183.0	91 ± 16/1676 ± 50	17.4	438.4 ± 32.8	194.8	1468.8	187.9 ± 8.4	27.7 ± 1.2
M_{fr}	253.7	183.4	54/1565	17.1	357.6	191.6	1924.1	170.1	27.7
I_{av}	227.0	171.9	89 ± 5/1340+−64	16.9	500.9 ± 17.9	177.6	1678.8	224.6 ± 8.6	24.4 ± 0.4
I_{fr}	230.6	170.2	98/1413	15.8	452.6	164.7	1745.9	187.8	26.2

Notes: m_{lost} = total loss of mass during burning; time to ignition/time of burning; time to ignition = time interval from insertion of the sample below the cone heater to ignition (piloted ignition by spark); time of burning = time of the flame burning of the sample until self-extinction; EHC = effective heat of combustion; peak HRR = maximum heat release rate corresponding to the second maximum of record; TOC = total oxygen consumed during combustion; av = average value; fr = sample coated with FR; MARHE = a maximum value of ARHE (average rate of heat emission) ARHE = $\frac{\int HRR*t*dt}{t1}$, with an integral in the nominator from 0 to t_1.

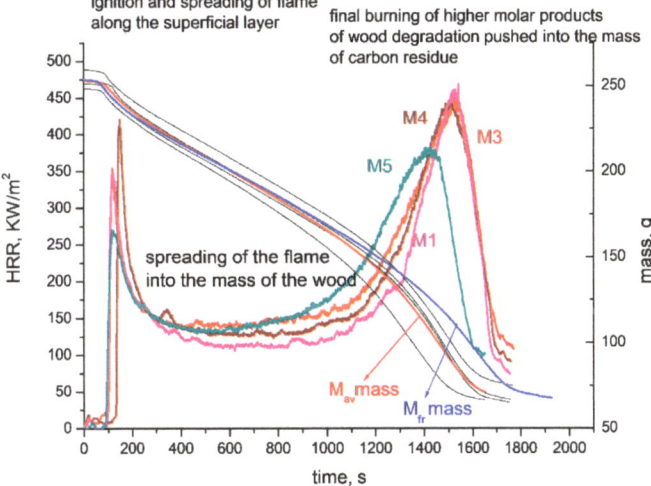

Figure 4. Merbau sample in four parallel runs of heat release rate (HRR) and mass with time. Cone radiancy is 35 kW/m^2; M_{av} mass is the average mass change of samples M_1, M_3, M_4, and M_5.

The most important parameters for similar experiments with spruce samples are in the Table 5.

As it may be seen from Figures 5 and 6, the peak HRR (the second maximum) of treated Cumaru is significantly lower than that of the untreated one; lower peak HRRs for treated samples were also observed for Kempas, Merbau, and Ipe, while Garapa gave a more intense peak of HRR. The difference is the highest for Cumaru.

All essential parameters read from the cone calorimetry experiments for both untreated and treated samples are presented in Table 4.

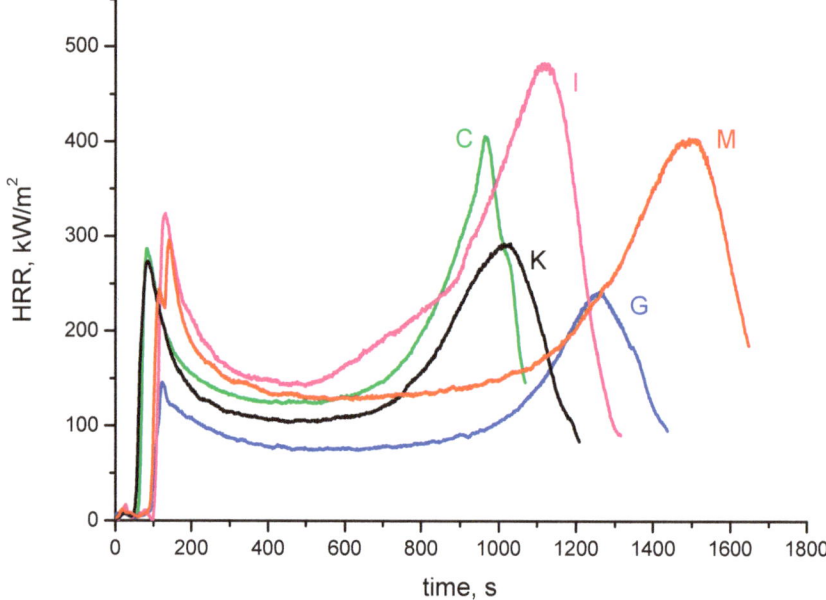

Figure 5. The average HRR–time runs for untreated samples at 35 kW/m^2; C—Cumaru, G—Garapa, I—Ipe, K—Kempas, M—Merbau.

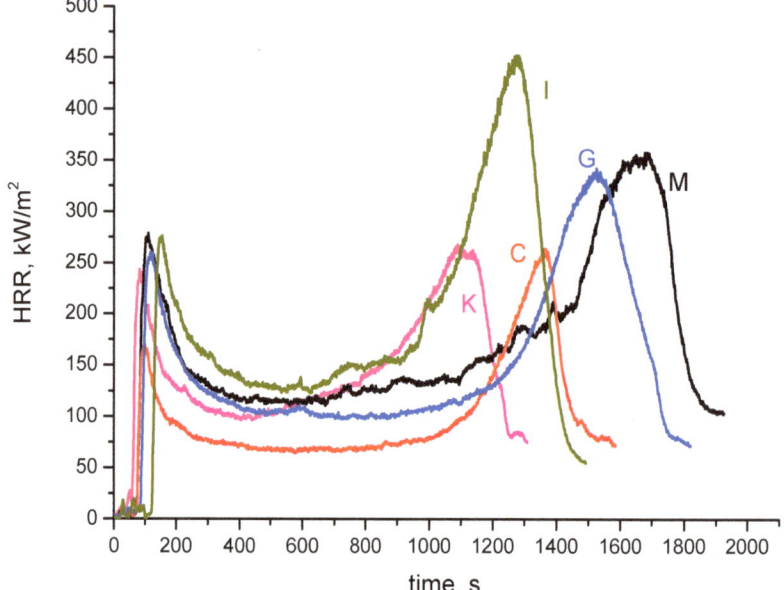

Figure 6. HRR of the samples coated with FR; C—Cumaru, G—Garapa, I—Ipe, K—Kempas, M—Merbau. Cone radiancy at 35 kW/m^2.

Table 5. Some parameters from the cone calorimeter after the burning of selected samples of spruce fir differing in initial density (dimensions of samples at 9 × 9 × 1 cm) with a cone radiancy of 35 kW/m².

Sample	Some Parameters from Cone Calorimeter after the Burning of Selected Samples						
	Density (kg/m³)	Initial Mass (g)	Time to Ignition (s)	EHC (MJ/kg)	TOC (g)	Peak HRR (kW/m²)	Carbon Residue (%)
Spruce Fir	561	55.02	47	15.2	41.6	227.0	17.2
	553	53.17	51	14.3	37.2	232.7	17.9
	445	43.64	47	14.4	31.7	234.9	15.1
	458	44.45	41	14.4	32.5	218.5	–
	385	37.37	31	15.3	30.1	224.2	12.6
	392	38.06	36	15.3	29.9	241.0	13.9

The ignition time t_{ig} (s) can be expressed as follows, as given in Equation (1):

$$t_{ig} = \pi k \rho c \left(\frac{T_{ig} - T_0}{2\delta I} \right)^2 \qquad (1)$$

where t_{ig} is ignition time in seconds, π is Pi, k is thermal conductivity in W/m K, ρ is density in kg/m³, c is specific heat in J/kg K, T_{ig} is the surface temperature of the material when ignited (in K), T_0 is the ambient temperature (in K), δ is emissivity, and I is radiation intensity (in kW/m²).

From this, it follows that the density and thus the initial mass at the identical dimensions of the sample is the decisive factor determining the time to ignition (Figure 7). On samples of the same quality (spruce), this is quite evident, while for tropical trees differing in structure and the amount of other additional compounds, the picture is rather complex. K_{av}, C_{av}, and I_{av} follow the straight line X approximately, but G_{av} and M_{av} decline. Figure 7 also shows the changes of the time to ignition when the samples were treated with the FR. The significant reduction of time to ignition occurs with samples Merbau and Garapa, while Cumaru, Kempas, and Ipe are affected only a little.

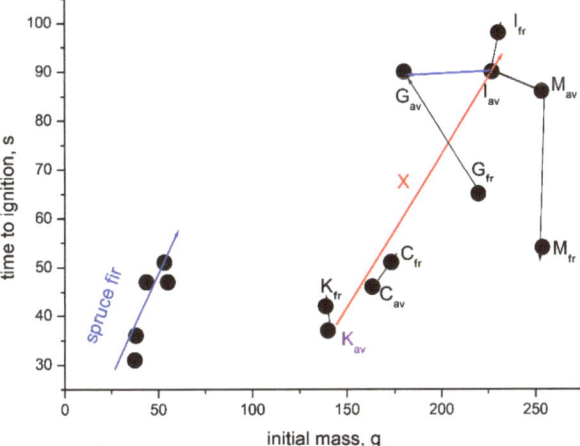

Figure 7. Time to ignition in dependence on the initial mass of the untreated and samples treated with FR.

4. Discussion

The main component of the fire retardant (FR) is secondary ferrous phosphate, which may provide antagonistic effects in retardation of burning as it comes to ignitability. This may represent a particular interest. Ferrous and subsequently ferric ions will promote the degradation of cellulose in a wood to lower molar mass products, while phosphate moiety may initiate crosslinking, and later on the increase of carbon residue, as it has a pronounced effect on the trapping of higher molar mass products of the degradation. This appears to be the case for Garapa wood (Table 2) where the prodegradation effect of ferrous ions seems to be suppressed. This may be documented by the higher peak HRR. We assumed that the carbonaceous residue is denser and keeps more degradation products, while in all other cases it is quite opposite. The reduction of peak HRR is the most pronounced in the case of Cumaru and Merbau. In the case of Kempas and Ipe, the effect is rather weak. The overview of the effect on the modified samples concerning the HRR and the time to ignition may be seen in Figure 8. The MARHE parameter may be considered as a scale of the samples' flammability (Figure 9). Except for Garapa, all other samples have a MARHE of lower flame retardancy (lower flammability) than untreated ones.

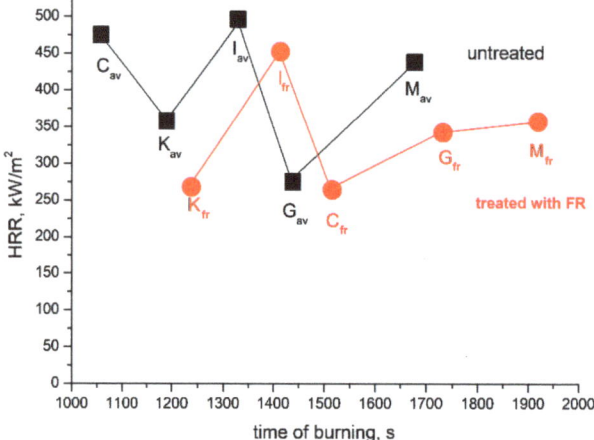

Figure 8. Changes of HRR with time of burning for tropical woods untreated and treated by FR (red line).

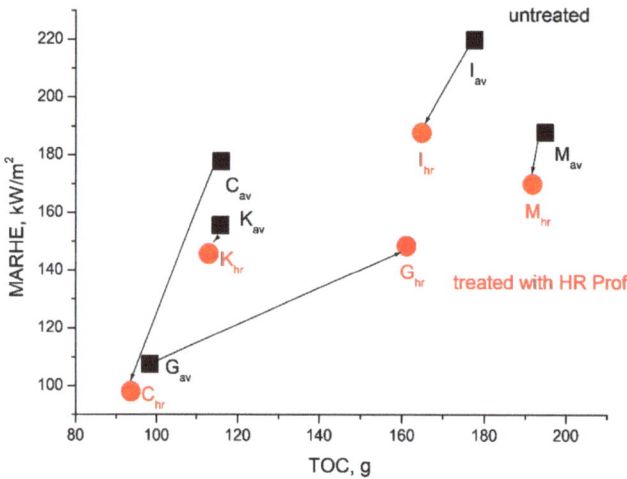

Figure 9. The correlation between MARHE and total oxygen consumed for untreated and treated samples of tropical woods (Table 2). The cone radiancy is 35 kW/m^2.

Figure 10 shows the universal relation between the effective heat of combustion (EHC) and the total oxygen consumed (TOC) divided by the mass lost during burning. The straight line received in both cases of tropical woods as well as spruce fir has the following shape: EHC = 1.008 + 15.46 × TOC/mass lost, where the slope around 15.5 kJ/g is higher than the parameter 13.1 kJ/g of consumed oxygen implemented in the cone calorimeter. The difference around 20% to 30% appears to be due to the formation of carbon residue in the case of woods. It may be of interest that the above correlation is valid also for spruce samples that are thinner than tropical woods.

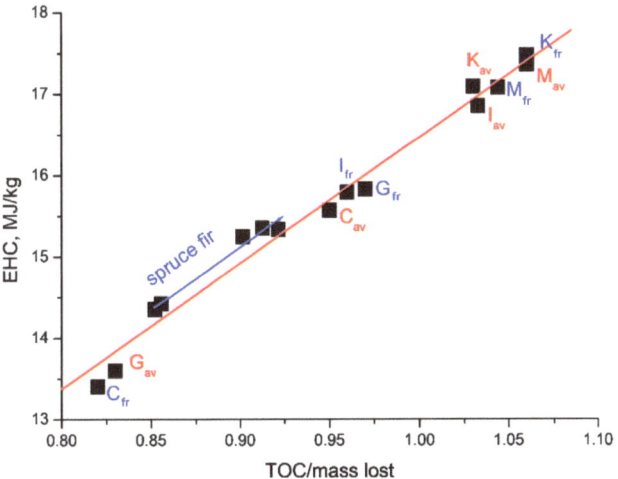

Figure 10. Universal correlation between effective heat consumption (EHC) and total oxygen consumed related to unity of mass lost during burning.

In Figure 11, we verified the assumption that the total smoke released should be inversely proportional to the carbonaceous residue formed. In the case of Cumaru and Garapa, after the modification with FR it appears to be true. The unmodified Garapa gave a relatively high carbon residue, while the total release of smoke is low; after modification, carbon residue is significantly reduced, and the total smoke increases. Unmodified Cumaru gave a higher smoke release and a lower amount of carbon residue, but in the case of modified samples it is vice versa. For other tropical woods, this is not unambiguous as the effect is rather low (Figure 11).

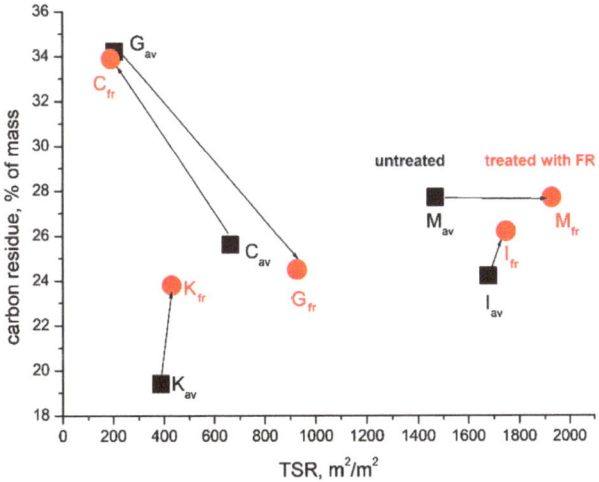

Figure 11. Correlation between carbon residue after sample extinction and total smoke release.

5. Conclusions

There is an unambiguous relation between the EHC and TOC related to mass lost during burning. The slope is, however, higher (about 15 kJ/g of consumed oxygen) than that implemented into the cone calorimeter (13.1 kJ/g of consumed oxygen). The data received from the spruce fall into this graph as well. This appears to be due to the compensation by carbon residue resting at the end of burning. This may contribute to a better understanding of the results obtained for the sample, as its residue leaves when burning. In the case of Cumaru and Garapa, there exists an inverse relation between the amount of smoke and carbon residue.

As it was also reported in [1,11,12], the time to ignition is affected by the initial mass of the sample. This is valid for the spruce and the Cumaru, Ipe, and Kempas, both treated and untreated with retardant, while the Garapa and Merbau were found to decline.

Cumaru and Garapa treated with the fire retardant show a significantly longer time of burning. The difference with the others was not so distinct. With the exception of Garapa, most of the samples performed a lower peak HRR. A similar effect may be found with a MARHE parameter that is lower (which indicates lower flammability) for treated samples, except again for Garapa wood.

Treated Garapa and Merbau also have a significantly lower time to ignition than untreated ones. This may be a demonstration of the antagonistic effect of the fire retarding agent—ferrous ions shorten the time to ignition, while phosphate moiety playing a role in the formation of carbon residue prolongs the overall time of sample burning.

Author Contributions: Conceptualization, L.M.O.; Formal analysis, P.K.; Investigation, L.M.O.; Methodology, L.M.O.; Resources, P.K.; Software, J.R.; Writing-original draft, L.M.O., J.R., and P.K.; Writing-review & editing, L.M.O.; supervision, L.M.O.; All authors have read and agreed to the published version of this manuscript.

Funding: This research received no external funding.

Conflicts of Interest: The authors declare no conflict of interest.

References

1. Momoh, M.; Horrocks, A.R.; Eboatu, A.N.; Kolawole, E.G. Flammability of tropical woods—I. Investigation of the burning parameters. *Polym. Degrad. Stab.* **1996**, *54*, 403–411. [CrossRef]
2. Elvira-Leon, J.C.; Chimenos, J.M.; Isabal, C.; Monton, J.; Formosa, J.; Haurie, L. Epsomite as flame retardant treatment for wood: Preliminary study. *Constr. Build. Mater.* **2016**, *126*, 936–942. [CrossRef]
3. Cekovska, H.; Gaff, M.; Osvald, A.; Kacik, F.; Kubs, J.; Kaplan, L. Fire resistance of thermally modified spruce wood. *BioResources* **2017**, *12*, 947–959. [CrossRef]
4. Koklukaya, O.; Carosio, F.; Grunlan, J.C.; Wagberg, L. Flame-retardant paper from wood fibers functionalized via layer-by-layer assembly. *ACS Appl. Mater. Interfaces* **2015**, *7*, 23750–23759. [CrossRef] [PubMed]
5. Östman, B.A.-L. Fire performance of wood products and timber structure. *Int. Wood Prod. J.* **2017**, *8*, 74–79. [CrossRef]
6. Jiang, J.; Li, J.; Hu, J.; Fan, F. Effect of nitrogen phosphorus flame retardants on thermal degradation of wood. *Constr. Build. Mater.* **2010**, *24*, 2633–2637. [CrossRef]
7. Ecochard, Y.; Decostanzi, M.; Negrell, C.; Sonnier, R.; Caillol, S. Cardanol and eugenol based flame retardant epoxy monomers for thermostable networks. *Molecules* **2019**, *24*, 1818. [CrossRef] [PubMed]
8. Maqsood, M.; Langensiepen, F.; Seide, G. The efficiency of biobased carbonization agent and intumescent flame retardant on flame retardancy of biopolymer composites and investigation of their melt-spinnability. *Molecules* **2019**, *24*, 1513. [CrossRef] [PubMed]
9. Sonnier, R.; Dumazert, L.; Livi, S.; Nguyen, T.K.L.; Duchet-Rumeau, J.; Vahabi, H.; Laheurte, P. Flame retardancy of phosphorus-containing ionic liquid based epoxy networks. *Polym. Degrad. Stab.* **2016**, *134*, 186–193. [CrossRef]
10. Sonnier, R.; Otazaghine, B.; Ferry, L.; Lopez-Cuesta, J.M. Study of the combustion effiency of polymers using a pyrolysis-combustion flow calorimeter. *Combust. Flame* **2013**, *160*, 2182–2193. [CrossRef]

11. Xu, Q.; Chen, L.; Harries, K.A.; Zhang, F.; Liu, Q.; Feng, J. Combustion and charring properties of five common constructional wood species from cone calorimeter tests. *Constr. Build. Mater.* **2015**, *96*, 416–427. [CrossRef]
12. Giraldo, M.P.; Haurie, L.; Sotomayor, J.; Lacasta, A.M.; Monton, J.; Palumbo, M.; Nazzaro, A. Characterization of the fire behaviour of tropical wood species for use in the construction industry. In Proceedings of the WCTE 2016: World Conference on Timber Engineering, Vienna, Austria, 22–25 August 2016; Technischen Universität Graz: Graz, Austria; pp. 5387–5395.
13. Gerad, J.; Guibal, D.; Paradis, S.; Cerre, J.C. *Tropical Timber Atlas Technological Characteristics and Uses*, 1st ed.; Editions Quæ: Versailles Cedex, France, 2017; pp. 600–602.
14. Wiemann, M.C. Characteristics and availability of commercially important woods. In *Wood Handbook: Wood as an Engineering Material*; Forest Products: Madison, WI, USA, 2010; pp. 2.1–2.45.
15. Cumaru—Specification Sheet. Available online: https://www.dlhwood.sk/wp-content/uploads/2017/01/60510_Tech_list_Cumaru_kor0.pdf (accessed on 13 November 2019).
16. Garapa—Specification Sheet. Available online: https://www.dlhwood.sk/wp-content/uploads/2017/01/60510_Tech_list_Garapa_kor0.pdf (accessed on 13 November 2019).
17. Ipe—Specification Sheet. Available online: https://www.dlhwood.sk/wp-content/uploads/2017/01/60510_Tech_list_Ipe_kor0-1.pdf (accessed on 13 November 2019).
18. Kempas—Spectification Sheet. Available online: https://www.dlhwood.sk/wp-content/uploads/2017/01/60510_Tech_list_Kempas_kor0.pdf (accessed on 13 November 2019).
19. Merbau—Specification Sheet. Available online: https://www.dlhwood.sk/wp-content/uploads/2017/01/60510_Tech_list_Merbau_kor0-1.pdf (accessed on 13 November 2019).
20. *ASTM E1354-17 Standard Test Method for Heat and Visible Smoke Release Rates for Materials and Products Using an Oxygen Consumption Calorimeter*; ASTM International: West Conshohocken, PA, USA, 2017. [CrossRef]
21. Cone Calorimeter. Available online: http://polymer.sav.sk/old/index.php (accessed on 22 December 2019).

© 2020 by the authors. Licensee MDPI, Basel, Switzerland. This article is an open access article distributed under the terms and conditions of the Creative Commons Attribution (CC BY) license (http://creativecommons.org/licenses/by/4.0/).

Review

A Review on the Effect of Wood Surface Modification on Paint Film Adhesion Properties

Jingyi Hang [1,2], Xiaoxing Yan [1,2,*] and Jun Li [1,2]

1 Co-Innovation Center of Efficient Processing and Utilization of Forest Resources, Nanjing Forestry University, Nanjing 210037, China; hangjingyi@njfu.edu.cn (J.H.); lijun0099@njfu.edu.cn (J.L.)
2 College of Furnishings and Industrial Design, Nanjing Forestry University, Nanjing 210037, China
* Correspondence: yanxiaoxing@njfu.edu.cn

Abstract: Wood surface treatment aims to improve or reduce the surface activity of wood by physical treatment, chemical treatment, biological activation treatment or other methods to achieve the purpose of surface modification. After wood surface modification, the paint film adhesion performance, gluing performance, surface wettability, surface free energy and surface visual properties would be affected. This article aims to explore the effects of different modification methods on the adhesion of wood coating films. Modification of the wood surface significantly improves the adhesion properties of the paint film, thereby extending the service life of the coating. Research showed that physical external force modification improved the hydrophilicity and wettability of wood by changing its surface structure and texture, thus enhancing the adhesion of the coating. Additionally, high-temperature heat treatment modification reduced the risk of coating cracking and peeling by eliminating stress and moisture within the wood. Chemical impregnation modification utilized the different properties of organic and inorganic substances to improve the stability and durability of wood. Organic impregnation effectively filled the wood cell wall and increased its density, while inorganic impregnation enhanced the adhesion of the coating by forming stable chemical bonds. Composite modification methods combined the advantages of the above technologies and significantly improved the comprehensive properties of wood through multiple modification treatments, showing superior adhesion and durability. Comprehensive analysis indicated that selecting the appropriate modification method was key for different wood types and application environments.

Keywords: paint film; adhesion performance; modification treatment; interface bonding; wood

Citation: Hang, J.; Yan, X.; Li, J. A Review on the Effect of Wood Surface Modification on Paint Film Adhesion Properties. *Coatings* **2024**, *14*, 1313. https://doi.org/10.3390/coatings14101313

Academic Editor: Marko Petric

Received: 7 August 2024
Revised: 9 October 2024
Accepted: 10 October 2024
Published: 14 October 2024

Copyright: © 2024 by the authors. Licensee MDPI, Basel, Switzerland. This article is an open access article distributed under the terms and conditions of the Creative Commons Attribution (CC BY) license (https://creativecommons.org/licenses/by/4.0/).

1. Introduction

Wood is widely used in the field of home decoration. In order to maintain its beauty for a long time, it is usually necessary to apply coating materials on its surface [1–3]. The paint film applied on the wood surface can isolate moisture in the air, effectively prevent the wood from decaying and mildewing, extend the service life of the wood, and play an important role in protecting the wood substrate [4]. However, the prerequisite for the coating to fully play the above role is good interfacial bonding between the coating and the attached wood substrate. The strength of this bonding is generally measured by the size of the adhesion [5,6]. The adhesion of the paint film is the ability of the coating to adhere to other coatings and between itself and the substrate.

At present, paint film adhesion is mainly divided into mechanical adhesion and chemical adhesion. From a macroscopic perspective, the principle of mechanical adhesion is that the surface of any substrate cannot be absolutely smooth. There will be certain pores and holes, and there will be a certain surface roughness. For example, the surfaces of substrates such as leather, wood, and paper are porous and have great roughness. When paint is applied to such substrates, the paint can penetrate into these pores and holes [7–9]. When the paint is cured into a film, the coating and the substrate are bonded together by this kind of mechanical adhesion method. The principle of chemical adhesion is explained from the perspective of microscopic physical and chemical reactions. It is believed that adhesion refers to the mutual attraction between molecules at the contact interface between wood and coating [10]. The components in the coating can react chemically with functional groups on the surface of the substrate to form covalent bonds or bonding structures, thereby enhancing the bonding between the coating and the substrate [11]. For example, polar groups such as carboxyl (–COOH) and ester (–COOR) in acrylic polymers and isocyanate (N–C–O) in polyurethane can form strong hydrogen bonds with hydroxyl groups in wood. These chemical reactions can produce a stronger bond between the coating and the substrate. Therefore, thorough research on the characteristics of wood and the performance of paint films will provide an important basis for improving the adhesion and overall durability of coatings.

1.1. Study on Wood Surface Properties

1.1.1. Wood Moisture Content and Fiber Hygroscopicity

If the moisture content of wood is too high, the room temperature is high or it is exposed to the sun, the evaporation of moisture in the wood will cause bubbling of the paint film, which may cause the paint film to fall off in severe cases and also lead to poor dimensional stability of the wood [12]. Generally, wood and wood products can only be painted when their moisture content is between 8% and 12% [13]. Too low a moisture content in wood leads to a decrease in the number of active hydroxyl groups in the wood, thereby reducing the bonding probability between the polar groups of the paint and the wood, resulting in a decrease in the adhesion ability of the paint film [14]. Wood generally has excellent hygroscopicity. Water absorption by wood fibers causes changes in the wood's dimensions, which in turn generates stress, weakens the adhesion of the paint film, and can easily lead to cracking of the paint film [15].

1.1.2. Wood Extracts and Surface Cleanliness

Wood extracts are a general term for substances extracted from wood using organic solvents such as ethanol, benzene, ether, acetone or dichloromethane, as well as water [16]. Wood contains a variety of extracts. Extracts precipitated from the wood surface and other impurities such as dust scattered on the wood surface will seriously reduce the wettability of the wood surface, affecting the adhesion of wood coatings to wood [17,18]. Therefore, before coating or bonding processes, the wood surface usually needs to be cleaned to remove these extractables and impurities to improve wettability and ensure good coating effect and durability.

1.1.3. Wood Surface Roughness

Surface roughness refers to microscopic geometric features composed of small spaces, peaks and valleys formed after cutting or pressure processing on the surface [19]. The roughness of the wood surface helps to improve the mechanical adhesion between it and the coating, expand the actual contact area of the wood, and enhance the effective bonding interface of the coating, thereby improving the adhesion of the coating to the wood [20]. If the roughness of the wood surface is too high, the continuity and integrity of its surface will be destroyed. Wood chips generated during the grinding process will fill the surface pores, which hinders the paint film from penetrating into these pores, thereby reducing the contact area between the paint film and the substrate, and ultimately leading to reduced adhesion of the paint film [21].

1.1.4. Wood Surface Wettability

The wettability of the wood surface is an important parameter that reflects the interface phenomenon between the coating and the substrate, and has a significant impact on the interface bonding quality between the paint film, the adhesive and the substrate [22]. Wettability is a property that describes the behavior of liquids on the surface of solid substrates. When a liquid (such as water, adhesives, coatings, diiodomethane, etc.) is dropped onto the surface of a substrate, wettability is manifested as the process and effect of the liquid wetting, spreading on and penetrating the surface of the substrate [23]. As shown in Figure 1, the contact angle is used to characterize wettability. 90° is a dividing point of the wettability of a solid surface. When the contact angle value is less than 90°, the wettability is good. When the contact angle value is higher than 90°, the wettability of the solid material is poor. The larger the contact angle, the more difficult it is for the liquid to wet the solid surface, which in turn has a negative impact on the coating and bonding abilities of the surface [24].

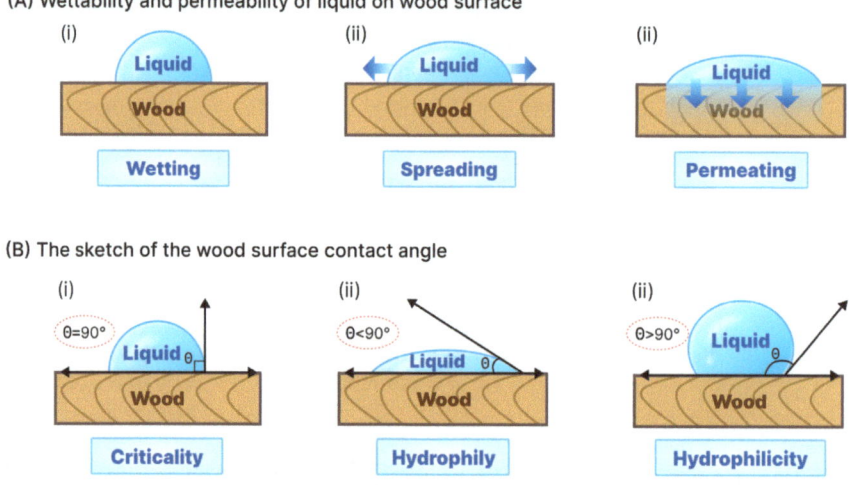

Figure 1. (**A**) Wettability and permeability of liquid on wood surface; (**B**) Sketch of the wood surface contact angle.

1.1.5. Wood Surface Free Energy

Wood is connected with and interacts with the outside world through its surface. Surface free energy is the interaction between the wood surface and external molecules, and is a further quantification of the wetting conditions of the substrate surface in terms of energy [25]. The surface free energy of wood is an important property that affects subsequent processing of the wood surface, such as gluing and painting [26].

After the wood is modified by some methods, its surface free energy is reduced, which is not conducive to surface coating performance. Wang et al. [27] grafted long-chain alkyl groups onto the cell walls of *Cunninghamia lanceolata*. Compared with the untreated material, the surface wettability of the modified *Cunninghamia lanceolata* was reduced, showing a lower surface free energy. Miklecic et al. [28] conducted a study on the thermal modification of beech wood. The results showed that heat-treated beech wood showed higher acidity and water contact angle, while the polar component of the surface free energy decreased. In addition, as the modification temperature increased, the contact angle of beech wood for water-based coatings also increased. Heat treatment increased the content of extracts soluble in hot water and organic solvents in beech wood, which reduced the wetting properties of the wood and weakened its adhesion.

1.2. Study on Paint Film Performance

1.2.1. Main Factors Affecting Coating Adhesion

Factors such as the film thickness, viscosity and compatibility with the surface of the coating have a crucial impact on the adhesion ability of the coating. Changes in any factor may significantly change the adhesion of the coating to the wood substrate [29]. For the technical parameters of coatings, film thickness is a key factor. Film thickness is an important indicator for detecting the uniformity and quality of the paint film [30]. Generally speaking, the larger the coating amount, the thicker the paint film, the smaller the stress between different paint films and the weaker the adhesion; this may cause cracks and peeling of the paint film. Under the condition of a certain surface roughness of the substrate, a coating that is too thin may not provide sufficient protection and wear resistance. According to the national standard GB/T1764-79, the film thickness of wood coatings is generally required to be between 60–120 μm [31].

At the same time, the choice of coating viscosity is also crucial. The viscosity will affect the leveling and uniformity of the coating. When the viscosity of the coating is low, its fluidity is enhanced, the penetration value becomes larger, and the coating penetrates faster into the surface of the substrate and can better fill the tiny gaps on the surface of the substrate to form closer contact. This close contact helps to enhance the mechanical bonding between the coating and the substrate, thereby improving the adhesion and durability of the coating. However, too low a viscosity may cause problems such as sagging or difficulty in spreading. Depending on the rheological properties of the coating, the viscosity of the coating is generally controlled between 90 and 120 KU. In addition, during the film-forming process, as the solvent evaporates, the viscosity of the coating increases, so the coating should be kept at low viscosity for a long enough time to fully penetrate the surface of the substrate. Long-term low viscosity can be achieved by using solvents with slow evaporation rates and maintaining a low cross-linking rate. The use of low molecular weight film-forming substances can also reduce the viscosity of the coating and improve adhesion. In addition, generally, baked paint has better adhesion than air-dried paint. One of the reasons is that a pre-cured coating at a high temperature has a lower viscosity and can easily penetrate the substrate to form mechanical bite adhesion.

The compatibility between the coating and the surface is also an important part of the technical parameters. The compatibility between the coating and the surface is a key factor in ensuring that the coating can be evenly applied to the wood surface to form a coating with uniform quality and stable performance. Good compatibility enables the coating to spread quickly on the wood surface and form a strong adhesion layer, thereby improving the adhesion and overall performance of the coating. Coatings with poor compatibility may result in poor wetting, and the coating cannot evenly cover the wood surface, thus affecting the coating quality and even causing problems such as peeling or blistering. The compatibility of the coating can be effectively improved by optimizing the coating formula, adjusting the wetting angle, or pretreating the wood surface. In order to improve the poor interfacial compatibility between bamboo fiber (BF) surface and polylactic acid surface, Zhang et al. [32] constructed a bioinspired polydopamine (PDA) functional coating on the BF surface and introduced the silane coupling agent KH560 as a bridge. The results showed that KH560 was successfully grafted onto the BF surface, and the PAD–KH560-modified BF enhanced the thermal stability of the PLA composite and improved the interfacial adhesion strength.

1.2.2. Coating Effectiveness Analysis Method

(1) Adhesion of paint film

Coating adhesion measurement is a commonly used test method to determine the adhesion strength of coatings and is one of the most important technical indicators for evaluating coatings or coating systems. There are many methods for evaluating the adhesion of paint films, among which the most commonly and widely used is the cross-cut test method. According to GB/T4893.4-2013 [33], the cross-cut method is to use a single-edged knife, a multi-edged knife or an instrument to cut the coating into a grid pattern. In this operation, all cuts should be cut to the substrate. After sticking pressure-sensitive adhesive tape to the cut area for about 5 s, hold one end of the tape at a 60-degree angle to the cut area and quickly tear off the tape. Visually or using a magnifying glass, inspect the cut area of the test coating, and judge the coating adhesion level by comparing it with the "Test Result Grading Table" in the standard (Table 1) based on the coating shedding situation. The adhesion level decreases from level 0 to level 5. It should be noted that the cross-cut test is not suitable for specimens with a film thickness exceeding 250 μm, and a multi-blade cutting tool is not suitable for specimens with a film thickness exceeding 120 μm. Therefore, before inspection, the film thickness of the specimen must be measured in accordance with relevant regulations. This method cannot determine the value of adhesion, but only assesses the resistance of the film to detaching from the wood. It belongs to the category of indirect measurement, but its operation is relatively simple, and it can assess the deformation of the film when subjected to external force, so it is widely used.

Table 1. Classification of test results for adhesion of paint film.

Classification	Illustrate	Surface Appearance of the Cross-Cut Area Where Peeling Occurred (Taking Six Parallel Cutting Lines as an Example)
0	Smooth cutting edges without any detachment	
1	A small amount of detachment at the cutting point; the detachment rate is no more than 5%	
2	The paint film peels off at the cutting point or in the grid, and the peeling rate is greater than 5% and not more than 15%	
3	The paint film peels off at the cutting point or in the grid, and the peeling rate is greater than 15% and not more than 35%	
4	The paint film peels off at the cutting point or in the grid, and the peeling rate is greater than 35% and not more than 65%	
5	The paint film peels off at the cut or inside the grid, and the peeling rate is greater than 65%	

(2) Abrasion resistance of paint film

The wear resistance of coatings evaluates the ability of coating surfaces to resist external forces such as friction and abrasion [34]. By simulating real usage conditions and measuring the wear resistance of coatings under a certain number of repetitions, the durability and performance of coatings can be evaluated. GB/T1768-2006 [35] provides detailed descriptions of the specimen size, test equipment, operating procedures, and evaluation methods for the wear resistance test.

First, the size of the test piece is (100 ± 1) mm × (100 ± 1) mm, with a suitable small hole in the center, generally with a diameter of 8 mm. In fact, it uses a test tool, namely a paint film abrasion tester, which is equipped with a pressure arm and a rubber grinding wheel for grinding the paint film. During test operation, the test piece is fixed on the workbench of the abrasion tester, and a weight and a rubber grinding wheel that meets the requirements are added to the pressure arm. Before a specific test, a test grinding of 50 turns is required to form a uniform and smooth grinding circle on the surface of the test piece, and the abrasion tester is readjusted to the specified number of turns [36]. Wear resistance is expressed by the mass loss of the paint film after a specified number of friction cycles, or by the number of cycles required to grind off the coating to the next coating or substrate. Good adhesion of the paint film means that the paint film cannot easily fall off of the substrate, so that when subjected to external friction, it can guarantee good wear resistance to a certain extent.

1.2.3. Effects of Various Varnish Products on Different Modified Wood

In order to gain a deeper understanding of the effects of different varnish products and different wood surface modification methods on paint film adhesion, Table 2 summarizes the important issues when finishing modified wood with different varnish products.

Table 2. Important issues when finishing modified wood with different varnish products.

Author	Wood Type	Modification Method	Coating Type	Adhesion (MPa)	Analysis
Xie et al. [37]	*Pinus sylvestris*	Chemical impregnation	Alkyd topcoat	2.5	Impregnation causes the wood cell walls to swell, increasing the contact area between the wood and the veneer, improving adhesion
Xiao et al. [38]	*Pinus sylvestris*	Chemical impregnation	Solvent-based alkyd resin	3.5	Solvent-based coatings have better penetration into wood substrates and better adhesion than water-based acrylic coatings
Herrera et al. [39]	*Fraxinus excelsior* L.	Heat treatment	UV coating material	1.1	The reduced acidity of the modified wood improves the adhesion of the coating to the surface
Altgen et al. [40]	*Picea abies* (L.) Karst.	Heat treatment	Waterborne acrylic paint	2.0	Thermal modification will significantly increase the brittleness of the paint film, which can lead to film cracking

In summary, the effects of different varnish products and wood surface modification methods on paint film adhesion show significant differences. Research shows that optimized treatment options need to take into account adhesion, material properties and possible defects to achieve the best coating results.

The development of wood surface coating technology has undergone significant changes in recent years. From traditional paints to modern high-performance coatings, researchers are constantly exploring new materials and technologies to improve the adhesion and durability of wood coatings. Research on wood surface modification treatments has increasingly become a cutting-edge topic in the field of wood protection and decoration [41,42]. Especially through physical and chemical modification, the physical and chemical properties of wood surfaces have been significantly improved, thereby improving the adhesion and weather resistance of coatings [43]. At the same time, the introduction of nanotechnology provides a new perspective on the optimization of coating performance. Research shows that the addition of nanoparticles can significantly enhance the adhesion and protective properties of the coating. These innovations are driving wood coating technology towards higher performance and wider applications. Sorting out the cutting-edge research not only provides a valuable reference for future wood application research, but also lays the foundation for multifunctional and high value-added development of wood.

2. Effect of Physical Modification of Wood Surface on the Adhesion Properties of Paint Film

Physical modification refers to the regulation of the wood surface structure through external forces (such as grinding, mechanical cutting, etc.) and high-temperature heat treatment [44]. Although some physical modification techniques can improve the adhesion of coatings by increasing the contact area and mechanical bite force between wood and coating, in certain cases these methods may also reduce the adhesion effect due to changes in the structure and wettability of the wood surface. Therefore, different physical modification techniques have different effects on the adhesion performance of wood coatings and must be analyzed based on specific circumstances.

2.1. Physical External Forces Modification

Surface roughness is the microscopic geometric morphology of the surface formed by the combined effects of mechanical processing and cutting, the wood's material, texture, porosity, etc. The surface roughness of wood is not only an inherent property of the wood, but also affected by the processing method. The surface roughness of the wood can also be affected by sanding the wood surface. After slicing or sanding, wood cells are exposed on the cut surface, and the arrangement of the cell tissue varies greatly depending on the anatomical structure of the wood [45]. Sanding the wood surface can not only reduce the impact of stains on the substrate surface on the coating performance, but also change the flatness of the substrate surface, increase the bonding area between the filler and the substrate, and improve the coating performance of the substrate surface [46].

Yu et al. [47] conducted experiments on six types of wood used for furniture production (*Pinus radiata*, *Pinus sylvestris*, Larch, Hemp oak, Catalpa, and Camphor), aiming to study the surface roughness, surface free energy and wetting of the wood. In the experiment, the researchers used sandpaper with different grits (P180, P240, P320, P400, P500) to polish the wood surface and tested the roughness, wettability and adhesion of the wood. Experimental results showed that hardwood exhibits slightly better coating adhesion than softwood. Hemp oak in particular shows good adhesion due to its porous structure and high wettability. The conclusion pointed out that the surface roughness and free energy of wood significantly affect the adhesion properties of the coating, and appropriate polishing treatment can improve the wettability and bonding quality of the wood surface.

Darmawan et al. [48] investigated the effect of roughness on the surface finishing performance by sanding longitudinal and transverse sections of sengon wood and jabon wood with 120#, 240#, and 360# sandpaper, respectively, and then coated the sanded surfaces with oil-based alkyd resin and water-based acrylic varnish. The results showed that as the roughness decreased, the mechanical interlocking between the substrate and the paint film was weakened, and the surface paint film adhesion became worse. There was no significant difference in paint film adhesion between jabon wood and sengon wood in the longitudinal and transverse directions. Compared to jabon wood, sengon wood has relatively good paint film adhesion due to its porous structure. Varnish type has a significant impact on wetting and bonding quality. Oil-based alkyds are easier to wet and produce better bonding qualities than water-based acrylics. This is because an oil-based alkyd varnish completes its polymerization reaction on the wood surface, forming a chemical bond with the wood and therefore creating a stronger adhesion to the wood surface. The water used as a solvent in water-based varnishes causes the wood fibers to swell, reducing permeability near the wood surface and resulting in a weaker interface between the wood material and the filler coating.

In general, rough surfaces can enhance intrinsic adhesion by providing a larger interfacial area and some mechanical interlocking mechanisms. It has been noted that rough surfaces provide paint with multiple possibilities for penetration and the formation of "resin fingers", which helps to form a strong joint. The anatomical structure of the wood will be one of the important factors affecting the quality of the adhesion of the coating to the wood surface. Wood with a porous structure is more wettable, increasing the absorption and penetration of the coating and providing better adhesion.

2.2. High-Temperature Heat Treatment Modification

High-temperature heat treatment of wood can greatly improve the quality of wood and its availability. High-temperature heat treatment can effectively reduce the hygroscopicity of wood [49]. After heat treatment, the moisture distribution and fiber structure inside the wood will become more uniform and stable, reducing the possibility of warping and deformation caused by humidity changes and thereby improving the dimensional stability, decay resistance and insect resistance of the wood, which is beneficial for painting [50].

Yu [51] subjected *Pometia pinnata* to high-temperature treatment and coated the surface with UV paint. It was found that after high-temperature treatment, the hydroxyl groups in the cellulose on the wood surface were dehydrated and turned into ether groups and ketone groups. As the number of hydroxyl groups decreased, the corresponding wettability of the wood surface also decreased. In addition, after high-temperature treatment, the wood became more brittle, and dust filled the pores of the wood during the sanding process. Although it was cleaned, the dust could not be completely cleaned out due to the large number of pores in the wood itself. This may also be a reason for the change in surface wettability. Due to the reduction in wettability, the penetration of UV coating on the wood surface was reduced, and the adhesion of the paint film was reduced, from level 1 to level 2.

Gurleyen et al. [52] investigated the effects of thermal modification on the properties of Turkish oak wood (*Quercus cerris* L.) coatings. They heat-treated Turkish oak wood at 190 °C for 2 h and 212 °C for 1 h and 2 h, respectively. Two coating methods were tested using a nano-coat system on treated and untreated wood surfaces. The first coating method applied a single coat of sealant at a dosage of 50 g/m^2, while the second coating method applied two coats of sealant at a dosage of 35 g/m^2 each. The results showed that the bond strength of both coated woods decreased with thermal treatment. The decrease was greater with longer treatment time and higher treatment temperature. This decrease may be due to the lower wettability and pH of the heat-treated wood surface.

Vidholdová et al. [53] thermally modified Turkish oak wood at 175 °C and 195 °C for 4 h. They then coated the heat-treated Turkish oak wood surfaces with linseed oil, a mixture of linseed oil and hard wax oil, or pure wax. The results showed that adhesion was affected by the type of wood substrate, with native wood having better adhesion than thermally modified wood. Among the coating materials tested, linseed oil showed the highest adhesion, while pure wax had the lowest adhesion. Therefore, when selecting oil-based or wax-based surface treatment agents for wood coating, their mechanical properties and durability must be comprehensively considered. It should be noted that these surface treatment agents perform differently on natural wood and thermally modified wood.

In general, although high-temperature heat treatment can effectively improve the dimensional stability, corrosion resistance and insect resistance of wood, its effect on coating adhesion is more complicated. Heat treatment usually causes reduced wettability by reducing the hygroscopicity of wood and changing its surface chemical structure, which in turn weakens the penetration and adhesion of coatings. The higher the treatment temperature and time, the greater the decrease in coating adhesion. In addition, different coating materials perform differently on thermally modified wood, so the appropriate coating must be selected according to the specific characteristics of the wood to ensure a balance between coating adhesion and durability.

3. Effect of Chemical Impregnation Modification on Wood Surface on Paint Film Adhesion Properties

Impregnation modification generally involves impregnating different types of chemical modifiers into the interior of the wood by means of a certain impregnation treatment process [54]. The modification mechanism of wood impregnation is shown in Figure 2. Some modifiers can undergo chemical cross-linking reactions with functional groups on the wood cell wall, thereby increasing the volume of the wood; other modifiers may not be able to penetrate the wood cell wall due to size limitations, and instead aggregate and fill the cell cavity, resulting in changes in the physical and chemical structure of the wood cell wall and cell cavity, thereby optimizing the performance of the wood. Since fast-growing wood species usually have lower density and larger pores, it is easier for modifiers to penetrate into the wood. Depending on the type of modifier, impregnation modification can be further divided into organic impregnation modification and inorganic impregnation modification [55,56].

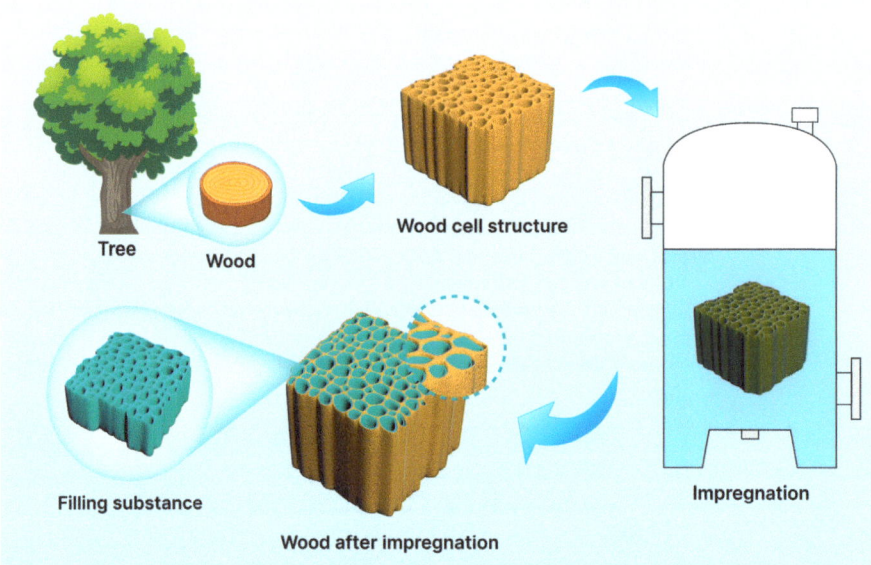

Figure 2. Modification mechanism of wood impregnation.

3.1. Organic Impregnation Modification

Organically modified wood is a process in which organic matter is impregnated into wood under certain conditions [57]. The organic modifier is deposited in the wood cell cavities or reacts chemically with the cell wall components, thereby becoming fixed in the wood and improving the dimensional stability, bacterial resistance, aging resistance, and so on [58].

Pepin et al. [59] conducted dipping experiments on eastern white pine (*Pinus strobus* L.) and white spruce (*Picea glauca* Moench (Voss)). In this experiment, wood samples were immersed in a treatment solution at a temperature of 65 °C for 15 s, followed by natural drying in a laboratory environment for one day and then in a conditioning chamber for six days. The treatment solution consisted of acrylic resin, amine oxide, borate buffer, and the organic biocide propiconazole. The treatment solution formula was adjusted according to the combination. Experiments showed that impregnating wood with amine oxide and propiconazole significantly improves paint film adhesion. Initially, the adhesion of the impregnated paint film is enhanced under wet conditions, benefiting from the fixation of the amine oxide and coating stability. However, as the aging cycle increases, the adhesion gradually weakens, especially after the seventh cycle. Although propiconazole inhibits microorganisms, it may affect the curing of the paint film and reduce adhesion. During the aging process, the permeability of amine oxides and propiconazole weakens, and the bonding force between wood and paint film decreases. Bleeding of resin from the wood may also interfere with coating curing, further weakening adhesion. Therefore, although amine oxide and propiconazole initially enhance adhesion, long-term aging leads to a decrease in adhesion. However, this acrylic paint is just a primer and more coats can be added to improve its performance.

Han et al. [60] used melamine-formaldehyde (MF) resin-coated transparent shellac–rosin microcapsule emulsion to impregnate ebony (*Microberlinla* sp.) substrates. The ebony board was immersed in the prepared microcapsule emulsion for 5 min, and then taken out and dried naturally. This process was recorded as one impregnation. The above steps were repeated for multiple impregnations of the ebony board. A water-based acrylic resin coating containing 3.0 wt.% transparent shellac microcapsules was applied to the surface of the ebony board. As the number of impregnations increased, the adhesion grade and impact strength of the ebony surface coating showed a slow upward trend. This may be because the number of microcapsules attached to the ebony surface increased, resulting in an increase in the roughness of the wood surface and a decrease in the adhesion of the coating. With the increase in the number of impregnations, the roughness of the ebony surface coating showed a trend of first increasing and then decreasing.

In summary, organic impregnation modification can represent an efficient and inexpensive single-step method for impregnating and infusing wood. Organic chemical impregnation treatment can improve the adhesion of the paint film in the early stage by increasing the roughness of the wood surface or changing the surface structure of the wood. However, the long-term performance of the paint film is affected by the stability of the treatment agent, the intrinsic properties of the wood, and the aging conditions. In order to meet the requirements of different coating systems, the impregnation treatment process needs to be further optimized to improve its adhesion and durability for long-term use.

3.2. Inorganic Impregnation Modification

Inorganic impregnation modification of wood is a method of immersing inorganic substances that can physically or chemically react with wood components into wood cells through infiltration, forming deposits in the wood cell cavities or even cell walls to play a filling role, or forming chemical connections with cell wall substances, thereby giving wood new properties and improving wood performance. Wood modified with inorganic substances combines the advantages of both wood and the inorganic substance, giving new functions to the inherent properties of wood. Since most of the modifiers are environmentally friendly, they have become a hot topic in wood modification research [61].

Among inorganic modifiers, silicates can not only significantly improve the physical mechanics and dimensional stability of wood, granting wood better flame retardant and heat resistance properties, but also maintain the environmental properties of wood. In particular, it has a good effect on improving the defects of Chinese fir, such as loose structure, low density and low strength [62].

Li et al. [63] used silicate to modify the surface of Chinese fir wood. The cured modifier achieved a physical filling effect in the pores of Chinese fir wood, and the silicate chemically interacted with the hydroxyl groups in the chemical composition of Chinese fir wood, forming a Si–O bond. This result reveals that although the modification treatment caused a slight reduction in the adhesion of the paint film on the Chinese fir wood surface, the degree of reduction was minimal. The modified Chinese fir wood has significantly improved bending strength, compressive strength and dimensional stability, showing superior mechanical properties.

Zhang et al. [64] used a sodium silicate/magnesium chloride composite to modify the surface of Chinese fir. After curing, the modifier formed a physical filling effect in the pores of the Chinese fir. The alkalinity of the sodium silicate solution partially removed wood hemicellulose and lignin, making it easier for the impregnant to penetrate. Due to the partial release of lignin and hemicellulose, the amount of pores in the Chinese fir increased and the permeability increased, promoting the modifier's penetration into the wood. Due to the decomposition and destruction of lignin and hemicellulose, more active –OH groups were exposed on the cellulose, which promoted the combination of sodium silicate and cellulose to form a Si–O–C structure, achieving a better impregnation effect. At the same time, its performance was also greatly improved. Therefore, sodium silicate modification of Chinese fir is an efficient, green, and environmentally friendly impregnation modification method.

Bao et al. [65] used poplar as the test material and sodium silicate as the modifier to obtain sodium silicate-impregnated poplar wood by vacuum pressure impregnation. Modified acrylic water-based paint and polyurethane oil-based paint were applied to the poplar surface for testing. The poplar specimens before and after modification were observed by scanning electron microscopy (SEM), as shown in Figure 3. As can be seen from Figure 3, after impregnation, sodium silicate mainly filled the poplar ducts. The study pointed out that in addition to physical filling, sodium silicate impregnation also leads to the formation of Si–O–C group chemical bonds with the wood structure. Sodium silicate modification increases the number of free hydroxyl groups and hydrogen bonding sites on the wood surface, promotes the combination of water molecules and wood, and improves the wettability and surface free energy of the poplar surface. Although sodium silicate impregnation makes the wood surface adhesion worse, there was no sample with an adhesion level greater than 3, indicating that the modification treatment would not excessively affect paint film adhesion on the poplar surface.

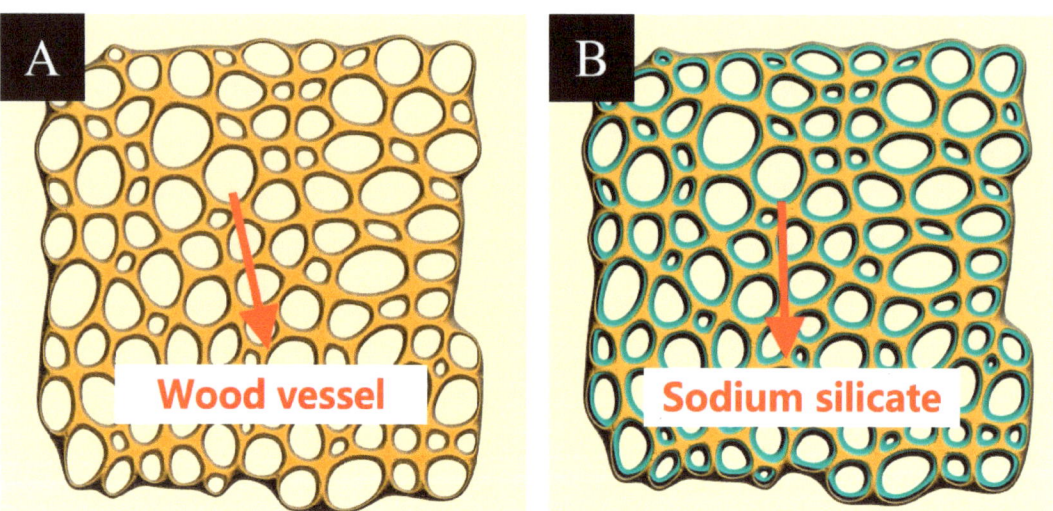

Figure 3. Cross-section schematic diagram of poplar wood before and after impregnation: (**A**) unmodified wood, (**B**) modified wood.

It is undeniable that inorganic impregnation modification can effectively improve the physical and mechanical properties and functionality of wood. However, impregnation treatments tend to affect coating adhesion due to changes in the chemical structure of the wood surface and physical filling effects. Although this treatment reduces the adhesion of the paint film, under reasonable modification conditions the decrease in adhesion is not significant, and the overall mechanical properties of the modified wood are greatly improved. Therefore, while inorganic impregnation modification improves the performance of wood, its impact on coating adhesion is still within a controllable range, and it is a promising wood modification method.

4. Effect of Composite Modified Wood Surface on Paint Film Adhesion Performance

Composite modification refers to the comprehensive use of two or more modification methods to improve the compatibility of the coating and substrate interface performance measures. Taghiyari et al. [66] impregnated common beech (*Fagus sylvatica* L.), black poplar (*Populus nigra* L.), and silver fir (*Abies alba* Mill.) with nanosilver and then subjected the impregnated specimens to high-temperature heat treatment (145 °C and 185 °C). The results showed a positive correlation between the density and pull-off strength of the solid wood species. Heat treatment at 145 °C resulted in an increase in the pull-off strength of the three species due to the formation of new bonds in the cell wall polymers. This increase was attributed to the formation of new bonds in the cell wall polymers by irreversible hydrogen bonds during water movement in the cell wall polymers. Thermal degradation of the polymers at 185 °C weakened the positive effect of the new bond formation, and impregnation with the silver nanosuspension reduced the pull-off strength of the beech samples. It is recommended to perform heat treatment at a lower temperature to increase the pull-off strength.

Zhang et al. [67] applied different silane coupling agents (KH550, KH560, KH570 and KH580) at a concentration of 8% on the surface of poplar wood modified with 29.5% silicate. The wood was then painted with a water-based sealing primer to study its surface properties. The results showed that the paint film adhesion of the poplar wood material was level 1, and the paint film adhesion of the poplar wood modified with silicate was level 3. This indicates that the paint film adhesion of poplar wood was significantly reduced after silicate modification. Silicate is an inorganic substance, and impregnation modification fills the voids in the poplar wood. The main component of water-based paint is organic resin. However, there is a problem of poor interfacial compatibility between organic and inorganic substances, reducing the adhesion of the organic paint film on the surface of silicate-modified poplar wood, which has an adverse effect on the subsequent processing and application of the poplar wood. After the silicate modified poplar wood (SMPW) surface was treated with KH550 and KH570, the paint film adhesion was level 1, which was significantly better than SMPW and the same as the level of unmodified poplar wood. After the SMPW surface was treated with KH560 and KH580, the adhesion of the surface paint film reached level 0, which was better than that of the paint film on SMPW and poplar. The Y group (–SH) in KH580 reacted strongly with the isocyanate group (R–N=C=O) and lipid group (–COOR) in water-based sealing varnish (WBSV), and KH580 self-condensed to form a cross-linked network, which improved the bonding strength between WBSV and the substrate; however, the silanol group generated by the hydrolysis of KH580 did not form a Si–O–C covalent bond with WBSV, which weakened the bonding strength between the water-based paint and the substrate to a certain extent. After KH560 was mixed with WBSV, the epoxy functional group (–OCH$_2$CH(O)CH$_2$) at the end of the Y group formed a hydroxyl group through a ring-opening reaction and combined with WBSV to form an unsaturated ether bond. The silanol group generated by the hydrolysis of KH560 formed a Si–O–C covalent bond with the water-based sealing primer, which made the bonding between the water-based paint and the substrate optimal. In conclusion, the silane coupling agent significantly improved the adhesion of the paint film on the SMPW surface.

Based on this experiment, Zhang et al. [68] studied the effect of different concentrations of silane coupling agent KH560 on the interfacial properties of 29.5% SMPW. KH560 solution was prepared according to the mass ratio of KH560: water of 1:1. Then, with anhydrous ethanol as the solvent, the concentration of KH560 was adjusted to 0.5%, 1.0%, 1.5%, 2.0%, 2.5% and 3.0%. The prepared solution was sprayed on the surface of the silicate modified poplar wood. The modified materials were all coated with water-based sealing primer and water-based bright varnish. The results showed that, except for the SMPW coating treated with 0.5% KH560, which had an adhesion grade of 2, SMPW coatings treated with 1.0%~3.0% KH560 had an adhesion grade of 1. SEM images showed that with the increase in KH560 concentration, the cracks between the water-based (water-based polyurethane modified acrylic) varnish and SMPW gradually decreased. The results of FI–TR and XPS showed that the characteristic peaks of Si–O–Si and the relative proportion of Si–O–Si groups in the treated SMPW were positively correlated with the concentration of KH560, indicating that KH560 was successfully grafted onto the substrate surface. After KH560 was composited with waterborne sealing primer (WBSP), the greater the concentration of KH560, the stronger the bonding force at 1730 cm^{-1}, indicating that the epoxy groups in KH560 reacted and connected with WBSP. After SMPW was wetted with 0.5%, 1.5%, and 2.5% KH560 solution, the contact angles of WBSP on the substrate surface were 62.55°, 60.10°, and 49.45°, respectively; the surface free energies of SMPW treated with 0.5%, 1.5%, and 2.5% KH560 were 45.87, 49.55, and 50.29 (10^{-3} J·m^{-2}), respectively. This indicates that increasing the KH560 concentration can improve wettability, increase the SMPW free energy value, and enhance film adhesion.

Through the above analysis, we can conclude that composite modification effectively improves the adhesion and interfacial compatibility of the wood surface by combining multiple modification methods. Compared with single modification, composite modification can more comprehensively improve the physical and chemical properties of wood, making the bond between the coating and the wood substrate stronger and the adhesion stronger. Especially when the properties of different modifiers are complementary, composite modification can significantly improve the durability and mechanical properties of wood, and has greater flexibility and potential in practical applications. Therefore, composite modification is an efficient and feasible technical approach to optimize wood surface properties.

5. Conclusions

This article comprehensively analyzed the latest research progress in wood surface modification technology, focusing on the application and effectiveness of physical modification, chemical impregnation modification and composite modification in improving wood properties. Physical external force modification significantly enhances the hydrophilicity and wettability of the wood by changing the surface structure of the wood, thereby improving the adhesion of the coating; high-temperature heat treatment effectively eliminates the internal stress and moisture of the wood and reduces the risk of coating cracking and peeling. Physical modification is simple to operate and does not use chemicals, meeting the requirements of sustainable development. New physical modification technologies, including plasma treatment and laser etching, may emerge in the future to further optimize wood properties. By selecting different impregnation materials and formulas, chemical impregnation modification can be effectively modified for specific application needs. In addition, new coating materials may be explored in the future to solve the stability problems of existing coatings under humidity and temperature changes. Composite modification shows great potential to achieve comprehensive performance improvements for different wood properties, especially significant improvements in coating adhesion. Future research will focus on composite modification technologies to achieve more efficient and environmentally friendly wood treatment solutions to meet the market demand for high-performance wood coatings. The future development of wood coatings will show a trend of diversification, environmental protection, intelligence and personalization. By exploring new coating materials and technologies, optimizing production processes, and strengthening cross-border integration and innovative cooperation, wood coatings will usher in more modified products and technical solutions with excellent performance and broad application prospects, promoting the upgrading and development of the entire industry chain.

Author Contributions: Conceptualization, methodology, validation, resources, data management, supervision, writing—review and editing, J.H.; formal analysis, X.Y.; investigation, J.L. All authors have read and agreed to the published version of the manuscript.

Funding: This project was partly supported by the Natural Science Foundation of Jiangsu Province (BK20201386).

Institutional Review Board Statement: Not applicable.

Informed Consent Statement: Not applicable.

Data Availability Statement: Not applicable.

Conflicts of Interest: The authors declare that there are no conflicts of interest.

References

1. Zhu, J.G.; Niu, J.Y. Green Material Characteristics Applied to Office Desk Furniture. *Bioresources* **2022**, *17*, 2228–2242. [CrossRef]
2. Wang, C.; Zhang, C.Y.; Zhu, Y. Reverse design and additive manufacturing of furniture protective foot covers. *Bioresources* **2024**, *19*, 4670–4678. [CrossRef]
3. Kaur, R.; Thakur, N.S.; Chandna, S.; Bhaumik, J. Sustainable Lignin-Based Coatings Doped with Titanium Dioxide Nanocomposites Exhibit Synergistic Microbicidal and UV-Blocking Performance toward Personal Protective Equipment. *ACS Sustain. Chem. Eng.* **2021**, *9*, 11223–11237. [CrossRef]
4. Yu, R.Z.; Liu, Y.; Konukcu, A.C.; Hu, W.G. A method of simulating seat load for numerical analysis of wood chair structure. *Wood Res.* **2024**, *69*, 432–444. [CrossRef]
5. Kamperidou, V.; Aidinidis, E.; Barboutis, I. Impact of Structural Defects on the Surface Quality of Hardwood Species Sliced Veneers. *Appl. Sci.* **2020**, *10*, 6265. [CrossRef]
6. Su, H.; Du, G.B.; Yang, H.X.; Wu, Y.C.; Liu, S.C.; Ni, K.L.; Ran, X.; Li, J.; Gao, W.; Yang, L. Novel ultrastrong wood bonding interface through chemical covalent crosslinking of aldehyde-amine. *Ind. Crops. Prod.* **2022**, *189*, 115800. [CrossRef]
7. Zhang, X.Y.; Xu, W.; Li, R.R.; Zhou, J.C.; Luo, Z.Y. Study on sustainable lightweight design of airport waiting chair frame structure based on ANSYS workbench. *Sustainability* **2024**, *16*, 5350. [CrossRef]
8. Zhong, Z.W. Surface roughness of machined wood and advanced engineering materials and its prediction: A review. *Adv. Mech. Eng.* **2021**, *13*, 16878140211017632. [CrossRef]
9. Liu, Y.; Hu, W.G.; Kasal, A.; Erdil, Y.Z. The State of the Art of Biomechanics Applied in Ergonomic Furniture Design. *Appl. Sci.* **2023**, *13*, 12120. [CrossRef]
10. Li, J.; Zheng, H.P.; Liu, L.; Meng, F.D.; Cui, Y.; Wang, F.H. Modification of graphene and graphene oxide and their applications in anticorrosive coatings. *J. Coat. Technol. Res.* **2021**, *18*, 311–331. [CrossRef]
11. Zou, W.R.; Fan, Z.Z.; Zhai, S.X.; Wang, S.W.; Xu, B.; Cai, Z.S. A multifunctional antifog, antifrost, and self-cleaning zwitterionic polymer coating based on poly (SBMA-co-AA). *J. Coat. Technol. Res.* **2020**, *17*, 765–776. [CrossRef]
12. Hanincová, L.; Procházka, J.; Novák, V.; Kopecky, Z. Influence of Moisture Content on Cutting Parameters and Fracture Characteristics of Spruce and Oak Wood. *Drv. Ind.* **2022**, *73*, 341–349. [CrossRef]
13. Hu, J.; Liu, Y.; Xu, W. Influence of Cell Characteristics on the Construction of Structural Color Layers on Wood Surfaces. *Forests* **2024**, *15*, 676. [CrossRef]
14. Wang, C.; Yu, J.H.; Jiang, M.H.; Li, J.Y. Effect of selective enhancement on the bending performance of fused deposition methods 3D-printed PLA models. *Bioresources* **2024**, *19*, 2660–2669. [CrossRef]
15. Hu, W.G.; Yu, R.Z. Mechanical and acoustic characteristics of four wood species subjected to bending load. *Maderas-Cienc. Tecnol.* **2023**, *25*, 39. [CrossRef]
16. Liu, J.B.; Cheng, W.D.; Jiang, X.Y.; Khan, M.U.; Zhang, Q.F.; Cai, H.Z. Effect of Extractives on the Physicochemical Properties of Biomass Pellets: Comparison of Pellets from Extracted and Non-extracted Sycamore Leaves. *BioResources* **2020**, *15*, 544–556. [CrossRef]
17. Hu, W.G.; Luo, M.Y.; Hao, M.M.; Tang, B.; Wan, C. Study on the effects of selected factors on the diagonal tensile strength of oblique corner furniture joints constructed by wood dowel. *Forests* **2023**, *14*, 1149. [CrossRef]
18. Hu, W.G.; Luo, M.Y.; Yu, R.Z.; Zhao, Y. Effects of the selected factors on cyclic load performance of T-shaped mortise-and-tenon furniture joints. *Wood Mater. Sci. Eng.* **2024**, 1–10. [CrossRef]
19. Hu, W.G.; Yu, R.Z.; Yang, P. Characterizing Roughness of Wooden Mortise and Tenon Considering Effects of Measured Position and Assembly Condition. *Forests* **2024**, *15*, 1584. [CrossRef]
20. Wu, W.; Xu, W.; Wu, S.S. Mechanical performance analysis of double-dovetail joint applied to furniture T-shaped components. *BioResources* **2024**, *19*, 5862–5879. [CrossRef]
21. Sundar, N.; Kumar, A.; Pavithra, A.; Ghosh, S. Studies on Semi-crystalline Poly Lactic Acid (PLA) as a Hydrophobic Coating Material on Kraft Paper for Imparting Barrier Properties in Coated Abrasive Applications. *Prog. Org. Coat.* **2020**, *145*, 105682.
22. Leggate, W.; McGavin, R.L.; Miao, C.; Outhwaite, A.; Chandra, K.; Dorries, J.; Kumar, C.; Knackstedt, M. The Influence of Mechanical Surface Preparation Methods on Southern Pine and Spotted Gum Wood Properties: Wettability and Permeability. *BioResources* **2020**, *15*, 8554–8576. [CrossRef]
23. Hu, W.G.; Luo, M.Y.; Liu, Y.Q.; Xu, W.; Konukcu, A.C. Experimental and numerical studies on the mechanical properties and behaviors of a novel wood dowel reinforced dovetail joint. *Eng. Fail. Anal.* **2023**, *152*, 107440. [CrossRef]
24. Wang, P.Y.; Cheng, S.; Cao, S.; Cai, J.B. Evaluation of Color Changes, Wettability, and Moisture Sorption of Heat-Treated Blue-Stained Radiata Pine Lumber. *BioResources* **2022**, *17*, 4952–4961. [CrossRef]
25. Hu, W.G.; Fu, W.J.; Zhao, Y. Optimal design of the traditional Chinese wood furniture joint based on experimental and numerical method. *Wood Res.* **2024**, *69*, 50–59. [CrossRef]

26. Wei, J.G.; Xu, Y.; Bao, M.Z.; Yu, Y.L.; Yu, W.J. Effect of Resin Content on the Surface Wettability of Engineering Bamboo Scrimbers. *Coatings* **2023**, *13*, 203. [CrossRef]
27. Wang, K.; Dong, Y.M.; Yan, Y.T.; Qi, C.S.; Zhang, S.F.; Li, J.Z. Preparation of Mechanical Abrasion and Corrosion Resistant Bulk Highly Hydrophobic Material Based on 3-D Wood Template. *RSC Adv.* **2016**, *6*, 98248–98256. [CrossRef]
28. Miklecic, J.; Jirous-Rajkovic, V. Influence of Thermal Modification on Surface Properties and Chemical Composition of Beech Wood (*Fagus sylvatica* L.). *Drv. Ind.* **2016**, *67*, 65–71. [CrossRef]
29. Hu, J.; Liu, Y.; Xu, W. Impact of cellular structure on the thickness and light reflection properties of structural color layers on diverse wood surfaces. *Wood Mater. Sci. Eng.* **2024**, 1–11. [CrossRef]
30. Wang, C.; Yu, J.H.; Jiang, M.H.; Li, J.Y. Effect of slicing parameters on the light transmittance of 3D-printed polyethylene terephthalate glycol products. *Bioresources* **2024**, *19*, 500–509. [CrossRef]
31. GB/T 1764-79; Film Thickness Measurement Method. Standardization Administration of the People's Republic of China: Beijing, China, 1979.
32. Zhang, K.Q.; Chen, Z.H.; Boukhir, M.; Song, W.; Zhang, S.B. Bioinspired Polydopamine Deposition and Silane Grafting Modification of Bamboo Fiber for Improved Interface Compatibility of Poly (Lactic Acid) Composites. *Int. J. Biol. Macromol.* **2022**, *201*, 121–132. [CrossRef] [PubMed]
33. GB/T 4893.4-2013; Test of Surface Coatings of Furniture. Part IV: Determination of Adhesion-Cross Cut. Standardization Administration of the People's Republic of China: Beijing, China, 2013.
34. Wang, C.; Zhang, C.Y.; Ding, K.Q.; Jiang, M.H. Immersion polishing post-treatment of PLA 3D printed formed parts on its surface and mechanical performance. *BioResources* **2023**, *18*, 7995–8006. [CrossRef]
35. GB/T 1768-2006; Paints and Varnishes—Determination of Resistance to Abrasion—Rotating Abrasive Rubber Wheel Method. Standardization Administration of the People's Republic of China: Beijing, China, 2006.
36. Wu, S.S.; Zhou, J.C.; Xu, W. A convenient approach to manufacturing lightweight and high-sound-insulation plywood using furfuryl alcohol/multilayer graphene oxide as a shielding layer. *Wood Mater. Sci. Eng.* **2024**, 1–8. [CrossRef]
37. Xie, Y.J.; Krause, A.; Militz, H.; Mai, C. Coating performance of finishes on wood modified with an N-methylol compound. *Prog. Org. Coat.* **2006**, *57*, 291–300. [CrossRef]
38. Xiao, Z.F.; Chen, H.O.; Mai, C.; Militz, H.; Xie, Y.J. Coating performance on glutaraldehyde-modified wood. *J. For. Res.* **2019**, *30*, 353–361. [CrossRef]
39. Herrera, R.; Muszynska, M.; Krystofiak, T.; Labidi, J. Comparative evaluation of different thermally modified wood samples finishing with UV-curable and waterborne coatings. *Appl. Surf. Sci.* **2015**, *357*, 1444–1453. [CrossRef]
40. Altgen, M.; Militz, H. Thermally modified Scots pine and Norway spruce wood as substrate for coating systems. *J. Coat. Technol. Res.* **2015**, *14*, 531–541. [CrossRef]
41. Ling, K.L.; Feng, Q.M.; Huang, Y.H.; Li, F.; Huang, Q.F.; Zhang, W.; Wang, X.C. Effect of Modified Acrylic Water-Based Paint on the Properties of Paint Film. *Spectrosc. Spect. Anal.* **2020**, *40*, 2133–2137.
42. Agarwal, C.; Csóka, L. Functionalization of wood/plant-based natural cellulose fibers with nanomaterials: A review. *Tappi J.* **2018**, *17*, 92–111. [CrossRef]
43. Hu, J.; Liu, Y.; Wang, J.X.; Xu, W. Study of selective modification effect of constructed structural color layers on European beech wood surfaces. *Forests* **2024**, *15*, 261. [CrossRef]
44. Li, X.R.; Liu, Y.X.; Yu, H.P.; Li, J. Evaluation of the Surface Roughness of Wood-based Environmental Materials and its Impact on Human Psychology and Physiology. *Adv. Mat. Res.* **2010**, *113-116*, 932–937. [CrossRef]
45. Hu, W.G.; Yu, R.Z. Study on the strength mechanism of the wooden round-end mortise-and-tenon joint using the digital image correlation method. *Holzforschung* **2024**, *78*, 519–530. [CrossRef]
46. Liu, Q.Q.; Gao, D.; Xu, W. Effect of Sanding Processes on the Surface Properties of Modified Poplar Coated by Primer Compared with Mahogany. *Coatings* **2020**, *10*, 856. [CrossRef]
47. Yu, Q.L.; Pan, X.; Yang, Z.; Zhang, L.; Cao, J.Y. Effects of the Surface Roughness of Six Wood Species for Furniture Production on the Wettability and Bonding Quality of Coating. *Forests* **2023**, *14*, 996. [CrossRef]
48. Darmawan, W.; Nandika, D.; Noviyanti, E.; Alipraja, I.; Lumongga, D.; Gardner, D.; Gerardin, P. Wettability and Bonding Quality of Exterior Coatings on Jabon and Sengon Wood Surfaces. *J. Coat. Technol. Res.* **2018**, *15*, 95–104. [CrossRef]
49. Suri, I.F.; Purusatama, B.D.; Kim, J.H.; Yang, G.U.; Prasetia, D.; Kwon, G.J.; Hidayat, W.; Lee, S.H.; Febrianto, F.; Kim, N.H. Comparison of physical and mechanical properties of *Paulownia tomentosa* and *Pinus koraiensis* wood heat-treated in oil and air. *Eur. J. Wood Wood Prod.* **2022**, *80*, 1389–1399. [CrossRef]
50. Li, Z.Z.; Luan, Y.; Hu, J.B.; Fang, C.H.; Liu, L.T.; Ma, Y.F.; Liu, Y.; Fei, B.H. Bamboo heat treatments and their effects on bamboo properties. *Constr. Build. Mater.* **2022**, *331*, 127320. [CrossRef]
51. Yu, J.H. Study on Coating Adhesion Improvment of Heat-Treated Wood. Master's Thesis, Chinese Academy of Forestry, Beijing, China, 2016.
52. Gurleyen, L.; Ayata, U.; Esteves, B.; Gurleyen, T.; Cakıcıer, N. Effects of thermal modification of oak wood upon selected properties of coating systems. *BioResources* **2019**, *14*, 1838–1849. [CrossRef]
53. Vidholdová, Z.; Slabejová, G.; Šmidriaková, M. Quality of Oil- and Wax-Based Surface Finishes on Thermally Modified Oak Wood. *Coatings* **2021**, *11*, 143. [CrossRef]

54. Cao, S.; Cai, J.B.; Wu, M.H.; Zhou, N.; Huang, Z.H.; Cai, L.P.; Zhang, Y.L. Surface Properties of Poplar Wood after Heat Treatment, Resin Impregnation, or Both Modifications. *BioResources* **2021**, *16*, 7561–7576. [CrossRef]
55. Li, W.Z.; Zhang, Z.; Yang, K.; Mei, C.T.; Van den Bulcke, J.; Van Acker, J. Understanding the effect of combined thermal treatment and phenol-formaldehyde resin impregnation on the compressive stress of wood. *Wood Sci. Technol.* **2022**, *56*, 1071–1086. [CrossRef]
56. Augustina, S.; Dwianto, W.; Wahyudi, I.; Syafii, W.; Gérardin, P.; Marbun, S.D. Wood Impregnation in Relation to Its Mechanisms and Properties Enhancement. *BioResources* **2022**, *18*, 4332–4372. [CrossRef]
57. Yu, F.J.; You, Z.Y.; Ma, Y.S.; Liu, H.Y.; Wang, Y.G.; Xiao, Z.F.; Xie, Y.J. Modification with carboxymethylation-activated alkali lignin/glutaraldehyde hybrid modifier to improve physical and mechanical properties of fast-growing wood. *Wood Sci. Technol.* **2023**, *57*, 583–603. [CrossRef]
58. Croitoru, C.; Roata, I.C. Ionic Liquids as Antifungal Agents for Wood Preservation. *Molecules* **2020**, *25*, 4289. [CrossRef] [PubMed]
59. Pepin, S.; Blanchet, P.; Landry, V. Interactions between a Buffered Amine Oxide Impregnation Carrier and an Acrylic Resin, and Their Relationship with Moisture. *Coatings* **2020**, *10*, 366. [CrossRef]
60. Han, Y.; Yan, X.; Tao, Y. Effect of Number of Impregnations of Microberlinla sp with Microcapsule Emulsion on the Performance of Self-Repairing Coatings on Wood Surfaces. *Coatings* **2022**, *12*, 989. [CrossRef]
61. Liu, Q.Q.; Du, H.J.; Lyu, W.H. Physical and Mechanical Properties of Poplar Wood Modified by Glucose-Urea-Melamine Resin/Sodium Silicate Compound. *Forests* **2021**, *12*, 127. [CrossRef]
62. Bi, X.Q.; Zhang, Y.; Li, P.; Wu, Y.Q.; Yuan, G.M.; Zuo, Y.F. Building bridging structures and crystallization reinforcement in sodium silicate-modified poplar by dimethylol dihydroxyethylene urea. *Wood Sci. Technol.* **2022**, *56*, 1487–1508. [CrossRef]
63. Li, P.; Zhang, Y.; Zuo, Y.F.; Lu, J.X.; Yuan, G.M.; Wu, Y.Q. Preparation and Characterization of Sodium Silicate Impregnated Chinese Fir Wood with High Strength, water Resistance, flame Retardant and Smoke Suppression. *J. Mater. Res. Technol.* **2020**, *9*, 1043–1053. [CrossRef]
64. Zhang, Y.; Bi, X.Q.; Li, P.; Wu, Y.Q.; Yuan, G.M.; Li, X.J.; Zuo, Y.F. Sodium silicate/magnesium chloride compound-modified Chinese fir wood. *Wood Sci. Technol.* **2001**, *55*, 1781–1794. [CrossRef]
65. Bao, X.D.; Zhang, M.Y.; Li, P.; Lu, J.X.; Yuan, G.M.; Zuo, Y.F. Investigating the surface wettability and surface free energy of sodium silicate-impregnated poplar wood. *Wood Mater. Sci. Eng.* **2023**, *18*, 141–150. [CrossRef]
66. Taghiyari, H.R.; Ilies, D.C.; Antov, P.; Vasile, G.; Majidinajafabadi, R.; Lee, S.H. Effects of Nanosilver and Heat Treatment on the Pull-Off Strength of Sealer-Clear Finish in Solid Wood Species. *Polymers* **2022**, *14*, 5516. [CrossRef] [PubMed]
67. Zhang, M.Y.; Lu, J.X.; Li, P.; Li, X.J.; Yuan, G.M.; Zuo, Y.F. Construction of high-efficiency fixing structure of waterborne paint on silicate-modified poplar surfaces by bridging with silane coupling agents. *Prog. Org. Coat.* **2022**, *167*, 106846. [CrossRef]
68. Zhang, M.Y.; Lyu, J.X.; Zuo, Y.F.; Li, X.G.; Li, P. Effect of KH560 concentration on adhesion between silicate modified poplar and waterborne varnish. *Prog. Org. Coat.* **2023**, *174*, 107267. [CrossRef]

Disclaimer/Publisher's Note: The statements, opinions and data contained in all publications are solely those of the individual author(s) and contributor(s) and not of MDPI and/or the editor(s). MDPI and/or the editor(s) disclaim responsibility for any injury to people or property resulting from any ideas, methods, instructions or products referred to in the content.

MDPI AG
Grosspeteranlage 5
4052 Basel
Switzerland
Tel.: +41 61 683 77 34

Coatings Editorial Office
E-mail: coatings@mdpi.com
www.mdpi.com/journal/coatings

Disclaimer/Publisher's Note: The title and front matter of this reprint are at the discretion of the Guest Editor. The publisher is not responsible for their content or any associated concerns. The statements, opinions and data contained in all individual articles are solely those of the individual Editor and contributors and not of MDPI. MDPI disclaims responsibility for any injury to people or property resulting from any ideas, methods, instructions or products referred to in the content.

www.ingramcontent.com/pod-product-compliance
Lightning Source LLC
LaVergne TN
LVHW072350090526
838202LV00019B/2514